Chemical Structure and Bonding

Roger L. DeKock
Calvin College

Harry B. Gray
California Institute of Technology

UNIVERSITY SCIENCE BOOKS
Sausalito, California

University Science Books
55D Gate Five Road
Sausalito, CA 94965
Fax: (415) 332-5393

Cover designer: Bob Ishi
Printer and binder: Maple-Vail Book Manufacturing Group

Library of Congress Catalog Number: 89-050820

ISBN 0-935702-61-X

Printed in the United States of America
10   9   8   7   6   5   4

Preface to the Second Edition

With this paperback edition, University Science Books becomes the publisher of *Chemical Structure and Bonding*. We thank the publisher, Bruce Armbruster, for his interest in chemistry and chemistry education.

Roger L. DeKock
Harry B. Gray

Preface to the First Edition

Chemical Structure and Bonding was written for courses devoted specifically to structure and bonding topics, and for courses in inorganic, physical, and quantum chemistry whose subject matter significantly involves these topics. We hope that for many undergraduate students, this textbook will be the first serious introduction to the concepts of structure and bonding. In addition, graduate students and others should find our treatment of these topics to be a helpful review. We have included numerous questions and problems at the end of each chapter and an Appendix with answers to most of the problems. The manuscript is without footnotes to the original literature, but for those who wish to pursue further reading and study, a "Suggestions for Further Reading" section is provided at the end of each chapter.

We have written *Chemical Structure and Bonding* so as to provide the flexibility of subject matter that instructors often require. For example, an instructor who wished to concentrate on molecular orbital theory could use mainly Chapters 4, 5, and 6. Alternatively, a course that was designed to cover several of the topics only briefly could omit those sections considered unnecessary or too advanced.

The coverage of material in this textbook is purposely broad. Chapter 1 starts with the elementary Bohr theory and proceeds to the quantum theory of Schrödinger and its application to the particle-in-a-box, the hydrogen atom, and multielectron atoms. Well prepared students could quickly read the first 10 sections of Chapter 1 and begin their serious study with the Schrödinger wave equation in Section 1-11. Our presentation of the effects of electron-electron repulsion in Section 1-14 has been an effective pedagogical tool in discussing the electronic configurations of transition metal atoms.

Chapter 2 begins with a thorough discussion of atomic radii, ionization potentials, and electron affinities. This is followed by a treatment of Lewis electron dot structures and the concept of resonance. The latter part of Chapter 2 introduces molecular geometry through the valence shell electron pair repulsion approach. Subsequently, molecular symmetry and polarity are introduced. Section 2-16 provides an insightful method for helping students to understand and categorize molecular topology.

Chapter 3 is devoted to the valence bond and hybrid orbital descriptions of chemical bonding. It has been our experience that students often feel that sp, sp^2, and sp^3 are the only possible hybridization schemes for s and p orbitals. By introducing the concept of hybridization with the H_2 molecule, we hope to dispel that notion and provide the student with a straightforward introduction to the mixing of orbitals.

Chapters 4 and 5 constitute the application of molecular orbital theory to diatomic and polyatomic molecules, respectively. Our approach is to tie the theory of molecular orbitals closely to experimental photoelectron spectroscopy (Section 4-5). The bonding in transition metal diatomic and triatomic molecules is treated in Sections 4-7, 4-10, and 5-10. In this way it should be clear that the bonding concepts needed in transition-metal chemistry are not distinct from those in main-group chemistry. Finally, the frontier orbital concept and the idea of "allowed" and "forbidden" reactions are discussed in Section 5-9.

The structure and bonding of transition metal complexes is treated in Chapter 6. A theoretical understanding of the shapes of transition metal complexes is covered in Section 6-12 by using the angular overlap model.

Chapter 7 provides an introduction to bonding in solids and liquids. This includes a discussion of van der Waals, metallic, and ionic bonding.

In the spring of 1975 Roger DeKock was teaching an introductory chemistry course at the American University of Beirut. One of the textbooks used in this course was *Chemical Bonds*, written by Harry B. Gray and published by Benjamin / Cummings in 1973. In the summer of 1975 DeKock wrote a letter to Gray suggesting that the pedagogy of molecular orbital theory would be enhanced by relating it closely to the experimental technique of photoelectron spectroscopy. Gray responded with enthusiasm and what resulted was a first draft written during the 1975–76 academic year. This first draft was revised by DeKock and edited by Jim Hall at the Aspen Writing Center during the summer of 1976. Gray class-tested the manuscript for three years (1976–78) and DeKock class-tested it during an interim course in January, 1979. After each class testing the manuscript underwent further revision and refinement until it reached its present status. We hope that this extensive class testing has removed many of the annoyances that students and instructors often find in first edition textbooks.

We acknowledge the able assistance of Jim Hall during the early stages of this project. We thank Mary Schwartz of Benjamin/Cummings who acted as editor and provided constant counsel and encouragement. Dick Palmer coordinated the production stages and managed to keep us on a time schedule. Thanks also go to Rich Huisman who did several of the initial drawings and assisted with proofreading and editorial comment. Finally, there are numerous individuals who worked at various stages on the typing of the manuscript, but special thanks go to Jan Woudenberg at Calvin College.

Roger L. DeKock   Harry B. Gray
Calvin College   California Institute of Technology
Grand Rapids, Michigan   Pasadena, California
February 1980   February 1980

Contents

Chapter 3 The Valence Bond and Hybrid Orbital Descriptions of Chemical Bonding 135

Chapter 4 The Molecular Orbital Theory of Electronic Structure and the Spectroscopic Properties of Diatomic Molecules 183

Contents   ix

1 Atomic Structure

The concept that molecules consist of atoms bonded together in definite patterns was well established by 1860. Shortly thereafter, recognition that the bonding properties of the elements are periodic led to widespread speculation concerning the internal structure of atoms themselves. The first major development in the formulation of atomic structure models followed the discovery, about 1900, that atoms contain electrically charged particles — negative electrons and positive protons. From charge-to-mass ratio measurements Sir Joseph John Thomson, a British physicist, realized that most of the mass of the atom could be accounted for by the positive (proton) portion. He proposed a jellylike atom with the small, negative electrons embedded in the relatively large proton mass. But from 1906 to 1909, a series of experiments directed by Ernest Rutherford, a New Zealand physicist working in Manchester, England, provided an entirely different picture of the atom.

1-1 Rutherford's Experiments and a Model For Atomic Structure

In 1906, Rutherford found that when a thin sheet of metal foil is bombarded with alpha (α) particles (He^{2+} ions), most of the particles penetrate the metal and suffer only small deflections from their original flight path. In 1909, at Rutherford's suggestion, Hans Geiger and Ernest Marsden performed an experiment to see if any α particles were deflected at a large angle on striking a gold foil. A diagram of their experiment is shown in Figure 1-1. They discovered that some of the α particles actually were deflected by as much as 90° and a few by even larger angles. They concluded:

> If the high velocity and mass of the α particle be taken into account, it seems surprising that some of the α particles, as the experiment shows, can be turned within a layer of 6×10^{-5} cm of gold through an angle of

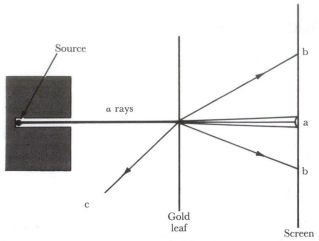

Figure 1-1 The experimental arrangement for Rutherford's measurement of the scattering of α particles by very thin gold foils. The source of the α particles was radioactive radium, which was encased in a lead block to protect the surroundings from radiation and to confine the α particles to a beam. The gold foil used was about 6×10^{-5} cm thick. Most of the α particles passed through the gold leaf with little or no deflection, at a. A few were deflected at wide angles, b, and occasionally a particle rebounded from the foil, c, and was detected by a screen or counter placed on the same side of the foil as the source.

90°, and even more. To produce a similar effect by a magnetic field, the enormous field of 10^9 absolute units would be required.

Rutherford quickly provided an explanation for this startling experimental result. He suggested that atoms consist of a positively charged nucleus surrounded by a system of electrons, and that the atom is held together by electrostatic forces. The effective volume of the nucleus is extremely small compared with the effective volume of the atom, and almost all of the mass of the atom is concentrated in the nucleus. Rutherford reasoned that most of the α particles passed through the metal foil because the metal atom is mainly empty space, and that occasionally a particle passed close to the positively charged nucleus, thereby being severely deflected because of the strong coulombic (electrostatic) repulsive force. The Rutherford model of an atom is shown in Figure 1-2, as is Thomson's earlier suggestion.

1-2 Atomic Number and Atomic Mass

Using the measured angles of deflection and the assumption that an α particle and a nucleus repel each other according to Coulomb's law of

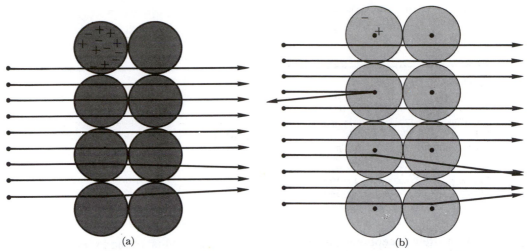

(a)   (b)

Figure 1-2 The expected outcome of the Rutherford scattering experiment, if one assumes (a) the Thomson model of the atom, and (b) the model deduced by Rutherford. In the Thomson model, mass is spread throughout the atom, and the negative electrons are embedded uniformly in the positive mass. There would be little deflection of the beam of positively charged α particles. In the Rutherford model all of the positive charge and virtually all of the mass is concentrated in a very small nucleus. Most α particles would pass through undeflected. But close approach to a nucleus would produce a strong swerve in the path of the α particle, and head-on collision would lead to its rebound in the direction from which it came.

electrostatics, Rutherford was able to calculate the nuclear charge (Z) of the atoms of the elements used as foils. In a neutral atom the number of electrons equals the positive charge on the nucleus. This number is different for each element and is known as its *atomic number.*

The mass of an atomic nucleus is not determined entirely by the number of protons in it. All nuclei, except for the lightest isotope of hydrogen, contain electrically neutral particles called neutrons. The mass of an isolated neutron is only slightly different from the mass of an isolated proton. The values of mass and charge for some important particles are given in Table 1-1.

1-3 Nuclear Structure

Rutherford's basic idea that an atom contains a dense, positively charged nucleus now is accepted universally. On the average, the electrons are far away from the nucleus, which accounts for an atom's relatively large effective diameter of about 10^{-8} cm. In contrast, the nucleus has a diameter of only about 10^{-13} cm.

TABLE 1-1. SOME LOW-MASS PARTICLES

Particle	Symbols	Mass (amu)*	Charge (esu × 10¹⁰)†	Relative Charge
Electron (β^- particle)	e^-, β^-, $_{-1}^{0}e$	0.00054858026	-4.803242	-1
Proton	p, $_1^1p$, $_1^1H$	1.007276470	-4.803242	$+1$
Neutron	n, $_0^1n$	1.008665012	0	0
α Particle	α, $_2^4He^{2+}$	4.001506191	$+9.606484$	$+2$
Deuteron	d, $_1^2d$, $_1^2p$	2.013553229	$+4.803242$	$+1$

* 1 amu = 1 atomic mass unit = $1.6605655 \times 10^{-24}$ g.
† 1 esu = 1 electrostatic unit = 1 $g^{1/2}$ $cm^{3/2}$ sec^{-1}.

With a few exceptions, nuclei are nearly spherical. Most of the exceptions occur in the rare-earth elements ($Z = 57$ to $Z = 71$), in which the shape is slightly ellipsoidal.

The distribution of mass and charge within the nucleus can be determined by various types of scattering experiments. In these experiments high energy particles (e.g., electrons, protons, neutrons, or α particles) are fired at nuclei. The interactions of these particles with nuclei reveal that all nuclei have approximately constant density. In extremely heavy nuclei there is evidence that the mutual electrostatic repulsion of the protons leads to a slightly lower central density. These experiments are only the beginning of the investigation of the internal structure of atomic nuclei.

1-4 Bohr Theory of the Hydrogen Atom

Immediately after Rutherford presented his nuclear model of the atom, scientists turned to the question of the possible structure of the satellite electrons. The first major advance was accomplished in 1913 by a Danish theoretical physicist, Niels Bohr. Bohr pictured the electron in a hydrogen atom as moving in a circular orbit around the proton. Bohr's model is shown in Figure 1-3, in which m_e represents the mass of the electron, m_n the mass of the nucleus, r the radius of the circular orbit, and v the linear velocity of the electron.

For a stable orbit to exist, the outward force exerted by the moving electron trying to escape its circular orbit must be opposed exactly by the forces of attraction between the electron and the nucleus. The outward force, F_0, is expressed as

$$F_0 = \frac{m_e v^2}{r}$$

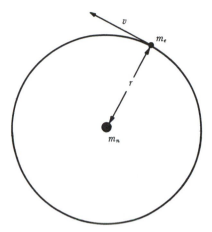

Figure 1-3 Bohr's picture of the hydrogen atom. A single electron of mass m_e moves in a circular orbit with velocity v at a distance r from a nucleus of mass m_n.

This force is opposed exactly by the sum of the two attractive forces that keep the electron in orbit—the electrostatic force of attraction between the proton and the electron, plus the gravitational force of attraction. Since the electrostatic force is much stronger than the gravitational force, we may neglect the latter. The electrostatic attractive force, F_e, between an electron of charge $-e$ and a proton of charge $+e$, is

$$F_e = -\frac{e^2}{r^2}$$

The condition for a stable orbit is that $F_0 + F_e$ equals zero:

$$\frac{m_e v^2}{r} - \frac{e^2}{r^2} = 0 \qquad \text{or} \qquad \frac{m_e v^2}{r} = \frac{e^2}{r^2} \tag{1-1}$$

Now we are able to calculate the energy of an electron moving in one of the Bohr orbits. The total energy, E, is the sum of the kinetic energy, KE, and the potential energy, PE:

$$E = KE + PE$$

in which KE is the energy due to motion,

$$KE = \tfrac{1}{2} m_e v^2$$

and PE is the energy due to electrostatic attraction,

$$PE = -\frac{e^2}{r}$$

Thus the total energy is

$$E = \tfrac{1}{2}m_e v^2 - \frac{e^2}{r} \tag{1-2}$$

However, Equation 1-1 can be written $m_e v^2 = e^2/r$, and if e^2/r is substituted for $m_e v^2$ into Equation 1-2, we have

$$E = \frac{e^2}{2r} - \frac{e^2}{r} = -\frac{e^2}{2r} \tag{1-3}$$

Now we only need to specify the orbit radius, r, before we can calculate the electron's energy. However, according to Equation 1-3, atomic hydrogen should release energy continuously as r becomes smaller. To keep the electron from falling into the nucleus Bohr proposed a model in which the angular momentum of the orbiting electron could have only certain values. The result of this restriction is that only certain electron orbits are possible. According to Bohr's postulate, the quantum unit of angular momentum is $h/2\pi$, in which h is the constant in Planck's famous equation, $E = h\nu$. (E is energy in ergs, $h = 6.626176 \times 10^{-27}$ erg sec, and ν is frequency in sec^{-1}). In mathematical terms Bohr's assumption was that

$$m_e vr = n\frac{h}{2\pi} \tag{1-4}$$

in which $n = 1, 2, 3, \ldots$ (all integers to ∞). Solving for v in Equation 1-4 we can write

$$v = n\left(\frac{h}{2\pi}\right)\frac{1}{m_e r} \tag{1-5}$$

Substituting the value of v from Equation 1-5 into the condition for a stable orbit (Equation 1-1) we obtain

$$\frac{m_e n^2 h^2}{4\pi^2 m_e^2 r^2} = \frac{e^2}{r}$$

or

$$r = \frac{n^2 h^2}{4\pi^2 m_e e^2} \tag{1-6}$$

Equation 1-6 gives the radius of the possible electron orbits for the hydrogen atom in terms of the quantum number n. The energy associated with each possible orbit now can be calculated by substituting the value of r from Equation 1-6 into the energy expression (Equation 1-3), which gives

$$E_n = -\frac{2\pi^2 m_e e^4}{n^2 h^2} = -\frac{k}{n^2} \qquad (1\text{-}7)$$

We have included the subscript n to indicate that the only variable in the energy expression is the quantum number n.

Exercise 1-1. Calculate the radius of the first Bohr orbit.

Solution. The radius of the first orbit can be obtained directly from Equation 1-6:

$$r = \frac{n^2 h^2}{4\pi^2 m_e e^2}$$

Assuming $n = 1$ and inserting the values of the constants from Appendix A we obtain

$$r = \frac{(1)^2 (6.626176 \times 10^{-27} \text{ erg sec})^2}{4(3.141593)^2 (9.109534 \times 10^{-28} \text{ g})(4.803242 \times 10^{-10} \text{ esu})^2}$$
$$= 0.529177 \times 10^{-8} \text{ cm} \approx 0.529 \text{ Å}$$

The Bohr radius for $n = 1$ is designated a_0.

Exercise 1-2. Calculate the velocity of an electron in the first Bohr orbit of a hydrogen atom.

Solution. From Equation 1-5,

$$v = n\left(\frac{h}{2\pi}\right)\frac{1}{m_e r}$$

Assuming $n = 1$ and substituting for $r = a_0 = 0.529177 \times 10^{-8}$ cm, we obtain

$$v = (1)(1.0545887 \times 10^{-27} \text{ erg sec})$$
$$\times \frac{1}{(9.109534 \times 10^{-28} \text{ g})(0.529177 \times 10^{-8} \text{ cm})}$$
$$= 2.18769 \times 10^8 \text{ cm sec}^{-1}$$

Exercise 1-3. Calculate the value of k in Equation 1-7 in (a) ergs, (b) reciprocal centimeters (cm^{-1}), and (c) electron volts (eV).

Solution. From Equation 1-7,

$$k = 2\pi^2 m_e e^4 / h^2$$

(a) Using the values for the constants found in Appendix A,

$$k = \frac{2(3.141593)^2 (9.109534 \times 10^{-28})(4.80324 \times 10^{-10})^4}{(6.626176 \times 10^{-27})^2}$$

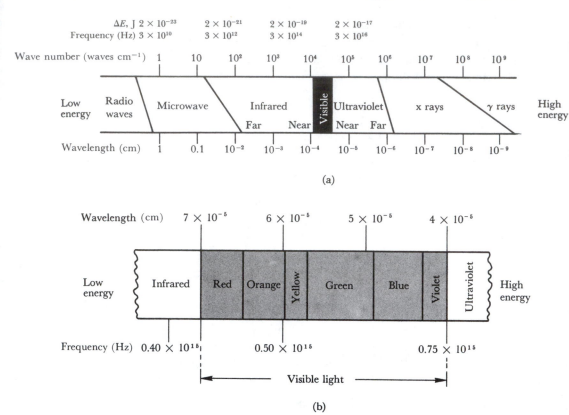

Figure 1-4 The spectrum of electromagnetic radiation. (a) The visible region is only a small part of the entire spectrum; (b) expanded view of visible region.

$$= 2.1799 \times 10^{-11} \text{ erg}$$

(b) $k = (2.1799 \times 10^{-11} \text{ erg})(8065.5 \text{ cm}^{-1}/1.6022 \times 10^{-12} \text{ erg})$
$$= 109{,}737 \text{ cm}^{-1}$$

(c) $k = (2.1799 \times 10^{-11} \text{ erg})(1 \text{ eV}/1.6022 \times 10^{-12} \text{ erg})$
$$= 13.606 \text{ eV}$$

1-5 Absorption and Emission Spectra of Atomic Hydrogen

All atoms and molecules absorb light of certain characteristic frequencies. The pattern of absorption frequencies is called an *absorption spectrum* and is an identifying property of any atom or molecule. Because frequency is directly proportional to energy $(E = h\nu)$, the absorption spectrum of an atom shows that the electron can have only certain energy values, as Bohr proposed.

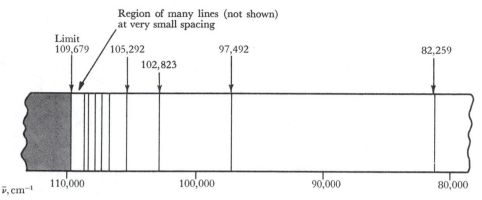

Figure 1-5 The electromagnetic absorption spectrum of hydrogen atoms. The lines in this spectrum represent ultraviolet radiation that is absorbed by hydrogen atoms as a mixture of all wavelengths is passed through a gas sample.

It is common practice to express positions of absorption in terms of the *wave number, $\bar{\nu}$,* which is the reciprocal of the wavelength, λ:*

$$\bar{\nu} = \frac{1}{\lambda}$$

Because λ is related to frequency (ν) by the relationship $\lambda\nu = c$ ($c =$ velocity of light $= 2.9979 \times 10^{10}$ cm sec^{-1}), we also have $\bar{\nu} = \nu/c$ and $E = h\bar{\nu}c$. Thus wave number and energy are directly proportional. If ν is in sec^{-1} units and c is in cm sec^{-1}, ν is expressed in (cm^{-1}) units. The spectrum of electromagnetic radiation is shown in Figure 1-4.

The absorption spectrum of hydrogen atoms is shown in Figure 1-5. The lowest-energy absorption corresponds to the line at 82,259 cm^{-1}. Notice that the absorption lines are crowded closer together as the limit of 109,679 cm^{-1} is approached. Above this limit absorption is continuous.

If atoms and molecules are heated to high temperatures, they emit light of certain frequencies. For example, hydrogen atoms emit red light when heated. An atom that possesses excess energy (an "excited" atom) emits light in a pattern known as its *emission spectrum.* A portion of the emission spectrum of atomic hydrogen is shown in Figure 1-6. The emission spectrum contains more lines than the absorption spectrum. The lines in the emission spectrum at 82,259 cm^{-1} and above occur at the same positions as the lines in the absorption spectrum, but the emission lines below 82,259 cm^{-1} do not appear in the absorption spectrum.

If we look more closely at the emission spectrum in Figure 1-6, we see that there are three distinct groups of lines. These three groups, or series, are named after the scientists who discovered them. The series that starts

*A complete list of Greek symbols appears in Appendix B.

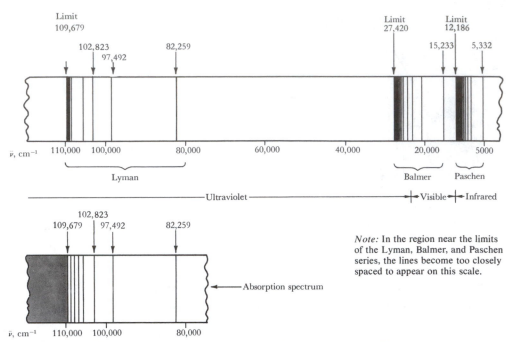

Figure 1-6 The emission spectrum from heated hydrogen atoms. The emission lines occur in series named for their discoverers, Lyman, Balmer, and Paschen. The Brackett and Pfund series are farther to the right in the infrared region. The lines become more closely spaced to the left in each series until they finally merge at the series limit.

at 82,259 cm⁻¹ and continues to 109,679 cm⁻¹ is called the *Lyman series* and is in the ultraviolet portion of the spectrum. The series that starts at 15,233 cm⁻¹ and continues to 27,420 cm⁻¹ is called the *Balmer series* and covers a small part of the ultraviolet and a large portion of the visible spectrum. The lines between 5,332 cm⁻¹ and 12,186 cm⁻¹ are called the *Paschen series* and fall in the near-infrared spectral region.

Although the emission spectrum of hydrogen appears to be complicated, Johannes Rydberg formulated a fairly simple mathematical expression that gives all the line positions. This expression, known as the *Rydberg equation,* is

$$\bar{\nu}_H = R_H\left(\frac{1}{n^2} - \frac{1}{m^2}\right)$$

In the Rydberg equation n and m are integers, with m greater than n; R_H is called the *Rydberg constant* and is known accurately from experiment to be 109,678.764 cm⁻¹.

Exercise 1-4. Calculate $\bar{\nu}_H$ for the lines with $n = 1$ and $m = 2, 3,$ and 4.

Solution.

$n = 1$, $m = 2$ line:
$$\bar{\nu}_H = 109{,}679\left(\frac{1}{1^2} - \frac{1}{2^2}\right) = 109{,}679\left(1 - \frac{1}{4}\right) = 82{,}259 \text{ cm}^{-1}$$

$n = 1$, $m = 3$ line:
$$\bar{\nu}_H = 109{,}679\left(\frac{1}{1^2} - \frac{1}{3^2}\right) = 109{,}679\left(1 - \frac{1}{9}\right) = 97{,}492 \text{ cm}^{-1}$$

$n = 1$, $m = 4$ line:
$$\bar{\nu}_H = 109{,}679\left(\frac{1}{1^2} - \frac{1}{4^2}\right) = 109{,}679\left(1 - \frac{1}{16}\right) = 102{,}824 \text{ cm}^{-1}$$

We see that the preceding wave numbers correspond to the first three lines in the Lyman series. Thus we can expect the Lyman series to correspond to lines calculated with $n = 1$ and $m = 2, 3, 4, 5, \ldots$ Let us check this by calculating the wave number for the line with $n = 1$, $m = \infty$:

$n = 1$, $m = \infty$ line:
$$\bar{\nu} = 109{,}679(1 - 0) = 109{,}679 \text{ cm}^{-1}$$

The wave number 109,679 cm^{-1} corresponds to the highest emission line in the Lyman series.

The wave number calculated for $n = 2$ and $m = 3$ is

$$\bar{\nu} = 109{,}679(\tfrac{1}{4} - \tfrac{1}{9}) = 15{,}233 \text{ cm}^{-1}$$

This corresponds to the first line in the Balmer series. Thus the Balmer series corresponds to the $n = 2$, $m = 3, 4, 5, 6, \ldots$, lines. You probably would expect the lines in the Paschen series to correspond to $n = 3$, $m = 4$, 5, 6, 7, 8, . . . , and they do. Now you should wonder where the lines are with $n = 4$, $m = 5, 6, 7, 8, \ldots$, and $n = 5$, $m = 6, 7, 8, 9, \ldots$. They are exactly where the Rydberg equation predicts they should be. The $n = 4$ series was discovered by Frederick Brackett and the $n = 5$ series was discovered by Pfund. The series with $n = 6$ and higher are located at very low frequencies and are not given special names.

The Bohr theory provided an explanation for the spectral absorption and emission lines in atomic hydrogen in terms of the orbits and energy levels shown in Figure 1-7. The orbit with $n = 1$ has the lowest energy; the one electron in atomic hydrogen occupies this level in its most stable state. The most stable electronic state of an atom or molecule is called the *ground state*. If the electron in a hydrogen atom can move only in certain orbits, it is easy to see why the atom absorbs or emits light only at specific wave

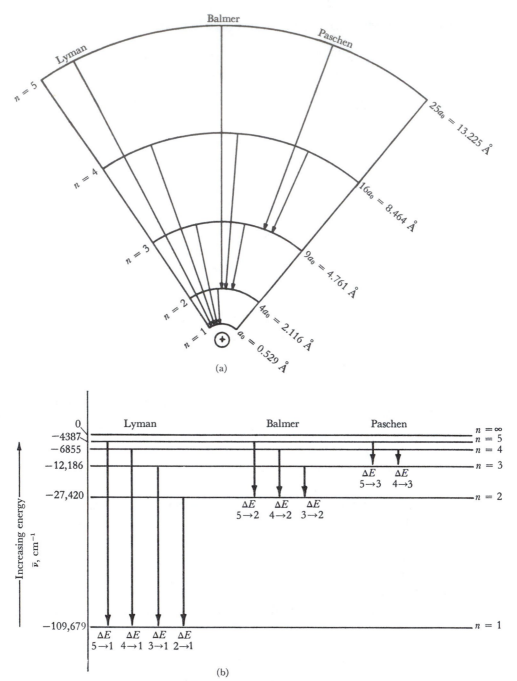

Figure 1-7 Orbits and energy levels of the hydrogen atom. (a) Hydrogen-atom orbits. Each arc represents a portion of the electron's circular path around the positive nucleus. Series for most energetic (excited) electrons dropping from outer levels to various inner levels are shown. (b) The energy changes for electrons dropping from excited energy states to less energetic levels. ΔE is determined by initial and final energies, which in turn are determined by the principal quantum number of the orbit, n.

numbers. The absorption of light energy allows the electron to jump to a higher orbit. An excited hydrogen atom, in which the electron is not in the lowest-energy orbit, emits energy in the form of light when the electron falls back into a lower-energy orbit. The following series of spectral lines are a consequence of electronic transitions:

1. The Lyman series of lines arises from transitions from the $n = 2, 3, 4, \ldots$ levels into the $n = 1$ orbit.

2. The Balmer series of lines arises from transitions from the $n = 3, 4, 5, \ldots$ levels into the $n = 2$ orbit.

3. The Paschen series of lines arises from transitions from the $n = 4, 5, 6, \ldots$ levels into the $n = 3$ orbit.

The Bohr theory also explains why there are more emission lines than absorption lines. Excited hydrogen atoms may have $n = 2, 3, 4, 5, \ldots$. Any transition to a lower level is accompanied by emission of a quantum unit of light of energy $h\nu$. Such a light quantum is called a *photon*. Since transitions are possible to all orbits below the occupied orbit in the excited atom, an excited hydrogen atom with an electron in the $n = 6$ orbit, for example, may "decay" to a less excited state by the transitions $n = 6 \rightarrow n = 5$, $n = 6 \rightarrow n = 4$, $n = 6 \rightarrow n = 3$, and $n = 6 \rightarrow n = 2$, or it may decay to the ground state by the transition $n = 6 \rightarrow n = 1$. In other words, for the excited hydrogen atom with an $n = 6$ electron, there are five possible modes of decay, each with a finite probability of occurrence. Thus five emission lines result which correspond to five emitted photon energies ($h\nu$). The lines in the absorption spectrum of hydrogen are due to transitions from the ground state ($n = 1$) orbit to higher orbits. Because almost no hydrogen atoms have n greater than one under typical absorption conditions, only the transitions $n = 1 \rightarrow 2$, $1 \rightarrow 3$, $1 \rightarrow 4$, $1 \rightarrow 5$, $\ldots$, and $1 \rightarrow \infty$ are observed. These are the absorption lines that correspond to the series of emission lines (the Lyman series) in which the excited states decay in one transition to the ground state.

Now we are able to derive the Rydberg equation from the Bohr theory. The transition energy (ΔE_H) of any electron jump in the hydrogen atom is the energy difference between an initial state, I, and a final state, II. That is,

$$\Delta E_H = E_{n_{II}} - E_{n_I}$$

In an absorption experiment $E_{n_{II}} > E_{n_I}$ and ΔE_H is positive. However, for emission we have $E_{n_{II}} < E_{n_I}$, because in the final state the electron is in a more stable orbit, and therefore ΔE_H is negative.

From Equation 1-7 the expression for the transition energy (whether for absorption or emission) becomes

$$\Delta E_H = -\frac{2\pi^2 m_e e^4}{n_{II}^2 h^2} - \left(-\frac{2\pi^2 m_e e^4}{n_I^2 h^2}\right)$$

or

$$\Delta E_H = \frac{2\pi^2 m_e e^4}{h^2}\left(\frac{1}{n_I^2} - \frac{1}{n_{II}^2}\right)$$

From conservation of energy we can write

$$\text{Absorption: } E_{n_I} + E_P = E_{n_{II}} \qquad \Delta E_H = E_P$$
$$\text{Emission: } E_{n_I} = E_{n_{II}} + E_P \qquad \Delta E_H = -E_P$$

In these equations $E_P = h\bar{v}c$ is the energy of the photon, whether absorbed or emitted. To compare directly with experiment, we wish to calculate $\bar{v}_H$. For absorption we have

$$\bar{v}_H = \frac{\Delta E_H}{hc} = \frac{2\pi^2 m_e e^4}{ch^3}\left(\frac{1}{n_I^2} - \frac{1}{n_{II}^2}\right) = R\left(\frac{1}{n_I^2} - \frac{1}{n_{II}^2}\right) \qquad (1\text{-}8)$$

in which $n_I < n_{II}$.

Equation 1-8 is equivalent to the Rydberg equation, with $n_I = n$, $n_{II} = m$, and $R = (2\pi^2 m_e e^4)/ch^3$. In the emission experiment we obtain

$$\bar{v}_H = -\frac{\Delta E_H}{hc} = -\frac{2\pi^2 m_e e^4}{n^2 h^2}\left(\frac{1}{n_I^2} - \frac{1}{n_{II}^2}\right) = R\left(\frac{1}{n_{II}^2} - \frac{1}{n_I^2}\right)$$

in which $n_I > n_{II}$. We can ensure that the calculated $\bar{v}_H$ will always be positive (as it must), no matter whether the transition is an absorption or an emission, by taking the absolute value

$$\bar{v}_H = R\left|\left(\frac{1}{n_I^2} - \frac{1}{n_{II}^2}\right)\right|$$

If we use the value of 9.109534×10^{-28} g for the rest mass of the electron, the Bohr theory value of the Rydberg constant is

$$R = \frac{2\pi^2 m_e e^4}{ch^3}$$
$$= \frac{2(3.141593)^2(9.109534 \times 10^{-28})(4.803242 \times 10^{-10})^4}{(2.997925 \times 10^{10})(6.626176 \times 10^{-27})^3}$$
$$= 109{,}737.3 \text{ cm}^{-1}$$

The value of R calculated from the Bohr theory assumed that the nucleus was stationary with respect to the orbiting electron. A careful analysis of the planetary motion that takes into account the slight orbiting effect of the nucleus reveals that the mass of the electron in the expression for R should be replaced by the *reduced mass*, μ_H, of the atom. The reduced mass is defined by the equation

$$\mu_H = \frac{m_e m_p}{(m_e + m_p)}$$

where m_p refers in general to the mass of the nucleus which for hydrogen consists only of a proton. Notice that for $m_p \gg m_e$ the reduced mass approaches m_e. The value of μ_H can be calculated from the data in Table 1-1; it is found to be $0.9994558 m_e$. The "corrected" theoretical value of R, which we now label R_H, is therefore $(109,737.3 \text{ cm}^{-1})(0.9994558) = 109,677.6 \text{ cm}^{-1}$. Recall that the experimental value of R_H is $109,678.764 \text{ cm}^{-1}$. This remarkable agreement between theory and experiment was a great triumph for the Bohr theory.

The previously calculated value of $R = 109,737.3 \text{ cm}^{-1}$ assumed that $\mu_H = m_e$, which can only occur for a nucleus of infinite mass. Consequently, this value of R is often referred to as R_∞.

1-6 Ionization Energy of Atomic Hydrogen

When light (photons) of energy $E = h\nu$ impinges on gaseous atomic hydrogen, it can cause the electron to be excited from the $n = 1$ orbit to some less stable orbit with $n > 1$. If the photon imparts sufficient energy to the atom, the electron will escape entirely from the proton. That is, it will be excited to the $n = \infty$ orbit with radius $r = \infty$ and $E_H = 0$. This process is called ionization. For the hydrogen atom the ionization process is

$$H(g) + (E = h\nu) \rightarrow H^+(g) + e^- (\tfrac{1}{2} m_e v^2)$$
$$\Delta E_H = E_{n_{II}} - E_{n_I} \tag{1-9}$$

This formulation shows that the ejected electron can have a kinetic energy $\tfrac{1}{2} m_e v^2$.

The *ionization energy, IE,* of an atom or molecule is the minimum energy required to remove an electron from the gaseous atom or molecule in its ground state, the ejected electron having zero kinetic energy and being infinitely distant from the proton. In 1905, Albert Einstein formulated a

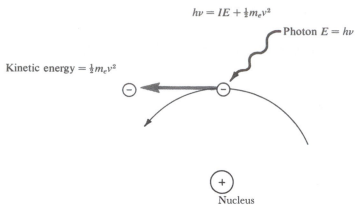

$$hv = IE + \tfrac{1}{2}m_e v^2$$

Photon $E = hv$

Kinetic energy $= \tfrac{1}{2}m_e v^2$

Nucleus

Figure 1-8 If a photon of sufficient energy impinges on the hydrogen atom, the result is ionization. The energetics of the ionization are governed by the Einstein photo-electric effect, which states that any photon energy not required to carry out the ionization process, *IE*, will be manifested as kinetic energy of the ejected electron $(\tfrac{1}{2}m_e v^2)$.

quantitative relationship between the photon energy, ionization energy, and kinetic energy of the ejected electron:

$$hv = IE + \tfrac{1}{2}m_e v^2 \tag{1-10}$$

Equation 1-10, known as the *Einstein photoelectric law,* shows that any photon energy in excess of that required to effect the ionization will be manifested as the kinetic energy of the ejected electron. This concept is illustrated in Figure 1-8. When ionization occurs as a result of photon impingement, the ionization process is called *photoionization* and the ejected electrons are called *photoelectrons.*

To calculate ionization energy for the hydrogen atom we first must recognize that when the hydrogen atom is ionized, its change in energy is simply the ionization energy, $\Delta E_{\mathrm{H}} = IE$. Then utilizing Equation 1-9 we obtain

$$\Delta E_{\mathrm{H}} = IE = E_{n_{\mathrm{II}}} - E_{n_{\mathrm{I}}}$$

For the ground state, $n_1 = 1$, and for the state in which the electron is removed completely from the atom, $n_{\mathrm{II}} = \infty$. Since $E_{\infty} = 0$ (Equation 1-7) we obtain

$$IE_{n_{\mathrm{I}}} = -E_{n_{\mathrm{I}}} \tag{1-11}$$

in which we have subscripted the ionization energy to indicate that the ionization is taking place from the energy state with $n = n_I$. Therefore taking into account the reduced mass effect,

$$IE_{n_I} = \frac{2\pi^2 \mu_H e^4}{h^2} = R_H hc$$
$$= (109{,}677.6)\,(6.626176 \times 10^{-27})\,(2.997925 \times 10^{10})$$
$$= 2.178721 \times 10^{-11} \text{ erg}$$

Ionization energies usually are expressed in electron volts. Since 1 eV $= 8065.479$ cm^{-1} (see Appendix A), we calculate

$$IE = (109{,}677.6 \text{ cm}^{-1})\,(1 \text{ eV}/8065.479 \text{ cm}^{-1}) = 13.5984 \text{ eV}$$

Thus, the calculated energy required to ionize one hydrogen atom is

$$IE = 13.5984 \text{ eV} = 109{,}677.6 \text{ cm}^{-1}$$

The experimental value of the ionization energy corresponds exactly to the Lyman series emission line $n = \infty \rightarrow n = 1$, which we stated previously to be 109,678.764 cm^{-1}. Using the conversion factor found in Appendix A, this corresponds to 13.5984 eV.

Equation 1-11 is easy to understand, because in order to remove an electron from its orbit, we must add sufficient energy to overcome the binding energy of the electron. This concept is illustrated in Figure 1-9.

1-7 General Bohr Theory for a One-Electron Atom

The problem of one electron moving around any nucleus of charge $+Z$ is very similar to the hydrogen-atom problem. Because the attractive force is $-Ze^2/r^2$ the condition for a stable orbit is

$$\frac{m_e v^2}{r} = \frac{Ze^2}{r^2}$$

Proceeding from this condition in the same way as with the hydrogen atom we find

$$r = \frac{n^2 h^2}{4\pi^2 m_e Z e^2}$$

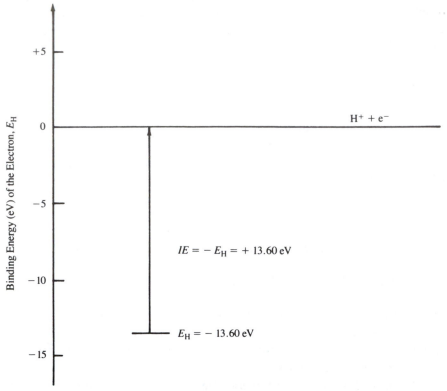

Figure 1-9 Illustration of the relationship between E_H and IE for the hydrogen atom. If the binding energy of the electron (E_H) is -13.60 eV, then the ionization energy (IE) must be $+13.60$ eV.

and

$$E = -\frac{2\pi^2 m_e Z^2 e^4}{n^2 h^2} = -\frac{kZ^2}{n^2}$$

Thus for the general case of nuclear charge $+Z$, for a transition from n_I to n_{II} we have

$$\Delta E = \frac{2\pi^2 m_e Z^2 e^4}{h^2}\left(\frac{1}{n_I^2} - \frac{1}{n_{II}^2}\right)$$

or simply

$$\Delta E = Z^2 \Delta E_H$$

The wave numbers of absorption and emission lines may be calculated from the formula

$$\bar{v} = Z^2 R_\infty \left| \left(\frac{1}{n_{\text{I}}^2} - \frac{1}{n_{\text{II}}^2} \right) \right|$$

In this equation we have ignored the small correction due to the reduced mass and simply employed R_∞. Recall from Equation 1-7 and Exercise 1-3 that a very useful expression for E is $E = -k/n^2$. For hydrogenlike atoms this becomes $E = -kZ^2/n^2$ in which $k = 2\pi^2 m_e e^4/h^2$.

Exercise 1-5. Calculate the third ionization energy of a lithium atom.

Solution. A lithium atom is composed of a nucleus of charge $+3$ ($Z = 3$) and three electrons. The *first ionization energy*, IE_1, of an atom with more than one electron is the energy required to remove one electron. For lithium

$$\text{Li}(g) \rightarrow \text{Li}^+(g) + e^- \qquad \Delta E = IE_1$$

The energy needed to remove an electron from the unipositive ion, Li^+, is defined as the *second ionization energy*, IE_2, of lithium,

$$\text{Li}^+(g) \rightarrow \text{Li}^{2+}(g) + e^- \qquad \Delta E = IE_2$$

and the *third ionization energy*, IE_3, of lithium is the energy required to remove the one remaining electron from Li^{2+}. For lithium, $Z = 3$ and $IE_3 = (3)^2(13.606 \text{ eV}) = 122.45 \text{ eV}$. The experimental value is also 122.45 eV.

The idea of an electron circling the nucleus in a well-defined orbit, analogous to the moon circling the earth, was easy to grasp and Bohr's theory gained wide acceptance. However, it soon was realized that this simple theory was not the complete description of the atom. One difficulty was the fact that an atom in a magnetic field has a more complicated emission spectrum than the same atom in the absence of a magnetic field. This phenomenon is known as the *Zeeman effect* and it cannot be explained by the simple Bohr theory. However, the German physicist Arnold Sommerfeld was able to rescue the simple theory temporarily by suggesting elliptical orbits, in addition to circular orbits, for the electron. The combined Bohr-Sommerfeld theory explained the Zeeman effect very well.

A more serious problem was the inability of even the Bohr-Sommerfeld theory to account for the spectral details of atoms having several electrons. The theory also failed completely as a means of interpreting the periodic properties of the chemical elements. Thus the "orbit" approach soon was abandoned in favor of a powerful new theory of electronic motion based on the methods of wave mechanics.

1-8 Matter Waves

In 1924, the French physicist Louis de Broglie advanced the hypothesis that all matter possesses wave properties. He postulated that for every moving particle there is an associated wave with a wavelength given by the equation

$$\lambda = \frac{h}{mv} = \frac{h}{p}$$

Mass times velocity, mv, is the momentum, p, of a particle and is a measure of inertia, or the tendency of the particle to remain in motion.

In 1927, Clinton Davisson and Lester Germer demonstrated that electrons are diffracted by crystals very much as x rays are. These electron-diffraction experiments dramatically supported de Broglie's postulate of matter waves. Thus modern experimental evidence indicates that "particle" and "wave" phenomena are not mutually exclusive, rather, they relate to different attributes of all matter. It is reasonable to expect that all things in nature possess both the properties of particles (discrete units) and the properties of waves (continuity). The particle aspect of matter is more important in describing the properties of the relatively large objects encountered in normal observations. Wave properties are more important in describing the characteristics of many of the extremely minute objects outside the realm of ordinary perception.

It is helpful to point out why the characteristics of wave motion are not apparent in the motion of easily observable objects. Everything from a baseball to a battleship has a wavelike nature associated with its movements. However, for relatively large objects the wavelengths are so small that we cannot perceive them. For example, consider a baseball with a mass of 200 g and a speed of 3.0×10^3 cm sec^{-1} From de Broglie's equation the wavelength of the associated wave is $6.63 \times 10^{-27}/(200)(3.0 \times 10^3)$, or approximately 10^{-32} cm. It is apparent that this wavelength is too small for ordinary observation. By contrast, an electron with a rest mass of approximately 10^{-27} g moving at the same speed would have a wavelength of about 2×10^{-3} cm, which is well within the realm of ordinary observation.

De Broglie's hypothesis of matter waves provided the foundation for an entirely new theory, one which firmly established the quantum properties of energy in physical systems. This theory, logically, is called *quantum mechanics*. Much of this book will be devoted to the application of quantum mechanics to the description of the electronic structures of atoms and molecules.

1-9 The Uncertainty Principle

One of the most important consequences of the dual nature of matter is the uncertainty principle, which was derived in 1927 by Werner Heisenberg. The *Heisenberg uncertainty principle* states that it is impossible to know simultaneously both the momentum and the position of a particle with certainty. This means that as the measurement of momentum or velocity of a particle is made more precisely, the measurement of the position of the particle is correspondingly less precise. Similarly, if the position of a particle is known precisely, the momentum must be less well known. Heisenberg showed that the lower limit of this uncertainty is Planck's constant divided by 4π. This relationship is expressed as

$$(\Delta p_x)(\Delta x) \geqq \frac{h}{4\pi} \tag{1-12}$$

in which Δp_x is the uncertainty in the momentum along the x direction and Δx is the uncertainty in the position. Equation 1-12 is only one of several ways of expressing the uncertainty principle. For example, we could write

$$(\Delta p_y)(\Delta y) \geqq \frac{h}{4\pi}$$

and

$$(\Delta p_z)(\Delta z) \geqq \frac{h}{4\pi}$$

The uncertainty principle can be clarified by considering an attempt to measure both the position and the momentum of an electron as sketched in Figure 1-10, where the position of the electron is to be pinpointed by using electromagnetic radiation as the measuring device. A photon must collide with and be reflected from the electron for that electron to be "seen." But since photons and electrons have nearly the same energy, collisions between them result in changes in the electron's velocity and consequently in its momentum.

The position of a particle at any instant can be observed with a precision comparable to that of the wavelength of the incident light. However, during that measurement the light photon interacting with the electron causes an alteration of the electron's momentum of the same magnitude as the momentum of the photon itself. To increase the precision in detecting the position

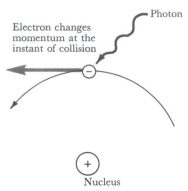

Figure 1-10 The position of an electron at an instant of time should be determinable by a "super microscope" with light of small wavelength, λ (x rays or γ rays). However, light photons of small λ have great energies and therefore very large momenta. A collision of one of these photons with an electron instantly changes the electronic momentum. Thus, as the position of the electron is better resolved, its momentum becomes more and more uncertain.

of the electron, one must decrease the wavelength of the incident light. But as the wavelength decreases the frequency of the light and the energy of the corresponding photons increase, thereby increasing the uncertainty in the momentum of the electron. Therefore uncertainty is inherent in the very nature of the measuring process.

Consider an electron moving with a velocity of 10^7 cm sec^{-1}. Suppose that an attempt to measure its position is made with visible light of frequency 10^{15} sec^{-1}. The energy of the electron is simply its kinetic energy, or energy of motion, which is $\frac{1}{2}mv^2$ or approximately 10^{-12} erg. The energy of light photons is given by $E = h\nu$. From this equation, with visible light of frequency 10^{15} sec^{-1}, the energy of a photon also is about 10^{-12} erg. Therefore, in a measurement collison the momentum of the electron will be changed instantaneously, with a correspondingly large uncertainty in the momentum at that instant.

The uncertainty principle need not be considered in measurements of the momentum and position of large bodies, because the measurements are not sufficiently precise to reveal any inherent uncertainty. Only for small particles does the uncertainty principle become important. The following comparison of a baseball and an electron illustrates this.

Recall the baseball weighing 200 g and traveling with a velocity of 3.0 $\times$ 10^3 cm sec^{-1}, mentioned in Section 1-8. This baseball has a momentum of 6.0×10^5 g cm sec^{-1}. Consider attempts to measure both the position and the momentum of the baseball. Suppose that it is possible to measure the momentum with a precision of one part in a trillion (10^{12}). This would mean that the momentum is known with an uncertainty no greater than 6.0×10^{-7} g cm sec^{-1}. The uncertainty principle asserts that the position then can be known with a precision of approximately 10^{-21} cm, which is much greater precision than typical physical measurements can achieve.

Now consider the electron with a mass of 10^{-27} g moving with the same velocity as the baseball. The electron would have a momentum of 3.0×10^{-24} g cm sec^{-1}. If it were possible to measure this momentum with the same precision as that for the baseball, Δp would be 3.0×10^{-36} g cm sec^{-1}, and the uncertainty in position, Δx, would be close to a billion centimeters. From these examples it is clear that the uncertainty principle is important only in considering measurements of the small particles that comprise an atomic system.

1-10 Atomic Orbitals

If electrons moved in simple orbits, then the momentum and position of an electron could be determined exactly at any instant. According to the uncertainty principle this situation does not really exist. Therefore in discussing the motion of an electron of known energy or momentum around a nucleus, it is necessary to speak only in terms of the probability of finding that electron at any particular position.

In view of the idea of the dual nature of matter we consider the motion of an electron as a wave. According to de Broglie's equation, the velocity (and consequently the energy) of the electron determines the wavelength and therefore the frequency of the associated wave. The amplitude of the wave in any region indicates the relative probability of finding the electron in that region.

To clarify the concept of electron probability it is helpful to do a hypothetical experiment in which we take a set of instantaneous pictures of an electron with a specific energy moving around a nucleus. If all these imaginary sequential pictures were superimposed, with the electron appearing as a small dot in each picture, a "cloud" similar to that shown in Figure 1-11(a) would result. This picture is called an electron-density or electron-cloud representation. The density of dots in a given spatial region is a pictorial representation of the *probability density* in that region.

If the electron cloud in Figure 1-11(a) is examined, the probability density is seen to be great near the nucleus and to decrease as the distance from the nucleus increases. Consequently an atom cannot be given a definite radius; rather, it is characterized by a fuzzy electron cloud possessing no definite boundaries. However, electronic motion with a poorly defined boundary is

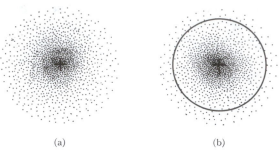

(a)   (b)

Figure 1-11 (a) Electron-density or electron-cloud representation of the motion of an electron around a positive nucleus. (b) Circular cross section of a spherical boundary surface enclosing 90–99 percent probability region of a $1s$ orbital. If an experimenter could make a large number of measurements of the position of the electron, 90–99 percent of the time he would find the electron within the sphere enclosed by the boundary surface.

unwieldy for efficient pictorial representation. Thus, as an arbitrary boundary for the electronic motion, it is convenient to set a volume with a surface along which the probability density is some constant value. This is called a *boundary surface of constant probability density*. It ensures that the shape of electron density will be represented accurately. It is common to define this surface as enclosing about 90–99 percent of the electron density.

For the lowest-energy orbital of the hydrogen atom the boundary surface is a sphere whose diameter is approximately 10^{-8} cm. In Figure 1-11(b) a circular cross-section of the boundary surface is superimposed over the electron-density picture. In place of the model using planetary orbits, the probability-density representation decribes the motion of the electron as being concentrated in a certain region with a certain shape around the nucleus. We will associate the name *electron orbital* with this type of model of electron motion. An orbital can be represented by a mathematical function, as well as pictorially by one of the probability-density representations that we have discussed.

In the following sections we introduce the mathematical equation that can be used to solve any quantum mechanical problem, including that of representing the electron orbitals discussed here. In the next section we apply the equation to the simple "particle-in-a-box" problem, then in Section 1-12 we return to a discussion of the hydrogen atom.

1-11 The Wave Equation and the Particle-in-a-Box Problem

In 1926, Erwin Schrödinger advanced the famous wave equation that relates the energy of a system to its wave properties. Because its application to the hydrogen atom is rather complicated, we shall first use the wave equation to solve the particle-in-a-box problem.

The Schrödinger Wave Equation

The Schrödinger equation expressed in three dimensions is

$$\frac{\partial^2 \psi}{\partial x^2} + \frac{\partial^2 \psi}{\partial y^2} + \frac{\partial^2 \psi}{\partial z^2} + \frac{8\pi^2 m}{h^2} (E - V)\psi = 0 \tag{1-13}$$

in which ψ = wave function or *eigenfunction*
x, y, z = coordinates in space
m = mass
h = Planck's constant
E = total energy or *eigenvalue*
V = potential energy

We can simplify Equation 1-13 by introducing the notation ∇^2 for the differential

$$\nabla^2 = \frac{\partial^2}{\partial x^2} + \frac{\partial^2}{\partial y^2} + \frac{\partial^2}{\partial z^2}$$

Thus Equation 1-13 can be rewritten as follows:

$$\nabla^2 \psi + \frac{8\pi^2 m}{h^2} (E - V)\psi = 0$$

This can be rearranged to give

$$\frac{-h^2}{8\pi^2 m} (\nabla^2 \psi) + V\psi = E\psi \tag{1-14}$$

It is convenient to introduce the *Hamiltonian operator, $\mathscr{H}$*, defined as

$$\mathscr{H} = \frac{-h^2}{8\pi^2 m} \nabla^2 + V$$

which permits us to rewrite Equation 1-14 as

$$\mathscr{H}\psi = E\psi$$

which is merely a shorthand version of the Schrödinger equation (1-13).

According to the de Broglie hypothesis any moving particle has an associated wavelength. Therefore, any moving particle can be described as a wave function, ψ, or $\psi(x, y, z)$. The latter form means that ψ is a function of coordinates x, y, and z. The wave function can be positive, negative, or imaginary. The probability of finding a particle in any volume element in space is proportional to $\psi^*\psi$, integrated over that volume of space.* This is the physical significance of the wave function. It should be emphasized that the probability of finding a particle in any volume element must be real and positive, and $\psi^*\psi$ always satisfies this requirement.

The Particle in a Box

The particle-in-a-box problem does not correspond to any real chemical system. Its usefulness in our context is that it illustrates several quantum mechanical features. To simplify the solution of the wave equation, we shall restrict ourselves to a one-dimensional box.

Our problem consists of a particle in a box as shown in Figure 1-12. In regions I and III (outside the box) the potential energy is taken to be infinite (the particle cannot escape the box) and the potential energy inside the box is taken to be zero. Under these conditions, classical mechanics predicts that the particle has an equal probability of being in any part of the box and the kinetic energy of the particle is allowed to have any value. As we shall see, solution of the wave equation leads to quite different results.

We can treat the problem in two steps:

1. Set up the wave equation for the problem. In our case the Schrödinger equation in one dimension is

$$\frac{d^2\psi}{dx^2} + \frac{8\pi^2 m}{h^2}(E - V)\psi = 0$$

In regions I and III of Figure 1-12 where $V = \infty$, we have

$$\frac{d^2\psi}{dx^2} + \frac{8\pi^2 m}{h^2}(E - \infty)\psi = 0 \tag{1-15}$$

In region II, $V = 0$ and

$$\frac{d^2\psi}{dx^2} + \frac{8\pi^2 m}{h^2}(E - 0)\psi = 0 \tag{1-16}$$

---

*The notation ψ^* (pronounced "sī star") refers to the *complex conjugate* of ψ. If ψ contains any imaginary term $i = \sqrt{-1}$, its sign is changed to $-i$ upon taking the complex conjugate.

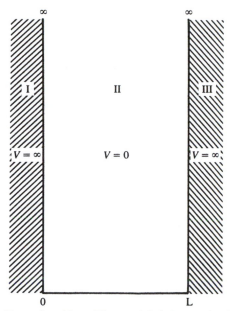

Figure 1-12 The one-dimensional box. The particle is constrained to move in region II.

2. Solve the Schrödinger equation to obtain the most general expression for ψ. In regions I and III, the only value of ψ that will satisfy Equation 1-15 is $\psi = 0$, because for any finite value of ψ the left side of Equation 1-15 will be infinite, not zero.

In region II, we must find a solution that satisfies Equation 1-16. We can rearrange Equation 1-16 as follows:

$$\left(\frac{-h^2}{8\pi^2 m}\right)\frac{d^2\psi}{dx^2} = E\psi \tag{1-17}$$

Equation 1-17 shows that the function ψ, when differentiated twice, gives the same function multiplied by E. The sine function possesses this behavior because

$$\frac{d^2(\sin ax)}{dx^2} = -a^2 \sin ax$$

Thus as a trial solution to Equation 1-17 we choose the function

$$\psi = A \sin ax, \tag{1-18}$$

in which A is a "normalization" constant. Substituting Equation 1-18 into Equation 1-17 we obtain

$$\frac{h^2 a^2}{8\pi^2 m}\psi = E\psi$$

or

$$E = \frac{a^2 h^2}{8\pi^2 m}$$

To obtain the value of a we must apply the *boundary conditions* to the wave function. The boundary conditions are that $\psi^*\psi$ must be *continuous*, *single-valued*, and *finite* everywhere. All of these conditions are only common sense. First, probability functions do not fluctuate radically from one place to another. Second, the probability of finding the particle in a given place cannot have two different values simultaneously. Third, because the probability of finding the particle somewhere must be 100 percent or 1.000, if the particle really exists, the probability at any one point cannot be infinite.

By choosing $\psi = A \sin ax$ we already guaranteed that our solution will be single-valued and finite. Now we wish to guarantee the continuity of the wave function. Recall that in regions I and III the solution was $\psi = 0$. Therefore, at the walls ($x = 0$ and $x = L$) we also must have $\psi = 0$. That is,

$$\psi = A \sin ax = 0$$

This means that $aL = n\pi$, in which $n = 1, 2, \ldots$, or rearranging to solve for a,

$$a = \frac{n\pi}{L}$$

Thus, in general, Equation 1-18 can be written as

$$\psi_n = A \sin \frac{n\pi x}{L}$$

and the corresponding energies are

$$E_n = \frac{n^2 h^2}{8mL^2} \tag{1-19}$$

We can determine the value of A by requiring that the wave function be *normalized*. Because the particle must be in the box, we write

$$\int_0^L \psi^*\psi \, dx = 1$$

but because ψ is real we have $\psi^* = \psi$ so

$$A^2 \int_0^L \sin^2 \frac{n\pi}{L} x \, dx = 1$$

The value of the integral is $L/2$, so

$$A^2\left(\frac{L}{2}\right) = 1 \qquad A = \left(\frac{2}{L}\right)^{1/2}$$

The final results for the particle in a one-dimensional box are

$$E_n = \frac{n^2 h^2}{8mL^2} \qquad \psi_n = \left(\frac{2}{L}\right)^{1/2} \sin \frac{n\pi x}{L}$$

Figure 1-13 shows the plots of the energy levels and the corresponding ψ's and ψ^2's.

We now wish to examine several properties of the energy levels and wave functions.

1. The energy of the particle does not vary continuously; rather, it only can have values of

$$\frac{h^2}{8mL^2}, \frac{h^2}{2mL^2}, \frac{9h^2}{8mL^2}, \cdots$$

2. The probability of finding the particle in the box is a function of distance x, and is not the same throughout the box. The positions of zero probability ($\psi = 0$ and therefore $\psi^2 = 0$) are called *nodes*.

3. The particle has a *zero-point* energy; that is, its lowest energy is not zero but

$$E_1 = \frac{h^2}{8mL^2}$$

4. Equation 1-19 shows that the spacing between energy levels increases as the mass of the particle decreases, and as the space to which the particle is confined decreases. Therefore, the effects of spacing of energy levels will be more important for systems of small mass confined to small regions of space. These are precisely the conditions we find in ordinary chemical systems, where the electron with a small mass is confined to a small region of space in an atom or molecule.

Although the particle in a box does not correspond to any real physical system, its description is of interest because it shows how the Schrödinger equation applies to several concepts that also will be involved in our discussion of the hydrogen atom. These are

1. the boundary conditions which result in the quantization of energy
2. the ideas of probability and nodes
3. the concept of normalization
4. the concept of zero-point energy

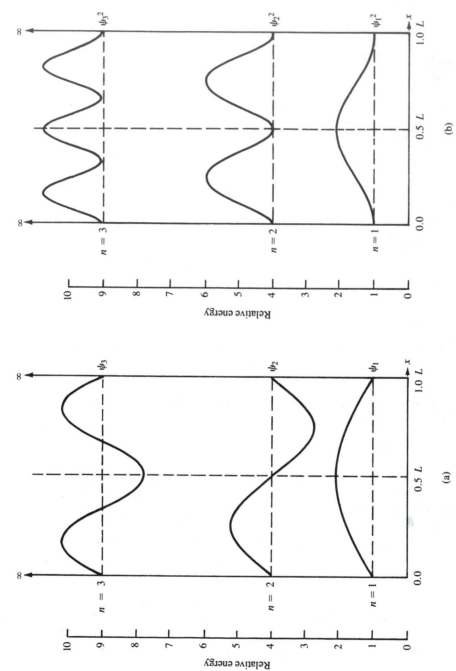

Figure 1-13 (a) Wave functions and (b) probability functions of a particle in a box. Notice that the energy is independent of x and is given by the dashed line for each value of n.

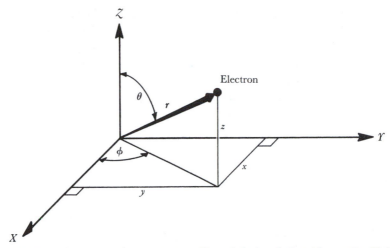

Figure 1-14 The polar coordinates θ, ϕ, and r and their relationships to the X, Y, and Z axes. It can be shown that $x = r \sin \theta \cos \phi$, $y = r \sin \theta \sin \phi$, and $z = r \cos \theta$.

1-12 The Wave Equation and the Hydrogen Atom

The steps required to solve the Schrödinger equation for the hydrogen atom are the same as those for the particle in a box. Because the hydrogen atom is three-dimensional we must use the Schrödinger equation in that form, Equation 1-13. In the hydrogen atom the potential energy is due to the electrostatic interaction, $-e^2/r$, in which r is the distance between the electron and the nucleus. This type of potential energy is referred to as a *central field*, for which the spherical polar coordinate system is more convenient than the Cartesian system, as illustrated in Figure 1-14. We shall not go into the details of transforming Equation 1-13 into the spherical polar coordinate system. Suffice it to say that the Schrödinger equation for the system of one proton and one electron (i.e., the hydrogen atom) can be solved exactly. A quantization results in which only certain orbitals (wave functions) and energies are possible.

The Quantum Numbers

These wave functions are specified by three quantum numbers (n, l, m_l) and the total wave function can be written as a product of the radial part R and the angular part Y:

$$\psi(n, l, m_l) = R_{n,l}(r)\ Y_{l,m_l}\ (\theta, \phi)$$

Solution of the Schrödinger equation imposes certain restrictions on the values that the quantum numbers can have. The first orbital quantum number is called the *principal quantum number.* It is given the symbol n and can have any positive integer value from one to infinity. The effective volume of an electron orbital depends on n. The orbital energy is determined by the value of n according to the expression

$$E = \frac{-k}{n^2}$$

in which $k = 2\pi^2 m_e e^4/h^2$, as in the Bohr theory. A very useful value for k is 13.6 eV.

The second orbital quantum number is designated l. The l quantum number determines the angular momentum of the electron but qualitatively may be called the *orbital-shape quantum number.* For any given value of n, $l = 0, 1, 2, \ldots, n-1$. These first two quantum numbers together determine the spatial properties of the electron orbital, as shown in Figure 1-15.

The third orbital quantum number, m_l, determines the orientation of a particular spatial configuration in relation to an arbitrary direction. The introduction of an external magnetic field most conveniently provides the arbitrary reference axis. We call the third quantum number the magnetic quantum number or the *orbital-orientation quantum number.* The m_l quantum number depends on the value of l and may have any integer value from $-l$ to $+l$:

$$m_l = -l, -l+1, \ldots, 0, \ldots, l-1, l$$

The preceding three orbital quantum numbers explicitly determine an atomic orbital.

An electron moving in a particular orbital has properties that can be explained by imagining that the electron is a tiny bar magnet with a north and a south pole. Its "bar magnet" behavior is described as *electron spin,* which can go in either of two directions in reference to an arbitrary axis. The *spin quantum number, m_s,* specifies the direction of spin in space with reference to such an arbitrary axis. The two orientations of electron spin in a magnetic field are shown schematically in Figure 1-16. Electron spin is quantized in half-integer units, and for each electron can have values of $+\frac{1}{2}$ or $-\frac{1}{2}$.

Exercise 1-6. An electron has the principal quantum number 4. What are the possible values of l, m_l, and m_S for this electron?

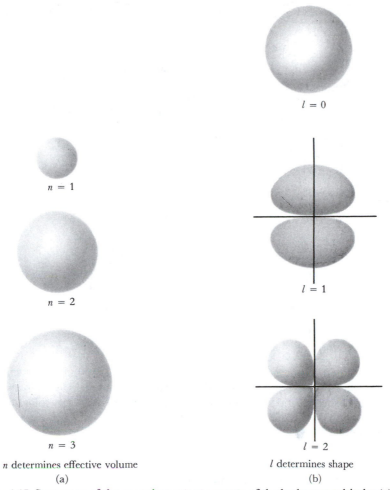

$l = 0$

$n = 1$

$l = 1$

$n = 2$

$n = 3$ $l = 2$

n determines effective volume l determines shape

(a)   (b)

Figure 1-15 Summary of the most important aspects of the hydrogen orbitals: (a) The principal quantum number, n, indicates approximately the effective volume of the orbital. (b) The orbital-shape quantum number, l, determines the general shape of the orbital.

Solution. With $n = 4$, l may have a value of 3, 2, 1, or 0.
 For $l = 3$ there are seven possible values for m_l: 3, 2, 1, 0, −1, −2, −3.
 For $l = 2$ there are five possible values for m_l: 2, 1, 0, −1, −2.
 For $l = 1$ there are three possible values for m_l: 1, 0, −1.
 For $l = 0$ there is only one possible value of m_l: 0.
 Since for each set of orbital quantum numbers the electron can have either $+\frac{1}{2}$ or $-\frac{1}{2}$ spin, there are 32 possible combinations of l, m_l, and m_s with $n = 4$.

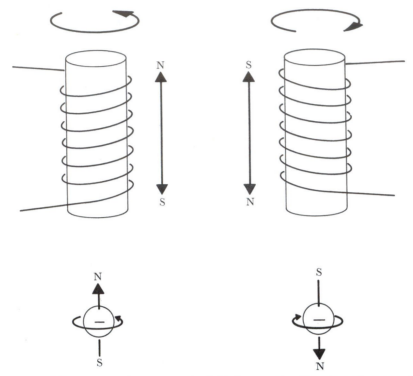

Figure 1-16 Electron spin in a magnetic field. Just as the direction of current flow around an iron bar determines the direction of the polarity of the magnet induced in the iron bar, so the direction of the spin of an electron determines its spin quantum number. Electron spin is quantized in half-integer units and for each electron can have values of $+\frac{1}{2}$ or $-\frac{1}{2}$.

The Schrödinger wave equation can be set up for any atom, but it can be solved exactly only for the hydrogen atom (or ions with one electron). Consequently the four quantum numbers, and the corresponding wave functions obtained by solving the Schrödinger equation of motion, actually apply only to "hydrogenlike" atoms. Nevertheless, the concept of quantum numbers and orbitals remains very useful. To describe many-electron atoms we assume orbitals that are analogous to hydrogenlike orbitals, but modify them somewhat because of the repulsive interactions of the electrons.

Quantum Number Specifications of Orbitals

The different n values for the hydrogen atom are called *energy levels* or *shells*. For each different shell (or n value) the l quantum number can have only certain discrete values, corresponding to orbitals of different shapes.

For historical reasons the 0, 1, 2, and 3 values of the l quantum number are designated by the letters s, p, d, and f, respectively. The combination of the principal quantum number and the appropriate letter corresponding to the value of the l quantum number is the shorthand notation for a particular orbital. For example, the combination $n = 1$, $l = 0$ is a 1s orbital, and that of $n = 3$, $l = 1$ is a 3p orbital. To specify this orbital completely, the value of m_l is also required. In the case of the 3p orbital there are three possibilities: $3p_{+1}$, $3p_0$, and $3p_{-1}$. For $m_l \neq 0$, the orbitals are imaginary functions. It is usually more convenient to deal with an equivalent set of real functions, which are linear combinations of the imaginary functions. The shorthand for the real hydrogen orbitals has an added subscript in Cartesian coordinates that now gives the angular dependency (for example, $3p_x$, $3p_y$, $3p_z$). The complete set of real orbitals for hydrogen through $n = 3$ is given in Table 1-2.

The wave function, ψ, which is a particular solution of the Schrödinger wave equation, mathematically describes the motion of an electron in an orbital. The amplitude of the wave function, indicated by its magnitude at various points in the region around a nucleus, gives the approximate probability of finding an electron in that orbital at any particular point. However, the precise value of the wave function squared, ψ^2, is a direct measure of the probability density of the electron at any point in space. In other words, the greater the amplitude or value of ψ in a given region, the greater the probability that the electron is in that region.

For the lowest-energy level or shell, $n = 1$. The rules relating the values of the quantum numbers require that the l and m_l quantum numbers be zero. Accordingly, there is only one orbital with $n = 1$ ($l = 0$, $m_l = 0$). This orbital is designated as 1s and is described by the following wave function, ψ:

$$\psi(1s) = N\left(\frac{1}{a_0}\right)^{3/2} e^{-r/a_0} \qquad (1\text{-}20)$$

in which

$$N = \frac{1}{\sqrt{\pi}}$$

In Equation 1-20 the symbol $\psi(1s)$ designates the wave function of the electron in the $n = 1$ orbital. The normalization constant, N, is fixed so that the probability of finding the electron somewhere in space is one.* The

---

*That is, $\int \psi^2 (1s)\, dv = 1$ where dv is the differential volume element and the integration is carried out over all space.

TABLE 1-2. IMPORTANT ORBITALS FOR THE HYDROGEN ATOM

Orbital Quantum Numbers			Orbital Designation	Radial Function* $R_{n,l}(r)$	Angular Function† $Y_{l,m_l}\left(\frac{x}{r}, \frac{y}{r}, \frac{z}{r}\right)$	Angular Function† $Y_{l,m_l}(\theta, \phi)$
n	l	m_l				
1	0	0	$1s$	$2e^{-r}$	$\dfrac{1}{2\sqrt{\pi}}$	$\dfrac{1}{2\sqrt{\pi}}$
2	0	0	$2s$	$\dfrac{1}{2\sqrt{2}}(r-2)e^{-r/2}$	$\dfrac{1}{2\sqrt{\pi}}$	$\dfrac{1}{2\sqrt{\pi}}$
2	1	(± 1)‡	$2p_x$	$\dfrac{1}{2\sqrt{6}}re^{-r/2}$	$\dfrac{\sqrt{3}(x/r)}{2\sqrt{\pi}}$	$\dfrac{\sqrt{3}(\sin\theta\cos\phi)}{2\sqrt{\pi}}$
2	1	0	$2p_z$	$\dfrac{1}{2\sqrt{6}}re^{-r/2}$	$\dfrac{\sqrt{3}(z/r)}{2\sqrt{\pi}}$	$\dfrac{\sqrt{3}(\cos\theta)}{2\sqrt{\pi}}$
2	1	(± 1)‡	$2p_y$	$\dfrac{1}{2\sqrt{6}}re^{-r/2}$	$\dfrac{\sqrt{3}(y/r)}{2\sqrt{\pi}}$	$\dfrac{\sqrt{3}(\sin\theta\sin\phi)}{2\sqrt{\pi}}$
3	0	0	$3s$	$\dfrac{2}{81\sqrt{3}}(27-18r+2r^2)e^{-r/3}$	$\dfrac{1}{2\sqrt{\pi}}$	$\dfrac{1}{2\sqrt{\pi}}$
3	1	(± 1)‡	$3p_x$	$-\dfrac{4}{81\sqrt{6}}(r^2-6r)e^{-r/3}$	$\dfrac{\sqrt{3}(x/r)}{2\sqrt{\pi}}$	$\dfrac{\sqrt{3}(\sin\theta\cos\phi)}{2\sqrt{\pi}}$

n	l	m_l		Radial	Angular	
3	1	0	$3p_z$	$-\dfrac{4}{81\sqrt{6}}(r^2-6r)e^{-r/3}$	$\dfrac{\sqrt{3}(z/r)}{2\sqrt{\pi}}$	$\dfrac{\sqrt{3}(\cos\theta)}{2\sqrt{\pi}}$
3	1	(±1)‡	$3p_y$	$-\dfrac{4}{81\sqrt{6}}(r^2-6r)e^{-r/3}$	$\dfrac{\sqrt{3}(y/r)}{2\sqrt{\pi}}$	$\dfrac{\sqrt{3}(\sin\theta\sin\phi)}{2\sqrt{\pi}}$
3	2	(±2)‡	$3d_{x^2-y^2}$	$\dfrac{4}{81\sqrt{30}}r^2e^{-r/3}$	$\dfrac{\sqrt{15}[(x^2-y^2)/r^2]}{4\sqrt{\pi}}$	$\dfrac{\sqrt{15}[\sin^2\theta\,(\cos^2\phi-\sin^2\theta)]}{4\sqrt{\pi}}$
3	2	(±1)‡	$3d_{xz}$	$\dfrac{4}{81\sqrt{30}}r^2e^{-r/3}$	$\dfrac{\sqrt{30}(xz/r^2)}{2\sqrt{2\pi}}$	$\dfrac{\sqrt{30}(\sin\theta\cos\theta\cos\phi)}{2\sqrt{2\pi}}$
3	2	0	$3d_{z^2}$	$\dfrac{4}{81\sqrt{30}}r^2e^{-r/3}$	$\dfrac{\sqrt{5}[(3z^2-r^2)/r^2]}{4\sqrt{\pi}}$	$\dfrac{\sqrt{5}(3\cos^2\theta-1)}{4\sqrt{\pi}}$
3	2	(±1)‡	$3d_{yz}$	$\dfrac{4}{81\sqrt{30}}r^2e^{-r/3}$	$\dfrac{\sqrt{30}(yz/r^2)}{2\sqrt{2\pi}}$	$\dfrac{\sqrt{30}(\sin\theta\cos\theta\sin\phi)}{2\sqrt{2\pi}}$
3	2	(±2)‡	$3d_{xy}$	$\dfrac{4}{81\sqrt{30}}r^2e^{-r/3}$	$\dfrac{\sqrt{15}(xy/r^2)}{2\sqrt{\pi}}$	$\dfrac{\sqrt{15}(\sin^2\theta\sin\phi\cos\phi)}{2\sqrt{\pi}}$

NOTE: Both the radial and the angular functions are normalized to one; r is in atomic units (that is, in units of a_0; see Exercise 1-1).
*To convert to a general radial function for a one-electron atom with any nuclear charge Z, replace r by Zr and multiply each function by $(Z)^{3/2}$.
†The last two columns present two equivalent ways of tabulating the angular functions. In the spherical polar coordinate system we have $x = r\sin\theta\cos\phi$, $y = r\sin\theta\sin\phi$, and $z = r\cos\theta$.
‡It is not correct to assign m_l values to the real functions x, y, $x^2 - y^2$, xz, yz, and xy. The real functions correspond to linear combinations of $+m_l$ and $-m_l$.

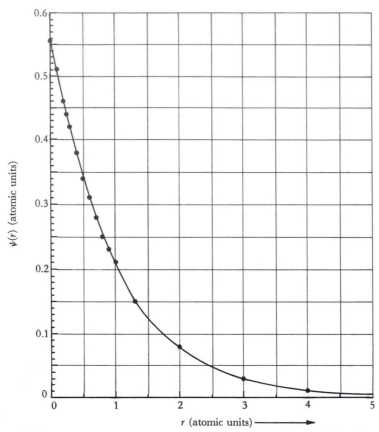

Figure 1-17 Plot of $\psi(1s) = Ne^{-r}$ for atomic hydrogen. The magnitude of the wave function, $\psi(r)$, gives approximately the chance of finding the $1s$ electron at any distance r from the nucleus. The distance r is measured in atomic units, that is, in units of a_0, the Bohr radius ($1a_0 = 0.529$ Å).

quantity e is the base of natural or Naperian logarithms, approximately 2.72. We can express $\psi(1s)$ in simpler form by giving the distance r from the nucleus in atomic units, that is, in units of a_0, the Bohr radius. The wave function then becomes

$$\psi(1s) = Ne^{-r}$$

The value of the wave function as a function of the radial distance from the center of the atom, r, is shown in Figure 1-17. The $\psi(r)$ function is for the hydrogen $1s$ electron orbital. The electron density associated with an electron orbital is obtained from the square of the wave function, $\psi^2(r)$, which gives the probability density, $P(r)$, for the electron at a given point in space:

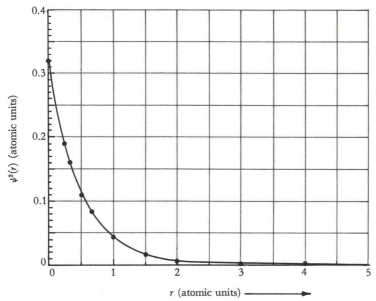

Figure 1-18 Plot of $\psi^2(r) = P(r) = N^2 e^{-2r}$ for atomic hydrogen. The precise value of the square of the wave function is a direct measure of the probability density of an electron at any distance r from the nucleus. The probability curve never reaches zero, even at $r = \infty$. However, the sphere around the nucleus that contains 99 percent of the probability [see Figure 1-11(b)] has a radius of 4.2 atomic units (2.2 Å).

$$P(r) = N^2 \left(\frac{1}{a_0}\right)^3 e^{-2r/a_0}$$

Notice that $P(r)$ has units of density. When r is expressed in units of a_0, the expression for probability density becomes

$$P(r) = N^2 e^{-2r} = \frac{1}{\pi} e^{-2r}$$

In Figure 1-18 the probability density for a $1s$ orbital is plotted as a function of r. For s orbitals the probability of finding an electron at the nucleus is finite (nonzero), whereas for all other orbitals the value of $\psi^2(r)$ at the nucleus is zero.

The electron density of a $1s$ orbital has been illustrated in Figure 1-11(a). As we have seen, the region of space referred to in the discussion of orbitals can be restricted by using a cross-sectional contour of constant probability density, as is shown in Figure 1-11(b) for the $1s$ orbital. Figure 1-19 shows two contours for each of several other hydrogen atomic orbitals.

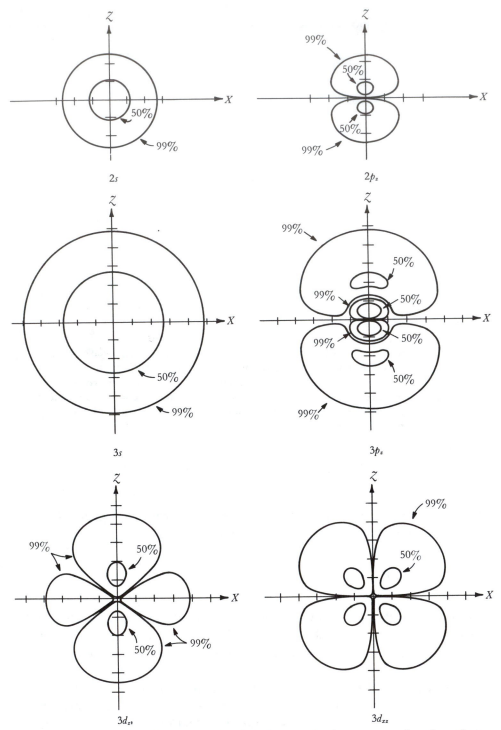

Figure 1-19 Contour diagrams in the XZ plane for hydrogen wave functions that show the 50 percent and 99 percent contours. X and Z axes are marked in intervals of 5 au. The $3p_z$ orbital differs from the $2p_z$ in having another nodal surface as a spherical shell around the nucleus at a distance of approximately 6 au.

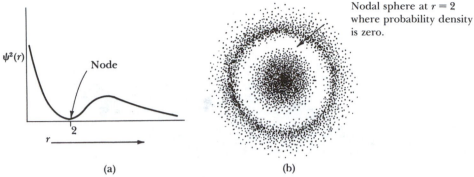

(a)   (b)

Figure 1-20 The hydrogen 2s orbital: (a) the graph of $\psi^2(r)$ against r; (b) a cross section through the probability function plotted in three dimensions. Probability density is represented by stippling.

When the principal quantum number is 2 the l quantum number is restricted to two values, 0 and 1. The orbital with $l = 0$ is the 2s orbital and can be described analytically as

$$\psi(2s) = \frac{1}{4\sqrt{2\pi}}(r - 2)e^{-r/2}$$

The important differences between the 2s and 1s orbitals for the hydrogen atom are that the 2s orbital is effectively larger than the 1s orbital, and that for $r = 2$ the 2s wave function is zero. Remember that a surface on which a wave function is zero is called a node (or nodal surface). Thus the 2s orbital of atomic hydrogen has a nodal sphere with a radius of two atomic units, as shown in Figure 1-20.

It is important to remember that all s orbitals are spherically symmetrical. Orbitals with $l = 0$ and different values of n differ only in "effective volume," and in the number of nodes.

In the second shell, with $n = 2$, an orbital with $l = 1$ is encountered. There are three orbitals with $l = 1$ because the m_l quantum number can have the values -1, 0, and 1. Unlike the s orbitals, which are spherically symmetrical, the three 2p orbitals have directional properties. Accordingly, the 2p orbital with $m_l = 0$ is specified in the polar coordinates of Figure 1-14 as

$$\psi(2p_z) = \frac{1}{4\sqrt{2\pi}}(\cos \theta)re^{-r/2} \tag{1-21}$$

The $2p_z$ orbital, which is described by Equation 1-21, has regions of greatest concentration of probability along the Z axis. The electron-density representation of this 2p orbital is shown in Figure 1-21(a). Examining this repre-

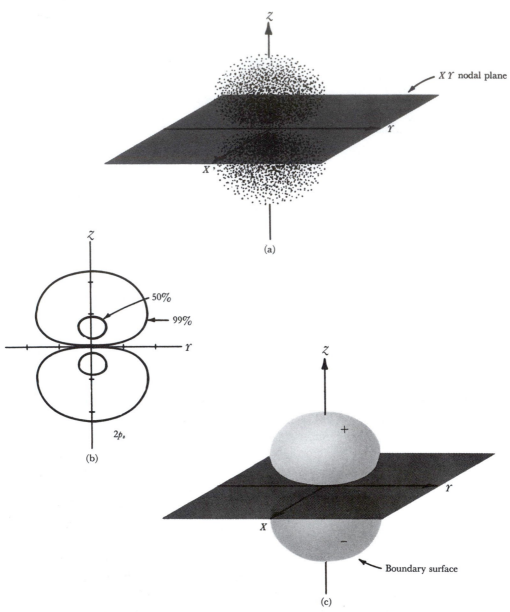

Figure 1-21 Three ways of representing the $2p_z$ atomic orbital of hydrogen: (a) ψ^2 represented by stippling. (b) contour diagram of the $2p_z$ orbital. The contours represent lines of constant ψ^2 in the YZ plane and have been chosen so that, in three dimensions, they enclose 50 or 99 percent of the total probability density. The $2p_z$ orbital is symmetrical around the Z axis. (c) the 99 percent probability shell portrayed as a surface. The plus and minus signs on the two lobes represent the relative signs of ψ and should not be confused with electric charge. Notice that there is no probability of finding the electron on the XY plane. Such a surface, which need not be planar, is a *nodal* surface.

sentation we see that the probability of finding the electron in the XY plane is zero [i.e., $\psi(2p_z) = 0$ when $\cos\theta = 90°$]. This nodal plane containing the atomic nucleus is a property of all p orbitals.

The $2p_z$ orbital also can be described with a contour diagram or a spatial representation. We can find the contour diagram by plotting lines of constant probability density, which correspond to constant $|\psi|$ or ψ^2. The square of the $\psi(2p_z)$ function is given by the equation

$$P(r) = [\psi(2p_z)]^2 = \frac{r^2 \cos^2\theta e^{-r}}{32\pi}$$

The resulting contours and a spatial representation for the $2p_z$ orbital are shown in Figure 1-21(b and c).

The signs of the two lobes in the spatial representation are a reminder that the $2p_z$ wave function is positive for positive Z values and negative for negative Z values; that is, positive for θ values between 0° and 90° and negative for θ values between 90° and 180°. All p functions change sign when inverted at the atomic nucleus, and are said to be antisymmetric. In contrast, s orbital functions are symmetrical because inversion does not generate a change in algebraic sign. The symmetry properties of orbitals are emphasized here because, as we shall see later, they are important in classifying bonds between atoms.

A complete set of p orbitals is shown in Figure 1-22. The three equivalent p orbitals differ only in their spatial orientations. They are designated p_x, p_y, and p_z, depending on their directional properties with respect to the X, Y, and Z axes.

When the principal quantum number is three, l can have three values: 0, 1, and 2. A $3s$ orbital has $l = 0$ and is similar to the s orbitals described previously. (It is a general result that the number of nodes equals $n - 1$; thus the $3s$ orbital has two nodes.) When $l = 1$ there are three $3p$ orbitals. The $2p$ and $3p$ orbitals with the same m_l qunatum number have the same angular dependence, although the boundary contours of the $3p$ orbitals are more complicated than those for the $2p$ orbitals, because of the presence of an additional node. However, the outer part of a $3p$ orbital looks like a $2p$ orbital (see Figure 1-19), and we usually represent all p orbitals by the three spatial pictures shown in Figure 1-22.

When $n = 3$ we encounter for the first time a value of two for l, thereby giving five possible values for the m_l quantum number. There are thus five d orbitals. Although their mathematical representations are more complicated (Table 1-2), pictorial representations of five energetically equivalent d orbitals are reasonably simple, as indicated in Figure 1-23. Notice that the d_{z^2} orbital, with $m_l = 0$, has a different shape than the others. Again, the

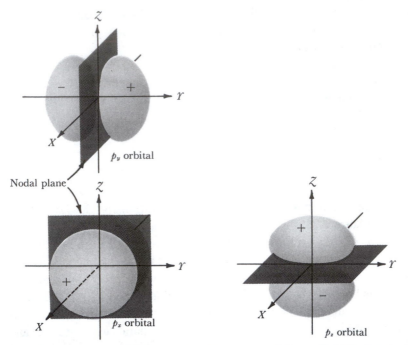

Figure 1-22 Boundary surfaces enclosing 99 percent of the probability for the $2p_y$, $2p_x$, and $2p_z$ orbitals of hydrogen. Note the nodal plane of zero probability density in each orbital.

signs labeling the various lobes indicate that the $3d$-orbital wave function is either positive or negative in that region. The d-orbital functions are symmetrical because inversion at the origin does not result in a change in algebraic sign.

There are orbitals of high l values in the shells with principal quantum numbers $n > 3$, but these orbitals are much more complicated and will not be introduced.

We already have indicated that each of the hydrogen atomic orbitals is normalized so that the probability of finding the electron somewhere in space is one. Besides being normalized, these wave functions are also *orthogonal*. Two wave functions, ψ_1 and ψ_2, are orthogonal if their *overlap integral* is zero

$$\int \psi_1 \psi_2 \, dv = 0$$

Applying this concept to the hydrogen atom, we present several examples of atomic orbitals that are orthogonal:

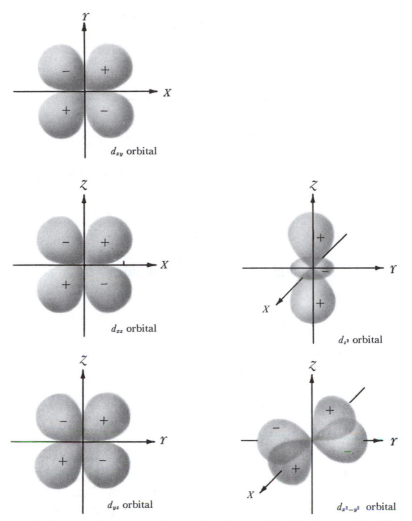

Figure 1-23 The five $3d$ atomic orbitals of hydrogen. The $4d$, $5d$, and $6d$ orbitals can be considered as essentially identical to these $3d$ orbitals except for an increase in size. Notice how the sign of the wave function changes from one lobe to the next in a given orbital. This change of sign is important when atomic orbitals are combined to make molecular orbitals.

$$\int \psi(1s)\psi(2s)\ dv = 0, \qquad \int \psi(2s)\psi(2p_z)\ dv = 0$$
$$\int \psi(2p_x)\psi(2p_z)\ dv = 0, \qquad \int \psi(2p_x)\psi(3s)\ dv = 0$$

We can illustrate the concept of orthogonality pictorially by using the $2s - 2p_z$ example of Figure 1-24. It is clear that equal positive and negative

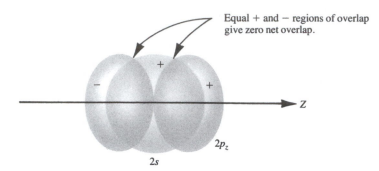

Equal + and − regions of overlap give zero net overlap.

Figure 1-24 Illustration of the orthogonality of the $2s$ and $2p_z$ wave functions. The net overlap of $2s$ and $2p_z$ is zero.

regions of overlap give zero net overlap, and therefore the functions are orthogonal. We will find the orthogonality property useful when we discuss chemical bonding in molecules in subsequent chapters.

1-13 Many-Electron Atoms

The Schrödinger equation can be set up for atoms with more than one electron, but it cannot be solved exactly in these cases. The second and subsequent electrons introduce the complicating feature of electron-electron repulsion, which is not present in the hydrogenlike atom. Nevertheless, with some modification the hydrogenlike orbitals account adequately for the electronic structures of many-electron atoms.

The key to the building or *Aufbau* process for many-electron atoms is called the *Pauli exclusion principle,* which states that no two electrons in an atom can have the same set of four quantum numbers. Thus two electrons in the ground state of atomic helium $(Z = 2)$ must possess the following quantum numbers:

$$n = 1, \quad l = 0, \quad m_l = 0, \quad m_s = +\tfrac{1}{2}$$

and

$$n = 1, \quad l = 0, \quad m_l = 0, \quad m_s = -\tfrac{1}{2}$$

In other words, the two electrons in the helium atom are placed in the $1s$ orbital with *opposite spins* to be consistent with the Pauli principle. We abbreviate the orbital electronic structure of helium

$$\textcircled{$\uparrow\downarrow$} \quad \text{or} \quad 1s^2$$

$1s$

in which $\uparrow$ stands for $m_s = +\frac{1}{2}$ and $\downarrow$ stands for $m_s = -\frac{1}{2}$.

It is a well-known result of quantum mechanics that, for multielectron atoms, *product* wave functions are solutions to the Schrödinger equation. By this we mean that the wave function describing the first electron must be *multiplied* by the wave function describing the second electron. The wave function that describes the ground state of the helium atom is a product wave function $\psi_{He}(1,2) = 1s(1)1s(2)$ in which the numbers 1 and 2 designate electrons number 1 and number 2. This equation describes only the *spatial* part of the wave function. If we also want to designate the *spin* of each electron, we must attach an additional spin "label." Typically, the symbol α represents a "spin-up" electron and β represents a "spin-down" electron. We immediately recognize two equally probable configurations for the helium atom: $[(1s(1)\alpha(1)] \, [1s(2)\beta(2)]$ or $\textcircled{$\uparrow\downarrow$}$ and $[1s(1)\beta(1)] \, [1s(2)\alpha(2)]$ or $\textcircled{$\downarrow\uparrow$}$. A full statement of the Pauli exclusion principle declares that the proper quantum mechanical wave function for helium must be *antisymmetric* with respect to the interchange of electron labels 1 and 2, That is, an interchange of labels causes the wave function to change sign. For helium, the antisymmetric wave function is obtained by taking the difference of the two equally probable configurations:

$$\psi_{He} = [1s(1)\alpha(1)] \, [1s(2)\beta(2)] - [1s(1)\beta(1)] \, [1s(2)\alpha(2)]$$
$$= 1s(1)1s(2)[\alpha(1)\beta(2) - \beta(1)\alpha(2)]$$

In discussing the properties of many-electron atoms such as helium, the concept of *effective nuclear charge*, Z_{eff}, is quite useful. Of course, the actual nuclear charge in a helium atom is +2, but the full force of this +2 charge is partially offset by the mutual repulsion of the two electrons. As far as either *one* of the electrons is concerned; Z_{eff} is *less* than +2, because of the "shielding" provided by the other electron. One atomic property which illustrates that Z_{eff} is less than +2 is the *IE* of helium, which is 24.59 eV. If there were no electron-electron repulsion, each electron would feel the full +2 nuclear charge. The *IE* in this hypothetical situation would be equal to the value for a hydrogenlike atom with $Z = +2$, or $IE = (2)^2(13.6$ eV$) = 54.4$ eV. Thus electron-electron repulsion drastically reduces the ionization energy of helium, and the utility of the concept of Z_{eff} is established.

Now consider the lithium atom $(Z = 3)$. Its $1s$ orbital is occupied fully by two electrons. The third electron must be placed in one of the orbitals with

$n = 2$. But which one? The decision is not important in a hydrogenlike atom because the $2s$ and $2p$ orbitals have the same energy. However, in a many-electron atom the shielding of any given electron from the nuclear charge depends on the l quantum number of the electron under consideration. For example, consider the $2s$ and $2p$ orbitals in the lithium atom. Both orbitals are shielded from the $+3$ nuclear charge by the $1s^2$ electrons, but because of its larger probability density very close to the nucleus (see Figure 1-20), the $2s$ orbital is not shielded as strongly as the $2p$ orbital. We say that the $2s$ orbital "penetrates" the inner $1s^2$ electron shell better than the $2p$ orbital does. Therefore the order of energies is $2s < 2p$, and the third electron in the lithium atom occupies the $2s$ orbital in the ground state:

$$\text{(↑↓)(↑)} \qquad \text{or} \qquad 1s^2 2s^1$$
$$1s \ \ 2s$$

The example of the lithium atom allows us to define the terms *core* and *valence electrons*. Core electrons are those corresponding to the electronic configuration of the previous noble gas element in the periodic table (He, Ar, Kr, Xe, or Rn). For lithium, the previous noble gas is helium, which has the electronic configuration $1s^2$. Therefore the core electrons for lithium are $1s^2$. Valence electrons then are defined as all electrons which are not core electrons. The lithium atom has only the $2s^1$ valence electron. We often use a shorthand notation to write the electronic configuration in which we designate the core electrons by using the symbol of the corresponding noble gas element.* Hence for lithium we could write $1s^2 2s^1$ or $[\text{He}]2s^1$.

A beryllium atom ($Z = 4$) has a filled $2s$ orbital, and a boron atom ($Z = 5$) has the fifth electron in a $2p$ orbital:

$$\text{Be: (↑↓)(↑↓)} \qquad\qquad\qquad \text{or} \qquad 1s^2 2s^2$$
$$1s \ \ 2s$$

$$\text{B: (↑↓)(↑↓)} \quad \text{(↑)(○)(○)} \qquad \text{or} \qquad 1s^2 2s^2 2p^1$$
$$1s \ \ 2s \qquad\quad 2p$$

For a carbon atom ($Z = 6$) we have a choice of electron placement, because there are several possible configurations for the second electron in a set of three $2p$ orbitals. For example, the three configurations

---

*Our definition of core electrons is too restricted for elements which occur after the transition series elements. For these elements the valence electrons are only those having the largest principal quantum number. For example, we could write the electronic configuration of Br as $[\text{Ar}]3d^{10}4s^2 4p^5$. The $3d^{10}$ electrons, however, should be considered as part of the core electrons.

TABLE 1-3. THE s, p, d, AND f ORBITAL SETS

Type of Orbital	Orbital Quantum Numbers	Total Orbitals in Set	Total Number of Electrons That Can Be Accommodated
s	$l = 0$; $m_l = 0$	1	2
p	$l = 1$; $m_l = 1, 0, -1$	3	6
d	$l = 2$; $m_l = 2, 1, 0, -1, -2$	5	10
f	$l = 3$; $m_l = 3, 2, 1, 0, -1, -2, -3$	7	14

all obey the Pauli principle. Which configuration most accurately represents the ground state of atomic carbon? The choice is made by invoking *Hund's rule*, which states that for any set of orbitals of equal energy the electronic configuration with the maximum number of parallel spins results in the lowest electron-electron repulsion. Thus the ground-state configuration of atomic carbon is

or $\quad 1s^2 2s^2 2p_x{}^1 2p_y{}^1$

The two electrons with parallel spins (same m_s value) are said to be "unpaired."

Now we are in a position to build the ground-state configuration of the atoms of all elements by filling the atomic orbital sets with electrons in order of increasing energy, making certain that the Pauli principle and Hund's rule are obeyed. The total number of electrons that the different orbital sets can accomodate is given in Table 1-3. The s, p, d, and f orbital sets usually are called *subshells*. As we noted previously, the group of subshells for any given n value is called a shell.

We have discussed the fact that the $2p$ orbitals have higher energy than the $2s$ orbital in terms of different degrees of shielding in a many-electron atom. Actually, the energy order was known from atomic spectral experiments long before a theoretical rationale was available. We also know both from atomic theory and from experiments that the complete order of increasing energy of the orbitals in the periodic table for many-electron, neutral atoms is $1s$, $2s$, $2p$, $3s$, $3p$, $4s$, $3d$, $4p$, $5s$, $4d$, $5p$, $6s$, $4f \approx 5d$, $6p$, $7s$, $5f \approx 6d$. This relative ordering is shown in Figure 1-25. The electronic configurations of the elements are presented in Table 1-4.

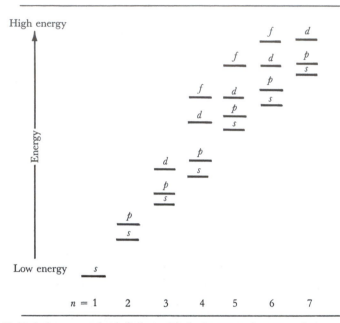

Figure 1-25 Relative energies of the orbitals in neutral, many-electron atoms. In building the periodic table for neutral, many-electron atoms, electrons are put in the available orbital of lowest energy. For example, the 4s orbital is filled before the 3d orbital, the 6s orbital is filled before the 4f or 5d orbitals, and so on.

TABLE 1-4. THE ELECTRONIC CONFIGURATIONS OF THE ELEMENTS

Z	Element	Electron Configuration	Z	Element	Electron Configuration
1	H	$1s$	19	K	$[Ar]4s$
2	He	$1s^2$	20	Ca	$[Ar]4s^2$
3	Li	$[He]2s$	21	Sc	$[Ar]3d4s^2$
4	Be	$[He]2s^2$	22	Ti	$[Ar]3d^24s^2$
5	B	$[He]2s^22p$	23	V	$[Ar]3d^34s^2$
6	C	$[He]2s^22p^2$	24	Cr	$[Ar]3d^54s$
7	N	$[He]2s^22p^3$	25	Mn	$[Ar]3d^54s^2$
8	O	$[He]2s^22p^4$	26	Fe	$[Ar]3d^64s^2$
9	F	$[He]2s^22p^5$	27	Co	$[Ar]3d^74s^2$
10	Ne	$[He]2s^22p^6$	28	Ni	$[Ar]3d^84s^2$
11	Na	$[Ne]3s$	29	Cu	$[Ar]3d^{10}4s$
12	Mg	$[Ne]3s^2$	30	Zn	$[Ar]3d^{10}4s^2$
13	Al	$[Ne]3s^23p$	31	Ga	$[Ar]3d^{10}4s^24p$
14	Si	$[Ne]3s^23p^2$	32	Ge	$[Ar]3d^{10}4s^24p^2$
15	P	$[Ne]3s^23p^3$	33	As	$[Ar]3d^{10}4s^24p^3$
16	S	$[Ne]3s^23p^4$	34	Se	$[Ar]3d^{10}4s^24p^4$
17	Cl	$[Ne]3s^23p^5$	35	Br	$[Ar]3d^{10}4s^24p^5$
18	Ar	$[Ne]3s^23p^6$	36	Kr	$[Ar]3d^{10}4s^24p^6$

TABLE 1-4. THE ELECTRONIC CONFIGURATIONS OF THE ELEMENTS (*continued*)

Z	Element	Electron Configuration	Z	Element	Electron Configuration
37	Rb	$[Kr]5s$	71	Lu	$[Xe]4f^{14}5d6s^2$
38	Sr	$[Kr]5s^2$	72	Hf	$[Xe]4f^{14}5d^26s^2$
39	Y	$[Kr]4d5s^2$	73	Ta	$[Xe]4f^{14}5d^36s^2$
40	Zr	$[Kr]4d^25s^2$	74	W	$[Xe]4f^{14}5d^46s^2$
41	Nb	$[Kr]4d^45s$	75	Re	$[Xe]4f^{14}5d^56s^2$
42	Mo	$[Kr]4d^55s$	76	Os	$[Xe]4f^{14}5d^66s^2$
43	Tc	$[Kr]4d^55s^2$	77	Ir	$[Xe]4f^{14}5d^76s^2$
44	Ru	$[Kr]4d^75s$	78	Pt	$[Xe]4f^{14}5d^96s$
45	Rh	$[Kr]4d^85s$	79	Au	$[Xe]4f^{14}5d^{10}6s$
46	Pd	$[Kr]4d^{10}$	80	Hg	$[Xe]4f^{14}5d^{10}6s^2$
47	Ag	$[Kr]4d^{10}5s$	81	Tl	$[Xe]4f^{14}5d^{10}6s^26p$
48	Cd	$[Kr]4d^{10}5s^2$	82	Pb	$[Xe]4f^{14}5d^{10}6s^26p^2$
49	In	$[Kr]4d^{10}5s^25p$	83	Bi	$[Xe]4f^{14}5d^{10}6s^26p^3$
50	Sn	$[Kr]4d^{10}5s^25p^2$	84	Po	$[Xe]4f^{14}5d^{10}6s^26p^4$
51	Sb	$[Kr]4d^{10}5s^25p^3$	85	At	$[Xe]4f^{14}5d^{10}6s^26p^5$
52	Te	$[Kr]4d^{10}5s^25p^4$	86	Rn	$[Xe]4f^{14}5d^{10}6s^26p^6$
53	I	$[Kr]4d^{10}5s^25p^5$	87	Fr	$[Rn]7s$
54	Xe	$[Kr]4d^{10}5s^25p^6$	88	Ra	$[Rn]7s^2$
55	Cs	$[Xe]6s$	89	Ac	$[Rn]6d7s^2$
56	Ba	$[Xe]6s^2$	90	Th	$[Rn]6d^27s^2$
57	La	$[Xe]5d6s^2$	91	Pa	$[Rn]5f^26d7s^2$
58	Ce	$[Xe]4f5d6s^2$	92	U	$[Rn]5f^36d7s^2$
59	Pr	$[Xe]4f^36s^2$	93	Np	$[Rn]5f^46d7s^2$
60	Nd	$[Xe]4f^46s^2$	94	Pu	$[Rn]5f^67s^2$
61	Pm	$[Xe]4f^56s^2$	95	Am	$[Rn]5f^77s^2$
62	Sm	$[Xe]4f^66s^2$	96	Cm	$[Rn]5f^76d7s^2$
63	Eu	$[Xe]4f^76s^2$	97	Bk	$[Rn]5f^97s^2$
64	Gd	$[Xe]4f^75d6s^2$	98	Cf	$[Rn]5f^{10}7s^2$
65	Tb	$[Xe]4f^96s^2$	99	Es	$[Rn]5f^{11}7s^2$
66	Dy	$[Xe]4f^{10}6s^2$	100	Fm	$[Rn]5f^{12}7s^2$
67	Ho	$[Xe]4f^{11}6s^2$	101	Md	$[Rn]5f^{13}7s^2$
68	Er	$[Xe]4f^{12}6s^2$	102	No	$[Rn]5f^{14}7s^2$
69	Tm	$[Xe]4f^{13}6s^2$	103	Lr	$[Rn]5f^{14}6d7s^2$
70	Yb	$[Xe]4f^{14}6s^2$			

1-14 Effects of Electron-Electron Repulsion in Many-Electron Atoms

In Section 1-13 we saw that electron-electron (e-e) repulsion had an important effect on the ionization energy of the helium atom and in determining the ground electronic configuration of the carbon atom. In this section we wish to show that e-e repulsion is also important in determining the electron occupation of the 3d and 4s orbitals of the transition metal atoms.

We must emphasize that Figure 1-25 gives the order of *filling* of orbitals only for neutral atoms. Orbital *energies* depend heavily on the atomic number and on the charge of the atom (ion), as is illustrated for the $3d$ and $4s$ orbital levels in the first transition series of elements. For example, Figure 1-25 indicates that the $4s$ orbital is filled before the $3d$ orbital, but chromium and copper, which are exceptions, have the electronic configurations $[Ar]3d^54s^1$ and $[Ar]3d^{10}4s^1$, respectively. Also, although the $4s$ orbital is occupied before the $3d$ orbitals for the scandium (Sc) atom, giving the electronic configuration $[Ar]3d^14s^2$, in the Sc^{2+} ion the $3d$ orbital is filled *before* the $4s$ orbital, with the resulting electronic configuration $[Ar]3d^1$.

We can use the case of the scandium atom to illustrate the importance of *e-e* repulsion in many-electron atoms. As we shall see, the sequence of energy levels ($4s < 3d$) shown in Figure 1-25 is not correct for the scandium atom. The simple relationship $IE_n = -E_n$ that we derived for the hydrogen atom (Equation 1-11) also can be applied qualitatively to many-electron atoms. Hence we can use the ionization energy of an electron to determine its orbital energy. The experimental results for scandium indicate that 6.62 eV is required to ionize a $4s$ electron, whereas 7.98 eV is required to ionize the $3d$ electron. That is,

$$Sc(3d^14s^2) \rightarrow Sc^+(3d^14s^1) + e^- \quad IE_{4s} = 6.62 \text{ eV}$$
$$Sc(3d^14s^2) \rightarrow Sc^+(3d^04s^2) + e^- \quad IE_{3d} = 7.98 \text{ eV}$$

from which we see that it is easier to ionize a $4s$ electron than a $3d$ electron.* Consequently we should employ the energy-level diagram

$$E \uparrow \quad \begin{array}{ll} 4s \,\textcircled{\tiny$\uparrow\downarrow$} & -6.62 \text{ eV} \\ 3d \,\textcircled{\tiny$\uparrow$} & -7.98 \text{ eV} \end{array}$$

to indicate that the $3d$ orbital is more stable than the $4s$ orbital. However, this energy-level diagram presents us with a dilemma, which can be stated as follows: "If the $3d$ orbital is more stable than the $4s$ orbital, why is the $4s$ orbital occupied in the ground state?" Alternatively, we could examine the experimental energy required to change one electron from the $4s$ orbital to the $3d$ orbital in forming the excited scandium atom with electron configuration $[Ar]3d^24s^1$. That is,

$$Sc(3d^14s^2) \rightarrow Sc(3d^24s^1); \qquad \Delta E = 2.03 \text{ eV}$$

*In the first transition series, Sc → Zn, the first ionization energy corresponds to the loss of a $4s$ electron.

The question now could be rephrased as follows: "If the $3d$ orbital is more stable than the $4s$ orbital, why is energy *required* to form the electronic configuration $[Ar]3d^2 4s^1$ from $[Ar]3d^1 4s^2$?" To answer this question we need to consider the effects of *e-e* repulsion.

Actually, in all many-electron atoms there are two effects to consider. The first is the *spin-pairing* energy; the second is the *e-e* repulsion. The latter results from the coulomb repulsion of like charges, and for the transition-metal valence electrons it is much larger than the spin-pairing energy, which we will neglect. If we assume that the interaction of core electrons with valence electrons is constant and that the energy of the core electrons is constant, we can use the following expression to represent the relative energy of an atom A with valence electronic configuration $a^m b^n$:

$$E(A, a^m b^n) = -mW(a) - nW(b) + \frac{m(m-1)}{2} J(a,a) + mnJ(a,b)$$

$$+ \frac{n(n-1)}{2} J(b,b) \tag{1-22}$$

in which W is the ionization energy of a system containing only one valence electron, and $J(a,b)$ represents the *e-e* repulsion between the charge densities ψ_a^2 and ψ_b^2. In our approach we will consider W and J as empirical parameters, that is, parameters to be obtained from experimental results.

To apply Equation 1-22 let us choose the specific example of the scandium atom discussed previously. We then obtain the following expressions for the relative energy of Sc^{2+}, Sc^+, and Sc:

$$E(Sc^{2+}, 3d^1 4s^0) = -W(d) \tag{1-23}$$
$$E(Sc^{2+}, 3d^0 4s^1) = -W(s) \tag{1-24}$$
$$E(Sc^+, 3d^1 4s^1) = -W(d) - W(s) + J(d,s) \tag{1-25}$$
$$E(Sc^+, 3d^0 4s^2) = -2W(s) + J(s,s) \tag{1-26}$$
$$E(Sc, 3d^1 4s^2) = -W(d) - 2W(s) + 2J(d,s) + J(s,s) \tag{1-27}$$
$$E(Sc, 3d^2 4s^1) = -2W(d) - W(s) + J(d,d) + 2J(d,s) \tag{1-28}$$

In Equations 1-23 through 1-28 the $-W(d)$ and $-W(s)$ terms represent the binding energy of the $3d^1$ and $4s^1$ electrons in Sc^{2+}, respectively. The J terms represent the *e-e* repulsion. For example, $J(d,s)$ represents the *e-e* repulsion between the $3d$ and $4s$ electrons. The number of coulomb repulsions depends on the number of electrons. In the $3d^1 4s^2$ electronic configuration of scandium there are two $J(d,s)$, one $J(s,s)$, and no $J(d,d)$ terms, as seen in Equation 1-27.

Using the experimentally determined ionization and excitation energies of scandium, we can obtain values for each of the W and J terms in Equations

1-23 through 1-28. The energy required to excite a $4s$ electron to a $3d$ orbital is known:

$$Sc(3d^14s^2) \rightarrow Sc(3d^24s^1)$$

$$\Delta E = E(Sc, 3d^24s^1) - E(Sc, 3d^14s^2) = 2.03 \text{ eV} \tag{1-29}$$

Also, the energy required to ionize a $4s$ electron or a $3d$ electron from the scandium atom has been determined:

$$Sc(3d^14s^2) \rightarrow Sc^+(3d^04s^2) + e^-$$

$$\Delta E = E(Sc^+, 3d^04s^2) - E(Sc, 3d^14s^2) = 7.98 \text{ eV} \tag{1-30}$$

and

$$Sc(3d^14s^2) \rightarrow Sc^+(3d^14s^1) + e^-$$

$$\Delta E = E(Sc^+, 3d^14s^1) - E(Sc, 3d^14s^2) = 6.62 \text{ eV} \tag{1-31}$$

from which we can see that it is easier to ionize a $4s$ electron than a $3d$ electron. Finally, the ionization energies of Sc^{2+} are known and they allow us to determine values for $W(d)$ and $W(s)$:

$$Sc^{2+}(3d^14s^0) \rightarrow Sc^{3+}(3d^04s^0) + e^-$$

$$\Delta E = E(Sc^{3+}, 3d^04s^0) - E(Sc^{2+}, 3d^14s^0) = 24.75 \text{ eV}$$

$$Sc^{2+}(3d^04s^1) \rightarrow Sc^{3+}(3d^04s^0) + e^-$$

$$\Delta E = E(Sc^{3+}, 3d^04s^0) - E(Sc^{2+}, 3d^04s^1) = 21.60 \text{ eV}$$

However, since we have assumed the core electrons $1s^22s^22p^63s^23p^6$ to be constant in energy, these last two equations allow us to write directly

$$W(d) = 24.75 \text{ eV}$$

and

$$W(s) = 21.60 \text{ eV}$$

Using the experimental values given in Equations 1-29 through 1-31 we are able to obtain empirical values for each of the terms $J(d,d)$, $J(d,s)$, and $J(s,s)$ since we have three equations and three unknowns. If we write the appropriate energy differences for each of the Equations 1-25 through 1-28 and combine these with the experimental results given in Equations 1-29 through 1-31 we obtain

$$E(\text{Sc}, 3d^24s^1) - E(\text{Sc}, 3d^14s^2)$$
$$= W(s) - W(d) + J(d,d) - J(s,s) = 2.03 \text{ eV} \tag{1-32}$$

$$E(\text{Sc}^+, 3d^04s^2) - E(\text{Sc}, 3d^14s^2)$$
$$= W(d) - 2J(d,s) = 7.98 \text{ eV} \tag{1-33}$$

$$E(\text{Sc}^+, 3d^14s^1) - E(\text{Sc}, 3d^14s^2)$$
$$= W(s) - J(d,s) - J(s,s) = 6.62 \text{ eV}$$

From Equation 1-33 we easily can solve for $J(d,s)$ since we know that $W(d)$ = 24.75 eV. In a similar fashion we can solve for the $J(d,d)$ and $J(s,s)$ terms. The values obtained for the J terms are

$$J(d,s) = 8.38 \text{ eV}$$
$$J(d,d) = 11.78 \text{ eV}$$
$$J(s,s) = 6.60 \text{ eV}$$

Now we are in a position to answer the question why the ground-state electronic configuration of the scandium atom is $[\text{Ar}]3d^14s^2$ and not $[\text{Ar}]3d^24s^1$. Let us examine each of the terms in Equation 1-32, which we repeat here:

$$E(\text{Sc}, 3d^24s^1) - E(\text{Sc}, 3d^14s^2)$$
$$= W(s) - W(d) + J(d,d) - J(s,s)$$
$$= 21.60 - 24.75 + 11.78 - 6.60 = 2.03 \text{ eV}$$

We see that although the relative values of $W(d)$ and $W(s)$ would tend to favor the $[\text{Ar}]3d^24s^1$ configuration, this effect is overcome by the larger e-e repulsion in the $3d$ orbitals as compared to that of the $4s$ orbital. In summary,

$$J(d,d) - J(s,s) > W(d) - W(s)$$

and the $[\text{Ar}]3d^24s^1$ electron configuration is at higher energy than $[\text{Ar}]3d^14s^2$ due to the large e-e repulsion, $J(d,d)$, in the $3d$ orbitals. The outcome, $J(d,d)$ > $J(s,s)$, results from the fact that the $3d$ orbitals have principal quantum number $n = 3$, whereas the $4s$ orbital has $n = 4$. This causes the $3d$ orbitals to be less diffuse than the $4s$ orbital.

Now we pose a second question about the electronic structure of the scandium atom: "Why does the $4s$ electron ionize at lower energy than the $3d$ electron in forming Sc^+?" If we subtract Equation 1-26 from Equation 1-25 to obtain the energy difference between $\text{Sc}^+(3d^04s^2)$ and $\text{Sc}^+(3d^14s^1)$ we obtain

$$E(Sc^+, 3d^04s^2) - E(Sc^+, 3d^14s^1)$$
$$= W(d) - W(s) - J(d,s) + J(s,s)$$
$$= 24.75 - 21.60 - 8.38 + 6.60 = 1.37 \text{ eV}$$

From this equation it is clear that

$$W(d) - W(s) > J(d,s) - J(s,s)$$

and we see that the difference between the one-electron terms, $W(d) - W(s)$, is more important than the difference in e-e repulsion, $J(d,s) - J(s,s)$. That is, the scandium $4s$ electron ionizes at lower energy than the $3d$ electron because $W(s) < W(d)$. This outcome results from the fact that the $4s$ orbital does not have a node at the nucleus whereas the $3d$ orbital has two nodes containing the nucleus (Figure 1-23). Consequently, the $4s$ orbital is able to better penetrate the core electrons and become stabilized relative to the $3d$ orbital.

Accurate quantum-mechanical calculations have been carried out on the scandium atom to determine the most stable electronic configuration. These calculations show that the $[Ar]3d^14s^2$ configuration is more stable than $[Ar]3d^24s^1$ for the reason that we discussed previously, namely that $J(d,d)$ is large. In addition, there is a large repulsive term between the core electrons and the $3d$ electrons. This "core-valence" e-e repulsion is much less for an electron in a $4s$ orbital than in a $3d$ orbital.

1-15 Atomic Energy States and Term Symbols

Due to the effects of e-e repulsion in many-electron atoms, it is possible for one electronic *configuration* to result in several atomic energy *states*. For example, in the carbon atom it is readily noticed that the following orbital arrangements will have different energies:

Each of these orbital arrangements has the same electronic configuration, namely, $1s^22s^22p^2$. In this section we wish to show how to determine the number of energy states that may arise for each electronic configuration and how to assign a *term symbol* to each energy state.

The most convenient classification of an atomic energy state is in terms of total orbital angular momentum L, total spin angular momentum S, and total angular momentum J. The term symbol for any atomic energy state that has L, S, and J specified is written as $^{2S+1}L_J$. In atomic-term symbols the capital letters S, P, D, F, G, . . . are employed for the total electronic orbital angular momentum quantum number, $L = 0, 1, 2, 3, 4, \ldots$, respectively. This notation agrees with the corresponding lower case letters that are used to designate the quantum number l, which determines the angular momentum of electrons in individual orbitals. The possible values of J that specify the total angular momentum can be determined by

$$J = L + S, L + S - 1, \ldots, |L - S|.$$

For a one-electron atom or ion, the total orbital angular momentum is that of the orbital occupied by the electron. Consequently the term symbols S, P, D, F, G, . . . correspond exactly to the orbital designations s, p, d, f, g, ... Since for a single electron $s = \frac{1}{2}$ we obtain $S = \frac{1}{2}$, so $2S + 1 = 2$. The value of $2S + 1$ gives the spin degeneracy or *multiplicity* for the atom. In the case of a one-electron atom we have two possibilities, $m_s = +\frac{1}{2}$ or $-\frac{1}{2}$.

The term symbol for the ground-state hydrogen atom electronic configuration $1s^1$ is designated $^2S_{1/2}$ (read "doublet s one-half") since we have $S = \frac{1}{2}$, $L = 0$, and $J = \frac{1}{2}$. The excited-state hydrogen atom with electronic configuration $2s^1$ also is designated $^2S_{1/2}$ but these two $^2S_{1/2}$ states do not have the same energy. Sometimes these two states are distinguished as $1s\,^2S_{1/2}$ and $2s\,^2S_{1/2}$.

The excited-state hydrogen atom that has an electronic configuration $2p^1$ may have two term symbols, $^2P_{3/2}$ and $^2P_{1/2}$. Notice that in this case *two* energy states arise from *one* electronic configuration. The energy difference or splitting between these states is a result of spin-orbit coupling. That is, the L and S quantum numbers can combine in two different ways, $L \pm S$, to give J values of $\frac{3}{2}$ and $\frac{1}{2}$. Figure 1-26 shows the relative energy of the states resulting from the $1s^1$, $2s^1$, and $2p^1$ electronic configurations of the hydrogen atom. Notice that the energies of the $^2P_{3/2}$ and $^2P_{1/2}$ states are only slightly different. This indicates that for the hydrogen atom the effect of spin-orbit coupling is small. Experimentally it is found that for many-electron atoms, the splitting in the valence shells is approximately proportional to Z^2. Consequently, spin-orbit coupling only becomes important for elements with relatively large atomic numbers.

Energy States in Many-Electron Atoms

The individual orbital and spin angular momenta in many-electron atoms may be combined by the Russell-Saunders scheme. For a system of n elec-

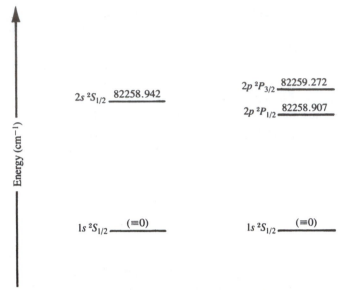

Figure 1-26 Relative energies (in reciprocal centimeters) of the $1s\ ^2S_{1/2}$, $2s\ ^2S_{1/2}$, and $2p\ ^2P_{3/2,1/2}$ states of the hydrogen atom. The $1s\ ^2S_{1/2}$ state is the ground state and therefore is assigned a relative energy of zero. Notice that the states derived from the $2s^1$ and $2p^1$ electronic configurations differ slightly in energy as a result of spin-orbit coupling in the $2p^1$ electronic configuration $(^2P_{3/2,1/2})$. The $^2P_{3/2,1/2}$ energy difference is greatly expanded in this figure.

trons, we define the total spin angular momentum as

$$S = s_1 + s_2 + \cdots + s_n, \ s_1 + s_2 + \cdots + s_n - 1,$$
$$s_1 + s_2 + \cdots + s_n - 2, \cdots$$
$$= \frac{n}{2}, \frac{n}{2} - 1, \frac{n}{2} - 2, \cdots, 0 \qquad (n \text{ even})$$
$$= \frac{n}{2}, \frac{n}{2} - 1, \frac{n}{2} - 2, \cdots, \frac{1}{2} \qquad (n \text{ odd})$$

From the preceding equations, we see that S goes in integer steps from a maximum that corresponds to all parallel spins to a minimum which is zero for an even number of electrons and $\frac{1}{2}$ for an odd number of electrons. The definition for the total orbital angular momentum is similar:

$$L = l_1 + l_2 + \cdots + l_n, \ l_1 + l_2 + \cdots + l_n - 1,$$
$$l_1 + l_2 + \cdots + l_n - 2, \ldots$$

If all l are equal, the minimum is zero; if one l is larger than the others, the minimum value is that given by orienting the other angular momenta to

TABLE **1-5.** TERMS ARISING
FROM THE CONFIGURATION
$2p^13p^1$

L	S	
	0	1
2	$^1D_2$	$^3D_{3,2,1}$
1	$^1P_1$	$^3P_{2,1,0}$
0	$^1S_0$	$^3S_1$

oppose it, subject to the condition that $L \geq 0$. Thus for an electronic configuration $2p^13p^1$ ($l_1 = 1$, $l_2 = 1$) we have the possible L and S values

$$L = 2, 1, 0 \qquad S = 1, 0$$

whereas for the electronic configuration $2p^13p^14f^1$ we have

$$L = 5, 4, 3, 2, 1 \qquad S = \tfrac{3}{2}, \tfrac{1}{2}$$

For any given energy state for which L and S have been specified, the J values can be determined readily. Thus for $L = 2$ and $S = 1$ we obtain $J = 3, 2, 1$.

In the discussion of energy states associated with a given electronic configuration, we have ignored the Pauli principle up to this point. This is because we have restricted our discussion to *nonequivalent* electrons, that is, electrons having different n or l quantum numbers. In these cases the Pauli exclusion principle will be obeyed automatically, and all of the combinations of L and S are possible. In Table 1-5 we tabulate these possibilities for the electronic configuration $2p^13p^1$.

Energy States in Many-Electron Atoms Containing Equivalent Electrons

We now wish to discuss a method by which the possible energy states of an atom containing *equivalent* electrons can be determined. By equivalent we mean that the electrons have the same n and l quantum numbers. To be sure that the Pauli exclusion principle is obeyed, we must use a method in which the m_l and m_s values of each electron are specified. Therefore we wish to know M_L and M_S for the total electronic system. In the Russell-Saunders scheme these are defined as

$$M_L = m_{l_1} + m_{l_2} + m_{l_3} + \cdots + m_{l_n} \qquad (1\text{-}34)$$
$$M_S = m_{s_1} + m_{s_2} + m_{s_3} + \cdots + m_{s_n} \qquad (1\text{-}35)$$

These relationships also exist between L and M_L, S and M_S:

$$M_L = L, L-1, L-2, \ldots, -L \tag{1-36}$$
$$M_S = S, S-1, S-2, \ldots, -S \tag{1-37}$$

Let us take the lithium atom as an illustrative example. The orbital electronic configuration of the ground state is $1s^2 2s^1$. The ground-state LSM_LM_S term is determined as follows:

1. Find the possible values of M_L:

$$M_L = m_{l_1} + m_{l_2} + m_{l_3}$$
$$m_{l_1} = m_{l_2} = m_{l_3} = 0 \text{ (all are } s \text{ electrons)}$$
$$M_L = 0$$

2. Find the possible values of L:

$$M_L = 0$$
$$L = 0$$

3. Find the possible values of M_S:

$$M_S = m_{s_1} + m_{s_2} + m_{s_3}$$
$$m_{s_1} = +\tfrac{1}{2}, \qquad m_{s_2} = -\tfrac{1}{2}, \qquad m_{s_3} = \pm\tfrac{1}{2}$$
$$M_S = +\tfrac{1}{2} \quad \text{or} \quad -\tfrac{1}{2}$$

4. Find the possible values of S:

$$M_S = +\tfrac{1}{2}, -\tfrac{1}{2}$$
$$S = \tfrac{1}{2}$$

In summary, the ground-state term for the lithium atom has $L = 0$, $S = \tfrac{1}{2}$, and $J = \tfrac{1}{2}$, designated $^2S_{1/2}$.

An excited electronic configuration for lithium would be $1s^2 2p^1$. For this configuration we find that $M_L = 1$, 0, or -1 ($L = 1$) and $M_S = \pm\tfrac{1}{2}$ ($S = \tfrac{1}{2}$). Therefore the term designation of this particular excited state is $^2P_{3/2}$ if $J = \tfrac{3}{2}$, and $^2P_{1/2}$ if $J = \tfrac{1}{2}$. The example of the lithium atom shows that in determining the number of energy states arising from a given electronic configuration, we may neglect filled shells and subshells since they always result in $M_L = 0$ ($L = 0$) and $M_S = 0$ ($S = 0$).

Admittedly, the lithium atom is a very simple example. To help find the energy states for more complicated electronic structures, we may construct a chart of the possible M_L and M_S values. This general procedure may be illustrated with the carbon atom, which has six electrons. Thus the electronic configuration of the ground state must be $1s^2 2s^2 2p^2$. It remains for us to determine how many energy states can arise from this one electronic configuration.

First we draw a chart as shown in Table 1-6. We need to consider only the electrons in incompletely filled subshells. For carbon the configuration $2p^2$ is important. Each of the two p electrons has $l = 1$ and therefore can have $m_l = +1, 0,$ or -1. Each of the two p electrons can have $m_s = +\frac{1}{2}$ or $-\frac{1}{2}$.

The next step is to write all allowable combinations of m_l and m_s values (called *microstates*) for the two p electrons, and to determine M_L and M_S values for each of these according to Equations 1-34 and 1-35. As shown in Table 1-6 there are 15 possible microstates.

According to Equations 1-36 and 1-37, a term corresponding to a given value of L and S has $(2L+1) \times (2S+1)$ combinations of M_L and M_S values associated with it. To find the L and S values that can be obtained from the $M_L - M_S$ combinations in Table 1-6, we begin with the maximum M_S and find the maximum M_L associated with it. From Table 1-6, this is $(M_S)_{max} = 1$, since $M_L = 2$ occurs only for $M_S = 0$. According to Equations 1-36 and 1-37, these microstates correspond to a term with

$$L = 1 \qquad M_L = 1, 0, -1 \qquad S = 1 \qquad M_S = 1, 0, -1$$

These values of M_L and M_S account for nine of the $M_L - M_S$ combinations in the table and leave

$$
\begin{array}{ccccccc}
M_S & 0 & 0 & 0 & 0 & 0 & 0 \\
M_L & -2 & 0 & 2 & -1 & 0 & 1
\end{array}
$$

From these combinations we again pick out the maximum M_S and M_L values (here $M_S = 0$, $M_L = 2$), and associate them with

$$L = 2 \qquad M_L = 2, 1, 0, -1, -2 \qquad S = 0 \qquad M_S = 0$$

Five more of the combinations have been accounted for, and there remains only $M_L = 0$ and $M_S = 0$, which must correspond to $L = 0$ and $S = 0$. Thus we find that, consistent with the Pauli principle, the $1s^2 2s^2 2p^2$ electronic configuration gives three energy states with

$$L = 1, S = 1 \qquad L = 2, S = 0 \qquad L = 0, S = 0$$

and the corresponding term symbols

$$^3P_{2,1,0} \qquad ^1D_2 \qquad ^1S_0$$

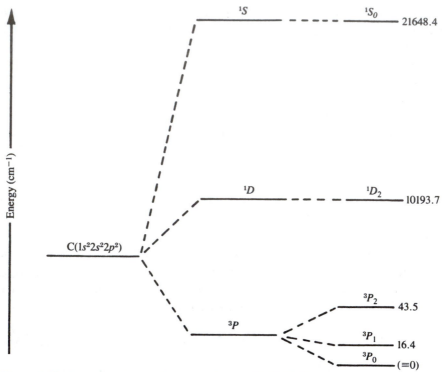

Figure 1-27 Term splitting in the ground-state $(1s^22s^22p^2)$ configuration of carbon. All energies are in cm^{-1}. The ^3P, ^1D, and ^1S terms are split as a result of electron–electron repulsion. The ^3P term is further split into $^3P_{2,1,0}$ as a result of spin-orbit coupling. The magnitude of the spin-orbit coupling is expanded in this figure.

Ground-State Terms for Many-Electron Atoms

Hund's rules allow us to decide which of several energy states will be lowest in energy. These rules can be formulated in three steps:

1. The ground state term always has maximum spin multiplicity. Therefore, for the carbon atom, the $^3P_{2,1,0}$ terms will be more stable than the $^1D_2$ and $^1S_0$ terms.

2. When comparing two states of the same spin multiplicity, the state with the higher L value is usually more stable. This is the case with the $^1D_2$ and $^1S_0$ terms of the carbon atom.

3. For given S and L values, the minimum J value is most stable if there is a single open subshell that is less than half-full, and the maximum J is most stable if the subshell is more than half-full. Thus for the carbon atom the $^3P_0$ state is the ground-state term.

The experimental relative energies of the energy states arising from the $1s^22s^22p^2$ electronic configuration of carbon are presented in Figure 1-27.

TABLE **1-6.** ALLOWED m_l AND m_s VALUES FOR THE p^2 CONFIGURATION

m_l															
$+1$		↑↓		↓	↑	↓	↑					↑	↓	↑	↓
0		↑↓	↓	↑			↑	↓	↑	↓			↑	↓
-1	↑↓		↑	↓	↑	↓			↑	↓	↑	↓	
M_S	0	0	0	0	0	0	0	0	0	1	-1	1	-1	1	-1
M_L	-2	0	2	-1	-1	0	0	1	1	-1	-1	0	0	1	1

It is clear that for a light atom such as carbon (small Z), the spin-orbit coupling is much smaller than the electrostatic interaction between the electrons.

When a subshell is more than half-full it is more convenient to work out the energy states and associated term symbols with respect to *electron holes* (i.e., vacancies in the various spin and orbital states) rather than electrons. For example, the electronic configurations for the carbon atom $(1s^2 2s^2 2p^2)$ and the oxygen atom $(1s^2 2s^2 2p^4)$ both give rise to the terms $^3P_{2,1,0}$, $^1D_2$, and $^1S_0$. However, in accordance with Hund's rule 3, the ground-state term symbol for the carbon atom is $^3P_0$, whereas for the oxygen atom it is $^3P_2$.

Determination of Only the Ground-State Term Symbol

For the carbon atom we used the 15 microstates presented in Table 1-6 to obtain the energy states $^3P_{2,1,0}$, $^1D_2$, and $^1S_0$. In many cases we wish to determine the ground-state term symbol $(^3P_0)$ without constructing all of the possible microstates. This can be done readily by writing the *one* microstate that satisfies Hund's rules. That is, we first choose the maximum M_S value that is consistent with the Pauli principle and then choose the maximum M_L value associated with this M_S value. For the carbon atom this gives

	⊘	⊘	↑	↑ ○	
	1s	2s		2p	
m_l	0	0	$+1$	0	-1
m_s	$\pm\frac{1}{2}$	$\pm\frac{1}{2}$	$+\frac{1}{2}$	$+\frac{1}{2}$	

$(M_S)_{max} = 1$ $(M_L)_{max} = 1.$

From Equations 1-36 and 1-37 we know that if $M_S = 1$, then $S = 1$, and if $M_L = 1$, then $L = 1$. The term symbol for $L = 1$ and $S = 1$ is ^3P, which was perviously determined to be the ground energy state.

Atomic Energy States and Valence-Orbital Ionization Energies

In our discussion of chemical bonding in subsequent chapters, we sometimes will require knowledge of the atomic orbital energies. We can obtain
the orbital energy from the ionization energy by applying the one-electron
equation $E_n = -IE_n$ (Equation 1-11). However, in many-electron atoms there
is often more than one energy state from which ionization can occur. Let us
return to the example of the carbon atom. In the ionization process we have

$$C\,(1s^2 2s^2 2p^2) \rightarrow C^+(1s^2 2s^2 2p^1) + e^-$$

in which the ground-state ionization process corresponds to $^3P_0 \rightarrow {}^2P_{3/2}$ and
requires 11.260 eV.

If we wish to know the *average* energy required to ionize a $2p$ electron
from the carbon atom we should consider ionization from all of the energy
states derived from the configuration $1s^2 2s^2 2p^2$. As shown in Figure 1-27
these are $^3P_0$, $^3P_1$, $^3P_2$, $^1D_2$, and $^1S_0$. Since C^+ has only one electron in its
incomplete subshell ($2p^1$), the terminal state produced in the ionization is
always $^2P_{3/2}$. To obtain the average ionization energy we simply take the
weighted average of the energy states arising from the carbon atom electronic configuration ($1s^2 2s^2 2p^2$). Each state has a degeneracy of $2J + 1$:

$$\begin{array}{cccccc} & {}^3P_0 & {}^3P_1 & {}^3P_2 & {}^1D_2 & {}^1S_0 \\ 2J+1 \rightarrow & 1 & 3 & 5 & 5 & 1 \quad \text{(degeneracy)} \end{array}$$

The average energy then is given as

$$\begin{aligned} E(1s^2 2s^2 2p^2)_{\text{average}} \\ = (\tfrac{1}{15})\,[E(^3P_0) + 3E(^3P_1) + 5E(^3P_2) + 5E(^1D_2) + E(^1S_0)] \end{aligned}$$

This energy is calculated relative to the ground state, that is, $E(^3P_0) \equiv 0$.

The *2p valence-orbital ionization energy, VOIE*, is defined as the energy
required to remove a $2p$ electron from the average energy state:

$$E(1s^2 2s^2 2p^2)_{\text{average}} \rightarrow E(1s^2 2s^2 2p^1)$$

Reference to Figure 1-28 shows that

$$VOIE\ (2p) = IE - E(1s^2 2s^2 2p^2)_{\text{average}}$$

in which *IE* refers to the ground-state ionization $^3P_o \rightarrow {}^2P_{3/2}$. Using the
experimental values given in Figure 1-28 and expressing the result in electron volts, we obtain

$$VOIE\ (2p) = 11.260 - 0.60 = 10.66 \text{ eV}$$

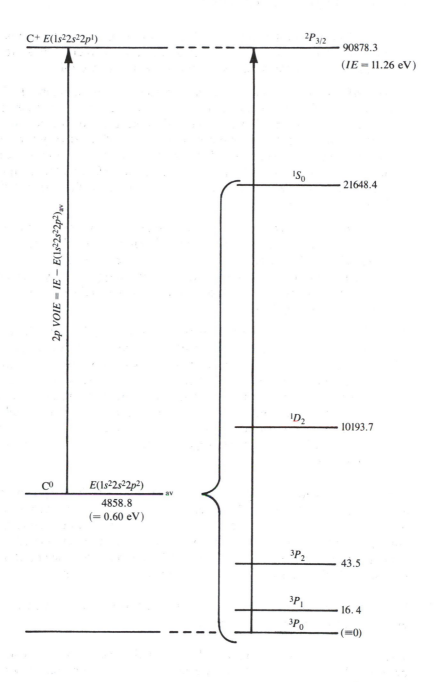

Figure 1-28 Evaluation of the $2p$ valence orbital ionization energy of carbon. Energies are expressed in reciprocal centimeters and in electron volts where indicated.

QUESTIONS AND PROBLEMS*

1. In each of the three following statements choose *one* of the four possibilities, (a), (b), (c), or (d), that most accurately completes the statement or answers the question. Read the statements carefully.

    A. Rutherford, Geiger, and Marsden performed experiments in which a beam of helium nuclei (α particles) was directed at a thin piece of gold foil. They found that the gold foil (a) severely deflected most of the particles of the beam directed at it; (b) deflected very few of the particles of the beam and deflected these only very slightly; (c) deflected most of the particles of the beam but deflected these only very slightly; (d) deflected very few of the particles of the beam but deflected these severely.

    B. From the results in (A) Rutherford concluded that (a) electrons are massive particles; (b) the positively charged parts of atoms are extremely small and extremely heavy particles; (c) the positively charged parts of atoms move with a velocity approaching that of light; (d) the diameter of an electron is approximately equal to the diameter of the nucleus.

    C. Which one of the following statements concerning the Bohr theory of the hydrogen atom is not true? The theory (a) successfully explained the observed emission and absorption spectra of the hydrogen atom; (b) requires that the greater the energy of the electron in the hydrogen atom the greater its velocity; (c) requires that the energy of the electron in the hydrogen atom can have only certain discrete values; (d) requires that the distance of the electron from the nucleus in the hydrogen atom can have only certain discrete values.

2. Consider two hydrogen atoms. The electron in the first hydrogen atom is in the $n = 1$ Bohr orbit. The electron in the second hydrogen atom is in the $n = 4$ orbit. (a) Which atom has the ground-state electronic configuration? (b) In which atom is the electron moving faster? (c) Which orbit has the larger radius? (d) Which atom has the lower potential energy? (e) Which atom has the higher ionization energy?

3. How much energy is required to ionize a hydrogen atom in which the electron occupies the $n = 5$ Bohr orbit?

4. Write an expression for the wavelength of the radiation that would be emitted by a He^+ ion when it decays from an excited state having

*Answers to selected problems appear in Appendix C.

principal quantum number $n = 4$ to a lower excited state having $n = 3$. Your expression should give the wavelength as a function of m_e, e, h, π, and c only. Calculate the numerical value of the wavelength of the emitted radiation. (You may ignore the small correction due to the reduced mass of He^+ and employ m_e in your expression.)

5. Calculate the wavelength of a photon of visible light with a frequency of 0.66×10^{15} sec^{-1}. What is the energy of the photon? What is the wave number?

6. Calculate the energy released when a hydrogen atom decays from the state having the principal quantum number 3 to the state having the principal quantum number 2.

7. Calculate the frequency of the light emitted when a hydrogen atom decays as in problem 6. Calculate the wave number of the light emitted in the decay of the hydrogen atom described.

8. The ground-state electronic configuration of lithium is $1s^2 2s^1$. When lithium is heated in a flame, it emits bright red light. The red color is due to light emission at a wavelength of 670.8 nm. No light emission is observed at longer wavelengths. (a) Suggest possible explanations for the emission at 670.8 nm. [*Hint:* The $1s^2$ electrons are not involved.] (b) What is the frequency of the light? (c) What is the wave number (cm^{-1}) of the light? (d) What is the energy of the process in kilocalories per mole?

9. Following are several electronic configurations that may be correct for the nitrogen atom $(Z = 7)$. Electrons are represented by arrows whose direction indicates the value of the spin quantum number, m_s. The three circles for the p orbitals indicate the possible values of the orbital-orientation quantum number, m_l. For each configuration choose the correct word: *excited,* if the configuration represents a possible excited state of the nitrogen atom; *ground,* if the configuration represents the ground state of the nitrogen atom; *forbidden,* if the configuration cannot exist.

10. Write the orbital electronic structures for the following atoms and ions and, where appropriate, show that you know Hund's rule: P ($Z = 15$); Na ($Z = 11$); As ($Z = 33$); C^- ($Z = 6$); O^+ ($Z = 8$).

11. Determine the number of unpaired electrons in the following atoms: C ($Z = 6$); F ($Z = 9$); Ne ($Z = 10$).

12. Draw spatial representations for the following orbitals and put in X, Y, and Z coordinates, if needed: $2p_z$; $3s$; $3d_{x^2-y^2}$; $3d_{xz}$; $n = 2$, $l = 1$.

13. An electron is in one of the $3d$ orbitals. What are the possible values of the orbital quantum numbers n, l, and m_l for the electron?

14. Write the orbital electronic structure of the ground state of (a) the calcium atom ($Z = 20$) and (b) the Mg^{2+} ion ($Z = 12$).

15. The Balmer series for atomic hydrogen occurs in the visible region of the spectrum. Which series in the emission spectrum of Be^{3+} has its lowest-energy line closest to the first line in the hydrogen Balmer series?

16. Suppose that you discovered some material from another universe that obeyed the following restrictions on quantum numbers:

$$n > 0$$
$$l + 1 \leq n$$
$$m_l = +l \text{ or } -l$$
$$m_s = +\tfrac{1}{2}$$

Assume that Hund's rule still applies. What would be the atomic numbers of the first three noble gases in that universe? What would be the atomic numbers of the first three halogens?

17. The spectrum of He^+ contains, along with many others, lines at 329,170 cm^{-1}, 390,125 cm^{-1}, 411,460 cm^{-1}, and 421,334 cm^{-1}. Show that these lines fit a Rydberg-type equation, $\bar{v} = R(1/n_I^2 - 1/n_{II}^2)$.

18. An atom is observed to emit light at 100.0 nm, 125.0 nm, and 500.0 nm. Theoretical considerations indicate that there are only two excited states involved. Explain why three lines are observed. How far above the ground state is the energy of the excited states?

19. Why do the $4s$, $4p$, $4d$, and $4f$ orbitals have the same energy in the hydrogen atom but different energies in many-electron atoms?

20. How can the same atom of hydrogen, in quick succession, emit a pho-

ton in the Brackett, Paschen, Balmer, and Lyman series? Can it emit them in the reverse order? Why, or why not?

21. A common photon source used for studying the ionization energies of atoms and molecules is the emission from excited helium atoms, which occurs at 58.4 nm:

$$He(1s^12p^1) \rightarrow He(1s^2)$$

(a) Calculate the frequency of light with wavelength of 58.4 nm. (b) What is the energy of the photon in ergs? In electron volts? (c) The ionization energy of the argon atom is 15.759 eV. If a photon from an excited helium atom strikes an argon atom, what will be the kinetic energy in electron volts of the ejected electron?

22. Consider the vanadium atom with the ground-state electronic configuration $[Ar]3d^34s^2$. Use the following interelectronic repulsion terms J and one-electron ionization energy terms W:

$$J(3d, 3d) = 17.36 \text{ eV}$$
$$J(3d, 4s) = 11.16$$
$$J(4s, 4s) = 8.68$$
$$W(3d) = 65.21$$
$$W(4s) = 46.86$$

(a) Calculate the excitation energy:

$$V(3d^34s^2) \rightarrow V(3d^5)$$

(b) Calculate the ionization energy for removal of a $3d$ electron:

$$V(3d^34s^2) \rightarrow V^+(3d^24s^2) + e^-$$

(c) Calculate the ionization energy for removal of a $4s$ electron:

$$V(3d^34s^2) \rightarrow V^+(3d^34s^1) + e^-$$

(d) From your results in parts (b) and (c), which electron, $3d$ or $4s$, has the lower ionization energy?

23. Find the ground-state term symbol for the following atoms: (a) Si, (b) Mn, (c) Rb, (d) Ni.

24. Find terms for the following orbital configurations, and in each case designate the term of lowest energy: (a) $2s$, (b) $2p^23s$, (c) $2p3p$, (d) $3d^2$, (e) $2p3d$, (f) $3d^9$, (g) $2s4f$, (h) $2p^5$, (i) $3d^24s$.

Suggestions for Further Reading

Berry, R.S., Advisory Council on College Chemistry Resource Paper, "Atomic Orbitals," *J. Chem. Educ.* 43: 283 (1966).

Birks, J.B., ed, *Rutherford at Manchester,* Menlo Park, Calif.: Benjamin, 1963.

Claydon, C.R., and Carlson, K.D., "Ground States, Configurations, and Truncated Orbital Bases of the Iron-Series Atoms," *J. Chem. Phys.* 49: 1331 (1968).

Cohen, I., and Bustard, T., "Atomic Orbitals; Limitations and Variations," *J. Chem. Educ.* 43: 187 (1966).

Devons, S., "Recollections of Rutherford and the Cavendish," *Physics Today* 24: 38 (1971).

Feynman, R.P., Leighton, R.B., and Sands, M., *The Feynman Lectures on Physics,* vol. 3, Reading, Mass.: Addison-Wesley, 1965.

Gray, H.B., *Electrons and Chemical Bonding,* Menlo Park, Calif.: Benjamin, 1965.

Gray, H.B., *Chemical Bonds,* Menlo Park, Calif.: Benjamin, 1973.

Hänsch, T.W., Schawlow, A.L., Series, G.W., "The Spectrum of Atomic Hydrogen," *Scientific American* (March, 1979).

Heisenberg, W., *The Physical Principles of Quantum Theory,* New York: Dover, 1930.

Jørgensen, C.K., "Photoelectron Spectra," in *Advances in Quantum Chemistry,* vol. 8, edited by P.-O. Löwdin, New York: Academic Press, 1974, p. 137.

Karplus, M., and Porter, R.N., *Atoms and Molecules: An Introduction for Students of Physical Chemistry,* Menlo Park, Calif.: Benjamin, 1970.

McGlynn, S.P., Vanquickenborne, L.G., Kinoshita, M., and Carroll, D.G., *Introduction to Applied Quantum Chemistry,* New York: Holt, Rinehart and Winston, 1972.

Ogryzlo, E.A., and Porter, G.B., "Contour Surfaces for Atomic and Molecular Orbitals," *J. Chem. Educ.* 40: 256 (1963).

Perlmutter-Hayman, B., "The Graphical Representation of Hydrogen-Like Functions," *J. Chem. Educ.* 46: 428 (1969).

Powell, R.E., "The Five Equivalent *d* Orbitals," *J. Chem. Educ.* 45: 1 (1968).

Weisskopf, V.F., "How Light Interacts with Matter," *Scientific American* (September, 1968).

2 Atomic and Molecular Properties

The analysis of orbital electronic structure guided by the Pauli principle clearly shows the simple basis for the periodic behavior of the elements. (A typical periodic table is found in Figure 7-9, page 426.) Generally, atoms with the same outer-orbital structure appear in the same column (group) of the periodic table. For example, atoms of the noble gas elements all have completely filled ns and np orbitals (closed-shell configurations). Metal atoms have very few electrons in the outermost s and p orbitals; thus they have a tendency to lose these electrons to achieve stable, closed-shell configurations. In contrast, nonmetals equal or exceed the s^2p^2 configuration (halfway to s^2p^6) in their outer-orbital structures, sometimes gaining electrons to achieve stable, closed-shell configurations.

The first-row transition metals are the elements scandium through zinc. These ten elements are the first to have orbital structures involving d electrons. Although the d electrons have important consequences for the spectroscopic properties of transition metal atoms, they do not grossly alter the chemical properties of these elements. The result is a "long period" of transition elements, all with similar properties.

2-1 Lewis Structures for Atoms

For all but a few atoms it is tedious to write the complete orbital electronic structure. It is also unnecessary, because only the valence electrons are important in chemical reactions. For example, in lithium the two $1s$ electrons are bound tightly to the nucleus of charge $+3$. Like the two electrons in helium, they are chemically unreactive. Thus we say that the valence electronic structure of lithium is $2s^1$, or Li·, in which the symbol Li represents the lithium nucleus and the two $1s$ electrons and the dot represent the valence

$2s$ electron. The shorthand "dot" notation is called a *Lewis structure,* after Gilbert N. Lewis, an American chemist. The Lewis notation vastly simplifies writing atomic structures.

Valence-orbital structures are so important in chemistry that all serious students should learn them for the main group elements. The learning task is made easy because the valence electronic structures are periodic. For example, oxygen, sulfur, and selenium atoms have the same valence structure ns^2np^4. Using sulfur as an example we write the Lewis formula for these atoms as $\cdot\ddot{S}:$. If we know that the valence electronic structure of atomic nitrogen is $\cdot\ddot{N}\cdot$, then we can write the Lewis formula for atomic phosphorus as $\cdot\ddot{P}\cdot$, because phosphorus is below nitrogen in the periodic table. Note that there is no distinction between the ns and np electrons in the dot notation.

Now consider the Lewis structures for the chlorine atom and the chloride ion. The closed-shell structure before chlorine is the neon structure, $1s^22s^22p^6$, and chlorine has seven electrons in addition to this closed-shell configuration. Thus the Lewis structure for chlorine has seven dots, $:\ddot{C}l:$, in which Cl represents the nucleus and the $1s^22s^22p^6$ electrons. When one electron is added to a chlorine atom to produce a chloride ion there are eight valence electrons, thereby giving the closed-shell configuration. The chloride ion has the structure $:\ddot{C}l:^-$, which shows the charge of -1.

2-2 Effective Atomic Radii in Molecules

Now we turn our attention to the relationship between atomic properties and valence-orbital structures. First we will consider the effective radius of an atom in a molecule. The *effective atomic radius* of an atom is defined as one half the distance between two nuclei of the element that are held together by a purely covalent single bond. (A covalent bond is a pair of electrons shared between two atoms.) For example, the separation of the two protons in the hydrogen molecule, H_2, is 0.74 Å. Therefore we assign each hydrogen atom in the H_2 molecule an atomic radius of 0.37 Å. The distance between lithium nuclei in Li_2 is 2.67 Å; thus the atomic radius of lithium is approximately 1.34 Å. The average effective radii of atoms of a selection of representative elements, shown in the periodic table arrangement of Figure 2-1, were determined from experimentally observed bond distances in many molecules. In most cases the atomic radius is compared with the size of the appropriate closed-shell positive or negative ion.

In terms of orbital structure the explanation of the shrinkage of atomic radii across a given row (or period) in Figure 2-1 is as follows. In any given

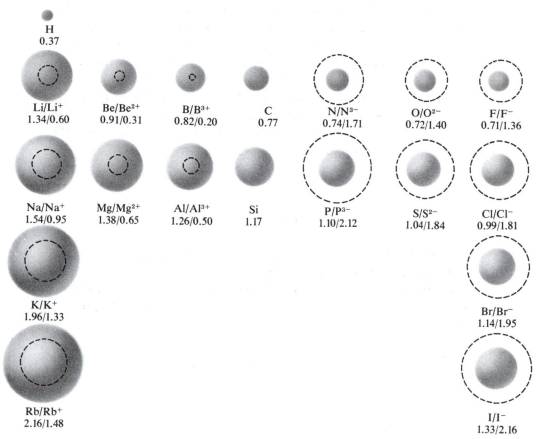

Figure 2-1 Relative atomic radii of some elements compared with the radii of the appropriate closed-shell ions. Radii are in angstroms. Solid spheres represent atoms, dashed circles represent ions. Notice that positive ions are smaller than their neutral atoms and that negative ions are larger.

period electrons are added to s and p orbitals, which are not able to shield each other effectively from the increasing positive nuclear charge. Thus an increase in the positive charge of the nucleus results in an increase in the effective nuclear charge Z_{eff}, thereby decreasing the effective atomic radius. This is the reason why a beryllium atom, for example, is smaller than a lithium atom.

From hydrogen to lithium there is a large increase in effective atomic radius because the third electron in a lithium atom is in an orbital that has a much larger effective radius than the hydrogen $1s$ orbital. According to the Pauli principle the third electron in lithium must be in an orbital with a larger principal quantum number, namely the $2s$ orbital. Seven more electrons can be added to the 2s and 2p orbitals, which have approximately the same radii. However, these electrons do not effectively shield each other from the positive nuclear charge as it increases; the result is an increase in Z_{eff} and a corresponding decrease in radius in the series lithium $(Z = 3)$ through neon $(Z = 10)$. After neon, additional electrons cannot be accommodated by the $n = 2$ level. Thus an eleventh electron must go into the $n = 3$ level, specifically, into the 3s orbital. Since the effective radius increases from the $n = 2$ to $n = 3$ valence orbitals, the effective size of an atom also increases with increasing atomic number within each group in the periodic table.

2-3 Ionization Energy and Orbital Configuration

As we explained in Section 1-6, the ionization energy, IE, of an atom is the energy required to remove an electron from the gaseous atom. The first ionization energy, IE_1, is the energy needed to remove one electron from the neutral gaseous atom to produce a unipositive gaseous ion. The process can be written

$$\text{atom}(g) + (\text{energy} = IE_1) \rightarrow \text{ion}^+(g) + e^- \qquad \Delta E = IE_1$$

For sodium we can write

$$\text{Na}(g) \rightarrow \text{Na}^+(g) + e^- \qquad IE_1 = 5.139 \text{ eV}$$

The first ionization energies for most of the elements are listed in Table 2-1, together with their ground-state orbital electronic configurations. For any atom, IE_1 is always the smallest ionization energy.

In all atoms except hydrogen, further ionizations are possible (Table 2-1). Ionization energy data provide the best evidence for the electronic structure of atoms. For example, the lithium atom, which has the electronic configuration $1s^2 2s^1$, has three ionization energies,

$$\text{Li}(g) \rightarrow \text{Li}^+(g) + e_{2s}^- \qquad IE_1 = 5.392 \text{ eV}$$
$$\text{Li}^+(g) \rightarrow \text{Li}^{2+}(g) + e_{1s}^- \qquad IE_2 = 75.638 \text{ eV}$$
$$\text{Li}^{2+}(g) \rightarrow \text{Li}^{3+}(g) + e_{1s}^- \qquad IE_3 = 122.45 \text{ eV}$$

We have indicated the type of electron that is ejected in each case. The large increase in the order $IE_1 < IE_2 < \ldots < IE_m$ is reasonable, because as successive electrons are lost, the ejected electron must be removed from an ion with greater positive charge. The first electron ejected from Li leaves behind a Li^+ ion, whereas the second ionized electron leaves the doubly charged Li^{2+} ion. In general, ionization of the mth electron produces an ion of charge $+m$.

Using Coulomb's law of electrostatics, we can develop a basis for interpreting IE_m. The energy by which a negative charge $-q_1$ is *attracted* to a positive charge $+q_2$ is

$$E_m = \frac{(-q_1)\ (+q_2)}{r_m} = \frac{-q_1 q_2}{r_m}$$

where r_m is the distance between the two charges. Therefore the energy required to *separate* the two charges for the ionization process is

$$IE_m = \frac{q_1 q_2}{r_m} \tag{2-1}$$

To interpret the ionization process using this simple model we employ $q_1 = e$, $q_2 = me$, and r_m as the average distance of the mth electron from the nucleus in the atom. The expression then becomes

$$IE_m = m\frac{e^2}{r_m}$$

Dividing this expression by m, we remove the simple Coulomb's law term e^2/r_m; thus the term we are more interested in for interpreting ionization energies is IE_m/m. Tabulating these values for lithium we have

$$Li(g) \rightarrow Li^+(g) + e^- \qquad IE_1/1 = 5.392 \text{ eV}$$
$$Li^+(g) \rightarrow Li^{2+}(g) + e^- \qquad IE_2/2 = 37.819 \text{ eV}$$
$$Li^{2+}(g) \rightarrow Li^{3+}(g) + e^- \qquad IE_3/3 = 40.82 \text{ eV}$$

The first electron is easy to remove and has an $IE_1/1$ value of about 5 eV. The other two are more difficult to remove and have IE_m/m values of about 40 eV. Referring to Equation 2-1 we see that the only term which can account for these differences in IE_m/m is the average distance between the mth electron and the nucleus, r_m. If the second and third electrons have a much larger IE_m/m value than the first electron, they must have an r_m value that is much smaller than the first electron's. This deduction is in good agreement with the electronic configuration $1s^2 2s^1$. That is, the two $1s$ electrons have their electron densities concentrated much closer to the nucleus than does the valence $2s$ electron. This effect explains the large difference in effective atomic radii shown in Figure 2-1 for Li (1.34 Å) and Li^+ (0.60 Å).

TABLE 2-1. IONIZATION ENERGIES OF THE ELEMENTS (eV)

Z	Element	IE_1	IE_2	IE_3	IE_4	IE_5	IE_6	IE_7	IE_8
1	H	13.598							
2	He	24.587	54.416						
3	Li	5.392	75.638	122.451					
4	Be	9.322	18.211	153.893	217.713				
5	B	8.298	25.154	37.930	259.368	340.217			
6	C	11.260	24.383	47.887	64.492	392.077	489.981		
7	N	14.534	29.601	47.448	77.472	97.888	552.057	667.029	
8	O	13.618	35.116	54.934	77.412	113.896	138.116	739.315	871.387
9	F	17.422	34.970	62.707	87.138	114.240	157.161	185.182	953.886
10	Ne	21.564	40.962	63.45	97.11	126.21	157.93	207.27	239.09
11	Na	5.139	47.286	71.64	98.91	138.39	172.15	208.47	264.18
12	Mg	7.646	15.035	80.143	109.24	141.26	186.50	224.94	265.90
13	Al	5.986	18.828	28.447	119.99	153.71	190.47	241.43	284.59
14	Si	8.151	16.345	33.492	45.141	166.77	205.05	246.52	303.17
15	P	10.486	19.725	30.18	51.37	65.023	220.43	263.22	309.41
16	S	10.360	23.33	34.83	47.30	72.68	88.049	280.93	328.23
17	Cl	12.967	23.81	39.61	53.46	67.8	97.03	114.193	348.28
18	Ar	15.759	27.629	40.74	59.81	75.02	91.007	124.319	143.456
19	K	4.341	31.625	45.72	60.91	82.66	100.0	117.56	154.86
20	Ca	6.113	11.871	50.908	67.10	84.41	108.78	127.7	147.24
21	Sc	6.54	12.80	24.76	73.47	91.66	111.1	138.0	158.7
22	Ti	6.82	13.58	27.491	43.266	99.22	119.36	140.8	168.5
23	V	6.74	14.65	29.310	46.707	65.23	128.12	150.17	173.7
24	Cr	6.766	16.50	30.96	49.1	69.3	90.56	161.1	184.7
25	Mn	7.435	15.640	33.667	51.2	72.4	95	119.27	196.46
26	Fe	7.870	16.18	30.651	54.8	75.0	99	125	151.06
27	Co	7.86	17.06	33.50	51.3	79.5	102	129	157
28	Ni	7.635	18.168	35.17	54.9	75.5	108	133	162
29	Cu	7.726	20.292	36.83	55.2	79.9	103	139	166
30	Zn	9.394	17.964	39.722	59.4	82.6	108	134	174
31	Ga	5.999	20.51	30.71	64				
32	Ge	7.899	15.934	34.22	45.71	93.5			
33	As	9.81	18.633	28.351	50.13	62.63	127.6		
34	Se	9.752	21.19	30.820	42.944	68.3	81.70	155.4	
35	Br	11.814	21.8	36	47.3	59.7	88.6	103.0	192.8
36	Kr	13.999	24.359	36.95	52.5	64.7	78.5	111.0	126
37	Rb	4.177	27.28	40	52.6	71.0	84.4	99.2	136
38	Sr	5.695	11.030	43.6	57	71.6	90.8	106	122.3
39	Y	6.38	12.24	20.52	61.8	77.0	93.0	116	129
40	Zr	6.84	13.13	22.99	34.34	81.5			
41	Nb	6.88	14.32	25.04	38.3	50.55	102.6	125	
42	Mo	7.099	16.15	27.16	46.4	61.2	68	126.8	153
43	Tc	7.28	15.26	29.54					
44	Ru	7.37	16.76	28.47					
45	Rh	7.46	18.08	31.06					
46	Pd	8.34	19.43	32.93					
47	Ag	7.576	21.49	34.83					
48	Cd	8.993	16.908	37.48					
49	In	5.786	18.869	28.03	54				
50	Sn	7.344	14.632	30.502	40.734	72.28			
51	Sb	8.641	16.53	25.3	44.2	56	108		
52	Te	9.009	18.6	27.96	37.41	58.75	70.7	137	

TABLE 2-1. IONIZATION ENERGIES OF THE ELEMENTS (eV) (*continued*)

Z	Element	IE_1	IE_2	IE_3	IE_4	IE_5	IE_6	IE_7	IE_8
53	I	10.451	19.131	33					
54	Xe	12.130	21.21	32.1					
55	Cs	3.894	25.1						
56	Ba	5.212	10.004						
57	La	5.577	11.06	19.177	49.95				
58	Ce	5.47	10.85	20.198	36.758				
59	Pr	5.422	10.55	21.624	38.98				
60	Nd	5.489	10.73	22.1	40.41				
61	Pm	5.554	10.90	22.3	41.1				
62	Sm	5.631	11.07	23.4	41.4				
63	Eu	5.666	11.241	24.9	42.6				
64	Gd	6.141	12.09	20.63	44.0				
65	Tb	5.85	11.52	21.91	39.8				
66	Dy	5.927	11.67	22.8	41.5				
67	Ho	6.02	11.80	22.84	42.5				
68	Er	6.10	11.93	22.74	42.6				
69	Tm	6.184	12.05	23.68	42.7				
70	Yb	6.254	12.18	25.03	43.7				
71	Lu	5.426	13.9	20.96	45.19				
72	Hf	6.6	14.9	23.3	33.33				
73	Ta	7.89							
74	W	7.98							
75	Re	7.88							
76	Os	8.7							
77	Ir	9.1							
78	Pt	9.0	18.563						
79	Au	9.225	20.5						
80	Hg	10.437	18.756	34.2					
81	Tl	6.108	20.428	29.83					
82	Pb	7.416	15.032	31.937	42.32	68.8			
83	Bi	7.289	16.69	25.56	45.3	56.0	88.3		
84	Po	8.42							
85	At								
86	Rn	10.748							
87	Fr								
88	Ra	5.279	10.147						
89	Ac	5.2							
90	Th	6.1							
91	Pa	5.9							
92	U	6.05							
93	Np	6.2							
94	Pu	6.06							
95	Am	5.99							
96	Cm	6.02							
97	Bk	6.23							
98	Cf	6.30							
99	Es	6.42							
100	Fm	6.50							
101	Md	6.58							
102	No	6.65							
103	Lr								

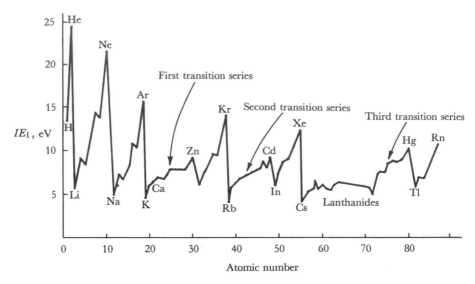

Figure 2-2 Variation of atomic ionization energy, IE_1, with atomic number. Notice that maximum ionization energies in a given row occur for the noble gases and that the ionization energies of the transition elements are similar.

Ionization Energies and Periodicity

Ionization energies of atoms exhibit periodic behavior, as illustrated for IE_1 in Figure 2-2. We see that in any given row in the periodic table the ionization energies generally increase with increasing atomic number, being smallest for the lithium family and largest for the helium family. However, there are irregularities; atoms with filled or half-filled subshells have larger ionization energies than might be expected. For example, beryllium $(2s^2)$ and nitrogen $(2s^2 2p^3)$ have larger ionization energies than boron and oxygen atoms respectively, as can be seen in Table 2-2.

The increase (although slightly irregular) in IE from lithium to neon is due to the steady increase in Z_{eff} with increasing atomic number. From lithium $(Z = 3)$ to neon $(Z = 10)$ all valence electrons are accommodated in $2s$ and $2p$ orbitals and are not able to shield each other completely from the increasing nuclear charge. Notice that the completely filled orbital structures in the helium family are especially stable. Electrons in helium and neon, which have the closed-shell $1s^2$ and $2s^2 2p^6$ structures, respectively, are attracted relatively closely to the nucleus, so the energies needed to remove an electron from these noble-gas atoms are correspondingly large.

TABLE 2-2. PERIODIC BEHAVIOR OF IONIZATION ENERGIES (IE)

General Increase of IE (eV) $\rightarrow$

Decrease of IE,	H·							He:
eV	13.60							24.59
$\downarrow$	Li·	Be·	·B·	·C̈·	·N̈·	·Ö:	:F̈:	:N̈e:
	5.39	9.32	8.30	11.26	14.53	13.62	17.42	21.56
	Na·							
	5.14							

Referring again to Figure 2-2, it can be seen that in any given family of elements the ionization energies decrease with increasing atomic number. For example, the IE_1 of sodium is less than the IE_1 of lithium. Recall that the $3s$ valence orbital in sodium has a larger effective radius than the $2s$ valence orbital in lithium. According to Coulomb's law, the net attraction between the $3s$ electron in a sodium atom and its effective nuclear charge is less than the net attraction between the $2s$ electron in a lithium atom and its effective nuclear charge. Thus it takes less energy to remove the $3s$ electron from the sodium atom than it does to remove the $2s$ electron from the lithium atom.

Ionization Energies of Core Electrons

Up to this point, we have considered ionization of core electrons only after all of the valence electrons have been removed. For example, if we wish to measure the energy required to remove a $1s$ electron from the lithium atom, it is the *second* ionization energy that is required:

$$Li^+(1s^2) \rightarrow Li^{2+}(1s^1) + e^-$$

However, to remove a $1s$ electron in beryllium, it is the *third* ionization energy that is required:

$$Be^{2+}(1s^2) \rightarrow Be^{3+}(1s^1) + e^-$$

There is another experimental method, however, by which we can measure ionization energies of core electrons without removing the valence electrons. This method requires monochromatic photons of sufficient energy ($E = h\nu$) to eject a core electron from the *neutral* atom. For example, the ionization of a $1s$ electron from the lithium atom corresponds to

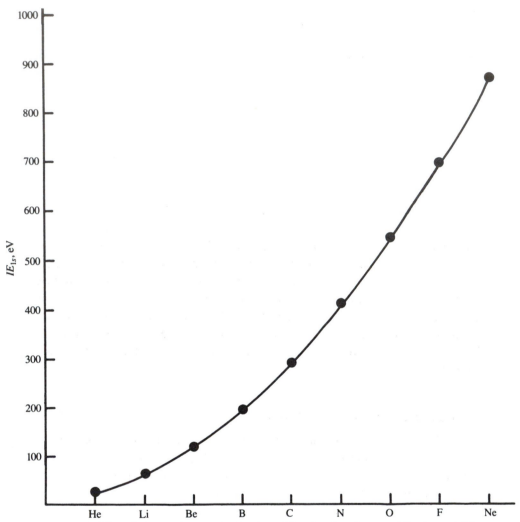

Figure 2-3 Plot of the ^1s electron ionization, IE_{1s}, for the elements He through Ne. The increase in IE_{1s} reflects the increase in effective nuclear charge, Z_{eff}, as Z increases.

$$\text{Li}(1s^2 2s^1) \rightarrow \text{Li}^+(1s^1 2s^1) + e^- \qquad IE_{1s} = 64.84 \text{ eV}$$

A plot of IE_{1s} versus atomic number is presented in Figure 2-3 for the elements helium through neon. Notice that IE_{1s} increases from 13.6 eV for H to 870.37 eV for Ne. These IE_{1s} results provide a ready illustration of the increase in Z_{eff} for the 1s electron as the atomic number, Z, increases.

Further insight into relative orbital energies may be obtained by comparing the following ionization energies of atomic lithium:

$$Li(1s^2 2s^1) \rightarrow Li^+(1s^2) + e^- \qquad IE_1 = 5.392 \text{ eV}$$
$$Li^+(1s^2) \rightarrow Li^{2+}(1s^1) + e^- \qquad IE_2 = 75.638 \text{ eV}$$
$$Li(1s^2 2s^1) \rightarrow Li^+(1s^1 2s^1) + e^- \qquad IE_{1s} = 64.84 \text{ eV}$$

Notice that IE_{1s} is larger than IE_1 because the core $1s$ electron is more difficult to remove than the valence $2s$ electron. In contrast, IE_{1s} is smaller than IE_2 although both correspond to removal of a $1s$ electron. This is a result of the fact that for IE_2, the $1s$ electron is ejected from the Li^+ cation, whereas for IE_{1s} it is ejected from the neutral Li atom.

2-4 Electron Affinity

The *electron affinity, EA,* of an atom is defined as the energy required to *remove* an electron from the negative ion:

$$\text{ion}^-(g) \rightarrow \text{atom}(g) + e^- \qquad \Delta E = EA$$

Using this definition it is easy to understand why the electron affinity sometimes is called the zeroth ionization energy, IE_0, since for a given atom A we have

$$A^-(g) + IE_0 \rightarrow A(g) + e^-$$
$$A(g) + IE_1 \rightarrow A^+(g) + e^-$$
$$A^+(g) + IE_2 \rightarrow A^{2+}(g) + e^-$$

If energy is *required* to remove an electron from a negative ion, the EA (IE_0) is positive. If energy is *released* for the reaction

$$A^-(g) \rightarrow A(g) + e^-$$

the EA (IE_0) is negative, meaning that A (g) is more stable than A^- (g). The electron affinity values for several atoms are listed in Table 2-3.

Atoms in the fluorine family have relatively large electron affinities (~ 3.5 eV) because an electron can be added to a valence p orbital relatively easily, thereby completing a closed-shell $s^2 p^6$ configuration. Atoms that already have closed shells or subshells often have small EA values. Examples are the noble gases (He, Ne, . . .), beryllium, magnesium, and zinc. Thus from the trends in both ionization energies and electron affinities it is apparent that closed shells or subshells represent particularly stable electronic arrangements.

TABLE 2-3. ATOMIC ELECTRON AFFINITIES (EA)

Atom	Orbital Electronic Configuration	EA (eV)	Orbital Electronic Configuration of Anion
H	$1s^1$	0.754	[He]
F	$[He]2s^22p^5$	3.34	[Ne]
Cl	$[Ne]3s^23p^5$	3.61	[Ar]
Br	$[Ar]4s^23d^{10}4p^5$	3.36	[Kr]
I	$[Kr]5s^24d^{10}5p^5$	3.06	[Xe]
O	$[He]2s^22p^4$	1.47	$[He]2s^22p^5$
S	$[Ne]3s^23p^4$	2.08	$[Ne]3s^23p^5$
Se	$[Ar]4s^23d^{10}4p^4$	2.02	$[Ar]4s^23d^{10}4p^5$
Te	$[Kr]5s^24d^{10}5p^4$	1.97	$[Kr]5s^24d^{10}5p^5$
N	$[He]2s^22p^3$	0.0 ± 0.2	$[He]2s^22p^4$
P	$[Ne]3s^23p^3$	0.77	$[Ne]3s^23p^4$
As	$[Ar]4s^23d^{10}4p^3$	0.80	$[Ar]4s^23d^{10}4p^4$
C	$[He]2s^22p^2$	1.27	$[He]2s^22p^3$
Si	$[Ne]3s^23p^2$	1.24	$[Ne]3s^23p^3$
Ge	$[Ar]4s^23d^{10}4p^2$	1.20	$[Ar]4s^23d^{10}4p^3$
B	$[He]2s^22p^1$	0.24	$[He]2s^22p^2$
Al	$[Ne]3s^23p^1$	(0.52)	$[Ne]3s^23p^2$
Ga	$[Ar]4s^23d^{10}4p^1$	(0.37)	$[Ar]4s^23d^{10}4p^2$
In	$[Kr]5s^24d^{10}5p^1$	(0.35)	$[Kr]5s^24d^{10}5p^2$
Be	$[He]2s^2$	≤ 0	$[He]2s^22p^1$
Mg	$[Ne]3s^2$	≤ 0	$[Ne]3s^23p^1$
Zn	$[Ar]4s^23d^{10}$	≤ 0	$[Ar]4s^23d^{10}4p^1$
Cd	$[Kr]5s^24d^{10}$	≤ 0	$[Kr]5s^24d^{10}5p^1$
Ti	$[He]2s^1$	0.62	$[He]2s^2$
Na	$[Ne]3s^1$	0.55	$[Ne]3s^2$
Cu	$[Ar]4s^23d^9$	1.28	$[Ar]4s^23d^{10}$
Ag	$[Kr]5s^14d^{10}$	1.30	$[Kr]5s^24d^{10}$
Au	$[Xe]4f^{14}5d^{10}6s^1$	2.31	$[Xe]4f^{14}5d^{10}6s^2$
He	$1s^2$	≤ 0	$1s^22s^1$
Ne	$[He]2s^22p^6$	≤ 0	$[He]2s^22p^63s^1$

SOURCE: E.N. Chen and W.E. Wentworth, "The Experimental Values of Atomic Electron Affinities," *J. Chem. Educ.* 52: 486 (1975). These values have been rounded to two significant figures after the decimal.

2-5 Covalent Bonding

A covalent bond forms when atoms that have electrons of similar or equal valence-orbital energies combine. For example, two atoms of hydrogen are joined by a covalent bond in the H_2 molecule. To understand covalent bond formation we now consider in detail the energy changes that occur if we allow two hydrogen atoms to come together from a large distance.

Each hydrogen atom consists of one electron and one proton. Because

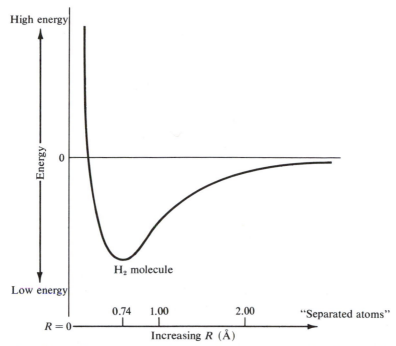

Figure 2-4 Potential energy curve for the H_2 molecule. As the distance between nuclei decreases the potential energy also decreases because of electron-nucleus attraction, and then increases again because of nucleus-nucleus repulsion. The point of minimum energy corresponds to the equilibrium internuclear separation, or bond distance, of H_2, which is 0.74 Å.

we want to focus our attention on the change in the energy of the system as we bring the two hydrogen atoms together, we will assume that the energy in the system is zero when the hydrogen atoms are isolated. In each isolated atom there is a single important force of attraction, which is the force between the electron and the proton. However, if we bring two hydrogen atoms together there are additional forces that we must consider. Two such forces are the *attractions* of the first electron for the second nucleus and the second electron for the first nucleus. There are also *repulsive* forces: the two electrons repel each other and the two protons repel each other. Thus there are four new electrostatic forces, two of them attractive and two repulsive.

   The most important force as the two hydrogen atoms come together is the force of attraction between the two electrons and the two protons. As the hydrogen atoms are brought together the energy of the system decreases, and it continues to decrease until the two hydrogen atoms are so close together that the nuclear repulsion — the repulsion between the protons — becomes significant, thereby causing the energy to increase. Figure 2-4

Electron–pair bond

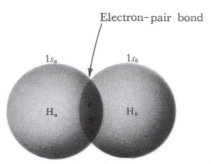

Figure 2-5 The two hydrogen $1s$ orbitals overlap to form an electron-pair covalent bond in H_2.

shows the energy curve for this process. Notice that there is a point—an equilibrium—at which the two hydrogen atoms are bound together in a stable configuration and the energy of the system is minimal. The separation of hydrogen nuclei at this position of minimum energy is the *equilibrium internuclear separation,* or the *bond distance* of H_2, 0.74 Å.

To separate two atoms in a diatomic molecule (e.g., H_2) requires approximately the energy difference between the minimum potential energy of the system and the zero energy of the isolated atoms. This is called the *bond energy* of a diatomic molecule. Bond energy usually is expressed in kilocalories per mole, that is, the number of kilocalories required to break one mole of bonds. The bond energy of H_2 is 103 kcal mole^{-1}.

The two electrons in H_2 are shared equally by the two hydrogen $1s$ orbitals. This in effect gives each hydrogen atom a stable, closed-shell (He) configuration. A simple orbital representation of the electron-pair bond in H_2 results from an *overlap* of the hydrogen $1s$ atomic orbitals as shown in Figure 2-5. Because the two electrons are shared equally between the nuclei, they are more stable than in the separated hydrogen atoms. This stability is shown clearly by a comparison of the ionization energies of H atoms and H_2 molecules:

$$H(g) \rightarrow H^+(g) + e^- \qquad IE = 13.60 \text{ eV}$$

$$H_2(g) \rightarrow H_2^+(g) + e^- \qquad IE = 15.46 \text{ eV}$$

Photon impingement on hydrogen at very low pressures forms the transient species H_2^+, which is the simplest of all molecules. This species, which is a combination of a proton and a hydrogen atom, is called the *hydrogen molecule-ion.* If a proton and a hydrogen atom come together, there is less net attractive force than when two hydrogen atoms approach each other. As a result the system's potential energy does not decrease as much, and the minimum energy occurs at a longer internuclear distance than for H_2. If we define one covalent bond as having two net bonding electrons, then H_2^+, with one electron, has one-half a bond. We find that H_2, with two electrons, has a bond length of 0.74 Å and that H_2^+, with one electron, has a bond length of 1.06 Å. The bond energy for H_2 is 103 kcal mole^{-1}, whereas the bond energy for H_2^+ is 61 kcal mole^{-1}. These comparisons illustrate the general fact that the more bonds there are between two atoms, the shorter the bond length and the stronger the bond. The number of bonds between two atoms is called the *bond order*.

2-6 Properties of H_2 and H_2^+ in a Magnetic Field

Most substances can be classified as either paramagnetic or diamagnetic, depending on their behavior in a magnetic field. A *paramagnetic substance* is attracted to a magnetic field. A *diamagnetic substance* is repelled by a magnetic field. Generally, atoms and molecules with unpaired electrons are paramagnetic, because there is a permanent magnetic moment associated with net electron spin. In many cases there is a further contribution to the permanent magnetic moment as a result of the movement of an electron in its orbital around the nucleus (or nuclei, in the case of molecules). In addition to the paramagnetic moment, magnetic moments may be induced in atoms and molecules by applying an external magnetic field. Such induced moments are opposite to the direction of the external magnetic field, thus repulsion occurs. The magnitude of this repulsion is a measure of the diamagnetism of an atom or molecule.

The paramagnetism of atoms and small molecules that results from unpaired electrons is larger than their induced diamagnetism, so such substances are attracted to a magnetic field. Atoms and molecules with no unpaired electrons, and therefore no paramagnetism due to electron spin, are diamagnetic and are repelled by a magnetic field. The H_2^+ ion, with one unpaired electron, is paramagnetic. The H_2 molecules, with two paired electrons, is diamagnetic.

2-7 Lewis Structures for Diatomic Molecules

The Lewis structures for two H atoms can be combined to give H:H, or H—H, in which the electron-pair bond is abbreviated with a single line. As we have already seen, the stability of H_2 is due to the tendency of each hydrogen atom to associate itself with two electrons, thereby achieving a closed shell configuration (He:). The Lewis structure of a fluorine atom is :$\ddot{\text{F}}\cdot$. Each fluorine atom lacks one electron to complete a closed-shell configuration. In the Lewis model two fluorine atoms achieve a closed-shell configuration by sharing two electrons, thereby forming an electron-pair bond (:$\ddot{\text{F}}$—$\ddot{\text{F}}$:). The electrons not involved in the bonding are called *nonbonding, unshared,* or *lone-pair* electrons.

The bond energy of F_2 is about 37 kcal mole^{-1}, compared to 103 kcal mole^{-1} for H_2. Although H_2 and F_2 each have a bond order of one, it takes much less energy to separate two fluorine atoms in a fluorine molecule than it does to separate two hydrogen atoms in a hydrogen molecule. In fact, the hydrogen molecule has an extraordinarily strong single bond, perhaps because there are no core electrons, and therefore a short H—H bond with strong interaction is possible. Most single bonds have a bond dissociation energy of 25–75 kcal mole^{-1}. A better comparison of single bond energies is provided by Li_2 and F_2, because both are second-row molecules and have a single bond according to the Lewis structure. The bond dissociation energy of Li_2 is only 26 kcal mole^{-1}, which is even less than that of F_2 (37 kcal mole^{-1}).

The H_2 and F_2 molecules are representative of many molecules in which electron-pair bonds are formed in such a way that each atom achieves a closed-shell configuration. For hydrogen to achieve a closed shell requires two electrons, which is the capacity of the $1s$ valence orbital. Each atom in the second row of the periodic table requires eight electrons (an octet) to achieve a closed shell, because the $2s$ and $2p$ orbitals have a total capacity of eight electrons. This requirement is commonly known as the *octet rule*. In the example of the fluorine molecule, each fluorine atom has eight electrons associated with it. For third-row atoms eight electrons usually are associated with each atom because of their $3s$ and $3p$ valence orbitals. However, these atoms also have $3d$ orbitals (although their energy is substantially higher than that of the $3p$ orbitals), so more than eight electrons may be associated with atoms in the third and higher rows.

2-8 Ionic Bonding

In a pure covalent bond (e.g., in H_2), electrons are shared equally between two atoms. A pure ionic bond is at the other extreme; that is, there is complete transfer of electrons from one atom to another, with no sharing. There probably is no diatomic molecule that has a completely ionic bond. However, alkali-metal halide molecules have such unequal sharing of electrons that they may serve as models for ionic bonding.

Consider a sodium chloride diatomic molecule. If two atoms of different elements are involved in a two-electron bond, there must be unequal sharing of electrons. Unequal sharing is caused by differences in the ionization energies and electron affinities of the two atoms. In NaCl the sodium atom has a small ionization energy of about 5 eV and a small electron affinity of about 0.5 eV. Therefore it easily loses an electron to form Na^+. The chlorine atom has a large ionization energy of more than 10 eV and a large electron affinity of almost 4 eV. Thus it does not easily lose an electron; rather, it can easily gain an electron. The preceding data provide a quantitative measure of the tendency for Cl to add an electron and for Na to lose an electron to achieve closed-shell configurations: a sodium atom and a chlorine atom can be assumed to form an ionic bond in which the one $3s$ valence electron in sodium is transferred to the one vacancy in the chlorine $3p$ orbitals.

We consider all ionic molecules as being composed of interacting ions. Thus sodium chloride contains the Na^+ ion, which has the inert neon configuration, and the Cl^- ion, which has the inert argon configuration. Therefore the correct Lewis structure for NaCl is

$[Na^+] [:\ddot{C}l:^-]$

Ionic or "ion-pair" structures are reasonable models of bonds involving alkali and alkaline-earth metals with oxygen or one of the halogens. Ionic bonds are appropriate in these molecules because the large differences in ionization energies and electron affinities must lead to extremely unequal sharing of electrons.

Interestingly, it is not the formation of closed shells that makes gaseous NaCl stable, since this stabilization process requires energy. We can see this by examining the following equations:

$$Na(g) \rightarrow Na^+(g) + e^- \qquad \Delta E = IE_{Na}$$
$$Cl^-(g) \rightarrow Cl(g) + e^- \qquad \Delta E = EA_{Cl}$$

Multiplying the second equation by -1 and adding the result to the first equation, we obtain

$$Na(g) + Cl(g) \rightarrow Na^+(g) + Cl^-(g) \qquad \Delta E = IE_{Na} - EA_{Cl}$$

in which $IE_{Na} - EA_{Cl} = (5.139 - 3.61) = 1.53$ eV. This equation shows that the formation of the gaseous ions from the neutral atoms *requires* energy. Therefore the stability of the gaseous sodium chloride molecule must result from the attraction of the oppositely charged ions.

A relatively simple calculation of the bond energy of sodium chloride will demonstrate that the ion-pair model is reasonable. As we have already noted, the standard bond energy, or bond dissociation energy, DE, is the energy required to dissociate a molecule into its component atoms. For sodium chloride the process is $NaCl(g) \rightarrow Na(g) + Cl(g)$. We can use the following thermochemical cycle to calculate the bond dissociation energy if we assume that the oppositely charged ions can be treated as point charges and are attracted according to Coulomb's law of energy, e^2/r:

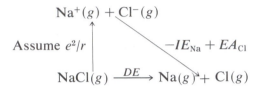

From this cycle it is clear that

$$DE = e^2/r - IE_{Na} + EA_{Cl} \tag{2-2}$$

We represent $NaCl(g)$ as a Na^+ ion and a Cl^- ion, separated by the experimental bond distance $r = 2.36$ Å. From Coulomb's law, an energy of e^2/r = 332 kcal mole^{-1} is required to dissociate oppositely charged bodies (each with unit charge) from a distance of $r = 1$ Å. The calculated energy for the process $NaCl(g) \rightarrow Na^+(g) + Cl^-(g)$ is $332/2.36$ or 141 kcal mole^{-1}. To calculate DE from Equation 2-2 we must convert the ionization energy and electron affinity from electron volts to kilocalories through the relationship 1 eV particle^{-1} = 23.06 kcal mole^{-1}. The ionization energy of sodium is 5.139 eV (119 kcal mole^{-1}) and the electron affinity of chlorine is 3.61 eV (83 kcal mole^{-1}). Thus the calculated bond dissociation energy is $DE = e^2/r$ = $IE_{Na} + EA_{Cl} = 141 - 119 + 83 = 105$ kcal mole^{-1}. The experimental value is 98 kcal mole^{-1}, so the ion-pair approximation allows us to calculate the bond dissociation energy within 7 percent of the experimental value.

Using the "point-charge" model we ignored the repulsions between filled electron shells. A better calculation, using the pure ionic model but taking the repulsion of the closed shells into account, gives the bond energy of NaCl within 2 percent of the experimental value. This and similar calculations indicate that an ionic model is justified for molecules formed from lithium- or beryllium-family atoms and either oxygen- or fluorine-family atoms. In addition, direct experimental evidence from electron-diffraction data shows that the electron distribution in a sodium chloride molecule is is approximately that given by the ionic model.

The calculation of bond dissociation energy indicates that the ionic bonding model is appropriate. The first ionization energy of the gaseous sodium chloride molecule also can be interpreted by the ionic model. The experimental ionization energy is 8.93 eV for

$$NaCl(g) \rightarrow NaCl^+(g) + e^-$$

If we treat NaCl in the ion-pair model as Na^+Cl^-, we can consider the ionization of NaCl as forming either $Na^{2+}Cl^-$ or Na^+Cl^0 ions. It is obvious, however, that the latter formulation is the correct one, since removal of an electron from Cl^- requires much less energy than from Na^+:

$$Cl^-(g) \rightarrow Cl(g) + e^- \qquad \Delta E = IE_0 = EA = 3.61 \text{ eV}$$

$$Na^+(g) \rightarrow Na^{2+}(g) + e^- \qquad \Delta E = IE_2 = 47.286 \text{ eV}$$

Let us now use the ion-pair model to calculate the ionization energy of NaCl by assuming that the ionization process is

$$Na^+Cl^- \rightarrow Na^+Cl^0 + e^-$$

The energy required to ionize this electron from NaCl is due to two factors: (1) The energy required to remove the electron from Cl^-, that is, the electron affinity of Cl, and (2) the energy required to remove the negatively charged electron from the positive electric field due to the Na^+ ion. We again assume the point-charge model, so the second term can be calculated from Coulomb's law of energy, e^2/r. The calculated ionization energy is

$$IE_{NaCl} = EA_{Cl} + e^2/r$$

To complete the calculation we must convert the term e^2/r to electron volts through the relationship 1 eV molecule^{-1} $\equiv$ 23.06 kcal mole^{-1}, and we obtain 332 kcal mole^{-1} = 14.40 eV per molecule. The calculated ionization energy for sodium chloride therefore becomes

$$IE_{NaCl} = EA_{Cl} + e^2/r = 3.61 + 14.40/2.36 = 3.61 + 6.10 = 9.71 \text{ eV}$$

This value should be compared with the experimental value of 8.93 eV. Just as in the calculation of the bond dissociation energy, use of the point-charge model results in an error, which for the ionization energy is about 9 percent. Nevertheless, the close agreement between calculation and experiment indicates that our model is essentially correct.

2-9 Electronegativity

Between the extremes of covalent and ionic molecules there are other molecules for which neither bond type is a sufficient description. A discussion of these intermediate molecules is facilitated by introducing the concept of electronegativity.

Electronegativity (EN) is a term that describes the relative ability of an atom to attract electrons to itself in a chemical bond. For example, in a sodium chloride molecule the chlorine atom has a large electronegativity and the sodium atom has a small electronegativity. The result is a virtually complete transfer of one electron from sodium to chlorine in the molecule. The American scientist Robert S. Mulliken has proposed that electronegativity be defined as proportional to the sum of the ionization energy and the electron affinity of an atom:

$$EN = c(IE + EA)$$

where c is a proportionality constant.

As we have seen, ionization energy is a measure of the ability of an atom to hold one electron, and electron affinity is a measure of the ability of an atom to attract an electron. It follows that an atom such as chlorine, which has both a large ionization energy and a large electron affinity, will have a large electronegativity. A quantitative Mulliken scale of atomic electronegativities can be obtained by assigning one atom a specific EN value, thereby fixing the proportionality constant, c. Unfortunately, not many atomic electron affinities are known accurately, thus only a few EN values can be calculated in this way.

A more widely applied quantitative treatment of electronegativity was introduced by the American chemist Linus Pauling in the early 1930s. The Pauling electronegativity value for a specific atom is obtained by comparing the bond energies of certain molecules containing that atom. If the bonding electrons were shared equally in a molecule AB, it would be reasonable to assume that the bond energy of AB would be the geometric mean of the bond energies of the molecules A_2 and B_2. However, the bond energy of an AB molecule almost always is greater than the geometric mean of the bond energies of A_2 and B_2. An example that illustrates this generalization is the HF molecule. The bond energy of HF is 135 kcal mole^{-1}, whereas the bond energies of H_2 and F_2 are 103 kcal mole^{-1} and 37 kcal mole^{-1}, respectively. The geometric mean of the latter two values is $(37 \times 103)^{1/2} = 62$ kcal mole^{-1}, which is much less than the observed bond energy of HF. This "extra" bond energy (designated Δ) in an AB molecule is assumed to be a consequence of the partial ionic character of the bond due to electronegativity differences

TABLE 2-4. ATOMIC ELECTRONEGATIVITIES

I	II	III	II	II	II	II	II	II	II	I	II	III	IV	III	II	I
H 2.20																
Li 0.98	Be 1.57											B 2.04	C 2.55	N 3.04	O 3.44	F 3.98
Na 0.93	Mg 1.31											Al 1.61	Si 1.90	P 2.19	S 2.58	Cl 3.16
K 0.82	Ca 1.00	Sc 1.36	Ti 1.54	V 1.63	Cr 1.66	Mn 1.55	Fe 1.83	Co 1.88	Ni 1.91	Cu 1.90	Zn 1.65	Ga 1.81	Ge 2.01	As 2.18	Se 2.55	Br 2.96
Rb 0.82	Sr 0.95	Y 1.22	Zr 1.33		Mo 2.16			Rh 2.28	Pd 2.20	Ag 1.93	Cd 1.69	In 1.78	Sn 1.96	Sb 2.05		I 2.66
Cs 0.79	Ba 0.89	La 1.10			W 2.36			Ir 2.20	Pt 2.28	Au 2.54	Hg 2.00	Tl 2.04	Pb 2.33	Bi 2.02		
		Ce 1.12	Pr 1.13 (III)	Nd 1.14 (III)		Sm 1.17 (III)		Gd 1.20 (III)		Dy 1.22 (III)	Ho 1.23 (III)	Er 1.24	Tm 1.25 (III)		Lu 1.27 (III)	
					U 1.38 (III)	Np 1.36 (III)	Pu 1.28 (III)									

NOTE: Roman numerals refer to the oxidation numbers of the atoms in the molecules used in the calculation of atomic electronegativity values.

between atoms A and B. In this model the electronegativity difference between two atoms A and B is defined as

$$EN_A - EN_B = 0.208\Delta^{1/2} \tag{2-3}$$

in which EN_A and EN_B are the electronegativities of atoms A and B, and Δ is the extra bond energy in kilocalories per mole. The extra bond energy is calculated from the equation

$$\Delta = DE_{AB} - [(DE_{A_2})(DE_{B_2})]^{1/2}$$

in which DE is the particular bond dissociation energy.

In Equation 2-3 the factor 0.208 converts kilocalories per mole to electron volts. The square root of Δ is used because it gives a more consistent set of atomic electronegativity values. Since only differences are obtained from Equation 2-3, one atomic electronegativity must be assigned a specific value: then the values of the electronegativities can be calculated easily. In a widely adopted version of the Pauling scale, the most electronegative atom, fluorine, is assigned an electronegativity of 3.98. A compilation of EN values based on this scale is given in Table 2-4.

Electronegativity is a concept useful for qualitatively describing the sharing of electrons in a bond between two atoms of different elements. For
sodium and chlorine the difference in the electronegativities of the two atoms
is so large that there is, in effect, complete transfer of the electron pair. However, there are many molecules in which the bond between dissimilar atoms
is better described as covalent, with some ionic character.

2-10 A Covalent Bond With Ionic Character: The HCl Molecule

The bond in a molecule composed of a hydrogen atom and a chlorine atom,
HCl, is neither purely covalent nor purely ionic. In the Lewis structure for
HCl an electron-pair bond between H· and ·C̈l: is formed, thereby associat-
ing two electrons with hydrogen and eight electrons with chlorine (H—C̈l:).
The covalent structure is a more accurate representation of the bonding in
HCl than is an ion pair such as we formulated for sodium chloride, because
the electronegativity of the hydrogen atom is much larger (and closer to that
of chlorine) than the electronegativity of the sodium atom. Although complete
transfer of the pair of electrons to the chlorine atom in H—C̈l: does not
occur, the electron density in the bond is more concentrated in the region of
the chlorine atom than in the region of the hydrogen atom. This unequal
charge distribution is illustrated by

$$\overset{\delta+}{H}\text{—}\overset{\delta-}{\overset{..}{\underset{..}{C}l}}:$$

There is a small, net positive charge associated with the hydrogen atom
because the electron pair is "pulled" toward the chlorine atom, which
acquires a small, net negative charge. Thus the HCl bond is said to have
ionic character.

A molecule such as HCl is polar. The measure of the tendency of a polar
molecule to become aligned in an electric field (Figure 2-6) gives a quantity
that is known as the electric *dipole moment,* which is related to the net
charge separation in the most stable electronic state of the molecule. An
HCl molecule has an electric dipole moment due to unequal sharing of the
two electrons in the bond. An H_2 molecule (covalent bond) has a zero dipole
moment, whereas a NaCl molecule (ionic bond) has a very large dipole
moment.

The dipole moment of a simple electrostatic system is equal to the product
of the charge, δ, and the distance of separation, r:

$$\text{dipole moment} = \mu = \delta \times r$$

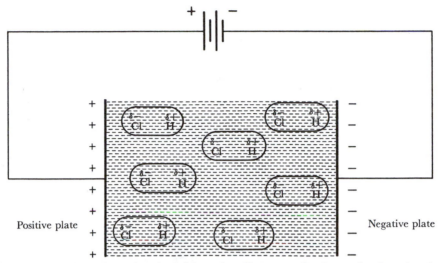

Figure 2-6 Effect of an electric field on the alignment of polar molecules. A polar molecule tends to align in an electric field, thereby maximizing the electrostatic attraction between the plates. In this illustration the preferred alignment of the polar molecule HCl is shown.

If the distance r is in centimeters and the charge δ is in electrostatic units, then μ is in electrostatic units times centimeters. Since the unit of electronic charge is $e = 4.8 \times 10^{-10}$ esu and bond distances are of the order of 10^{-8} cm (1 Å), dipole moments are of the order of 10^{-18} esu cm. It is convenient to express μ in *debye* units, D, in which $1\ D \equiv 10^{-18}$ esu cm.

Because it is possible to measure dipole moments we have an experimental method of estimating the partial ionic character of heteronuclear diatomic molecules. The experimental bond length of HCl is $r = 1.27$ Å (or 1.27×10^{-8} cm). Therefore the completely ionic structure H^+Cl^- (assuming $\delta = 1$ and the point-charge model) has a calculated dipole moment of 6.09 D (4.8×10^{-10} esu $\times 1.27 \times 10^{-8}$ cm). The experimental dipole moment is only 1.08 D. The separated partial charge calculated from the data is 1.08/6.09 = 0.18, which represents a partial ionic character of 18 percent.

Dipole moments for several diatomic molecules are given in Table 2-5.

2-11 Lewis Structures For Polyatomic Molecules

Communication among chemists commonly involves a language filled with Lewis structures for the molecules under discussion. It is important to develop considerable facility in formulating these "line and dot" structures

TABLE 2-5. DIPOLE MOMENTS OF
SOME DIATOMIC MOLECULES.

Molecule	Dipole Moment (D)
LiH	5.88
HF	1.82
HCl	1.08
HBr	0.82
HI	0.44
O_2	0
CO	0.112
NO	0.153
ICl	0.65
BrCl	0.57
ClF	0.88
BrF	1.29
KBr	10.41
KCl	10.27
KF	8.60
KI	11.05

for all types of molecules. In the following pages we discuss several representative polyatomic molecules.

Methane, Ammonia, and Water

The Lewis structures of CH_4, NH_3, and H_2O are as follows

Methane   Ammonia   Water

These three molecules are *isoelectronic;* that is, they have the same number of electrons.* The eight valence electrons around the central atom illustrate the octet rule. In CH_4, all eight electrons are involved in bond pairs, whereas the other molecules contain both nonbonding and bonding electron pairs.

*The term *isoelectronic* is often employed more generally to refer to molecules having the same number of *valence electrons.*

The usefulness of the octet rule is that it allows us to predict which molecules will be stable under ordinary conditions of temperature and pressure. For example, all of the following carbon hydrides are known: CH, CH_2, CH_3, CH_4. Only CH_4, however, obeys the octet rule and is stable under ordinary conditions.

Hydrides of Beryllium and Boron

The molecules BeH_2 and BH_3 represent further examples which do not obey the octet rule:

H—Be—H

Beryllium hydride

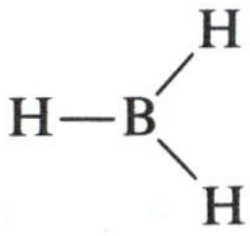

$$H-B\begin{matrix} H \\ \\ H \end{matrix}$$

Boron hydride

At normal temperatures and pressures both BeH_2 and BH_3 use their empty valence orbitals to form larger molecular aggregates. Beryllium hydride is a solid in which hydrogen atoms share electrons with adjacent beryllium atoms in "bridge" bonds, which may be represented as

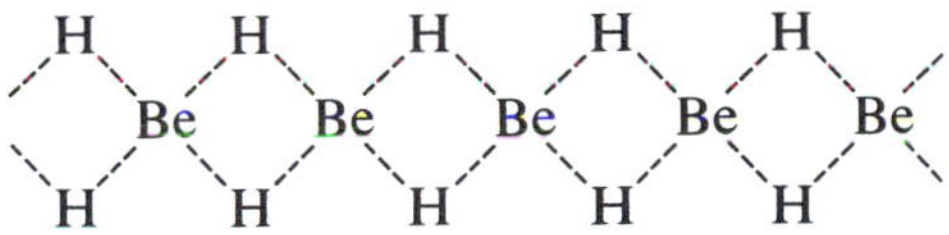

In a sense, each Be shares eight electrons in the solid, thereby achieving a closed valence shell.

Under normal conditions the compound of empirical formula BH_3 has the molecular formula B_2H_6 and is called *diborane*. The experimentally determined structure of B_2H_6 reveals two types of bonding for the hydrogen atoms, as shown in Figure 2-7. Two BH_2 units are held together through two B—H—B bridges, or *three-center,* bonds. In the B_2H_6 structure the regular (or terminal) B—H bond length is shorter than the B—H distance in the bridge bonds.

The solid state structure of BeH_2 and the structure of B_2H_6 represent species for which ordinary Lewis structures cannot be drawn. Molecules that do not have enough electrons for each pair of atoms to be bonded by

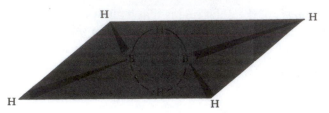

Figure 2-7 Bridging bonds in diborane. The arc from B through H to B represents a three-center electron-pair bond — one bonding molecular orbital spread over three nuclei with a capacity for two electrons. The B—H distance in the bridge bonds is 1.334 ± 0.027 Å, as compared to the terminal B—H bond length of 1.187 ± 0.030 Å.

two electrons are said to be *electron deficient*. For example, in B_2H_6 there are a total of 8 B—H bonds but only 12 electrons, not 16 as would be required by ordinary electron-pair bonds. The usual description of the bonding in B_2H_6 ascribes two electrons to each of the four terminal B—H bonds, which accounts for 8 of the 12 electrons. Of the remaining 4 electrons, two reside in each of the B—H—B bridge bonds. We shall consider the bonding in B_2H_6 more fully in Section 3-8.

The BH_3 molecule can react in other ways to achieve an octet around the boron atom. One example is the reaction of NH_3 and BH_3 to form the addition compound NH_3BH_3:

$$
\begin{array}{c}
\text{H} \\
| \\
\text{H—N:} \\
| \\
\text{H}
\end{array}
+
\begin{array}{c}
\quad \text{H} \quad \text{H} \\
\quad \diagdown \; \diagup \\
\quad \text{B} \\
\quad | \\
\quad \text{H}
\end{array}
\rightarrow
\begin{array}{c}
\text{H} \quad \text{H} \\
| \quad\quad | \\
\text{H—N}^+\text{—}^-\text{B—H} \\
| \quad\quad | \\
\text{H} \quad \text{H}
\end{array}
$$

In this compound, nitrogen, which has a lone electron pair, donates both the electrons of the covalent bond. The charges designated on the nitrogen and boron atoms are called *formal charges*. Formal charges are assigned by assuming that, in a bond involving two atoms joined by an electron pair, each atom "owns" one of the electrons. Thus each hydrogen atom in NH_3BH_3 owns one electron. But each neutral hydrogen atom has one electron to begin with, so it has zero formal charge in the molecule. However, the nitrogen atom owns only four electrons in NH_3BH_3. Since this is one less than the five electrons in atomic nitrogen, we assign a formal charge of $+1$ to the nitrogen atom in NH_3BH_3. A similar calculation results in a formal charge of -1 on the boron atom. Notice that the sum of the formal charges for a neutral molecule must equal zero. For a molecular ion the total formal charge of all the atoms must equal the total charge on the ion.

Magnesium Chloride, an Ionic Molecule

A magnesium atom has a small electronegativity and atomic chlorine has a large electronegativity; thus magnesium chloride requires a structure showing an ionic bond. One electron is transferred from the magnesium atom to each of the chlorine atoms in $MgCl_2$. The correct structures indicate the charges:

$$[:\ddot{\underset{\cdot\cdot}{Cl}}:^-][Mg^{2+}][:\ddot{\underset{\cdot\cdot}{Cl}}:^-]$$

There are two large bond dipoles ($\overset{\longleftarrow\ \longrightarrow}{\underset{-\ +\ +\ -}{Cl\ Mg\ Cl}}$) in opposite directions and of equal magnitude (since the chlorine ions are equivalent). These two bond dipoles cancel, because the geometrical structure of the molecule is linear.

Ammonium Chloride Molecule

The ammonium chloride molecule, NH_4Cl, contains NH_4^+ and Cl^- ions. The Lewis structure for the nitrogen atom is $:\dot{N}\cdot$; for each hydrogen atom it is H·. However, the NH_4^+ ion has one positive charge, which means that one of the nine electrons has been lost. Since all the hydrogen atoms in the ion are equivalent we give the nitrogen atom the positive charge, $\cdot\dot{N}\cdot^+$, and write

$$\begin{array}{c} H \\ | \\ H-N^+-H \\ | \\ H \end{array}$$

Thus the correct Lewis structure of the NH_4^+ ion has four single bonds and no unshared electrons. Notice that the ammounium ion is isoelectronic with methane. A formal charge of +1 is assigned to the nitrogen atom.

Now we assign the electron that was removed from NH_4 (to give NH_4^+) to the chlorine atom, giving the ion-pair structure for the ammonium chloride molecule:

$$\begin{array}{c} H \\ | \\ H-N^+-H \quad [:\ddot{\underset{\cdot\cdot}{Cl}}:^-] \\ | \\ H \end{array}$$

In short, the N—H bonds in NH_4^+ are covalent with some ionic character, whereas the NH_4^+ ion is attached to Cl^- by an ionic bond.

2-12 Molecules With Double and Triple Bonds

Now we consider molecules in which more than one electron pair is involved in a bond between two atoms. An example is the ethylene molecule, C_2H_4, in which the four hydrogen atoms are attached to the two carbon atoms, and the two carbon atoms are attached to each other. First we can draw the bonds to the hydrogen atoms:

$$
\begin{array}{cc}
H & H \\
\diagdown\;\diagup \\
C\!: \quad :\!C \\
\diagup\;\diagdown \\
H & H
\end{array}
$$

Then by hypothesizing a single bond between the carbon atoms and substituting a line for the bonding pair we obtain

$$
\begin{array}{cc}
H & H \\
\diagdown\;\diagup \\
\ddot{C}\!-\!\ddot{C} \\
\diagup\;\diagdown \\
H & H
\end{array}
$$

One unshared valence electron remains on each carbon atom. Counting the electrons we find that there are seven valence electrons associated with each carbon atom and two electrons with each hydrogen atom. If we suppose an additional electron-pair bond between the two carbon atoms, then each carbon atom has the required eight electrons:

$$
\begin{array}{cc}
H & H \\
\diagdown\;\diagup \\
C\!=\!C \\
\diagup\;\diagdown \\
H & H
\end{array}
$$

The bond between the two carbon atoms involves two electron pairs and is called a *double bond*. Therefore the carbon-carbon *bond order* in ethylene is two.

There are many compounds that require Lewis structures with double or triple bonds.

$\diagdown\quad\diagup$ $C\!\!=\!\!C$ $\diagup\quad\diagdown$	carbon-carbon double bond
$-C\!\equiv\!C-$	carbon-carbon triple bond
$-\ddot{N}\!=\!\ddot{N}-$	nitrogen-nitrogen double bond
$:N\!\equiv\!N:$	nitrogen-nitrogen triple bond
$:\ddot{O}\!=\!\ddot{O}:$	oxygen-oxygen double bond

$$\begin{array}{c}\diagdown\\ {}/\end{array}C\!=\!\ddot{N}\!-\!\quad\quad\text{carbon-nitrogen double bond}$$

$$-C\!\equiv\!N\!:\quad\quad\text{carbon-nitrogen triple bond}$$

$$\begin{array}{c}\diagdown\\ {}/\end{array}C\!=\!\ddot{O}\!:\quad\quad\text{carbon-oxygen double bond}$$

Among the preceding structures is that for the nitrogen molecule, N_2, which has a bond order of three. Nitrogen constitutes about 80% of the Earth's atmosphere, and it generally is regarded as being almost entirely unreactive. It is so inert because atoms of the element are very strongly bound together in diatomic molecules.

Carbon monoxide is isoelectronic with N_2, and the Lewis structures of both molecules are represented by a triple bond:

$$:N\!\equiv\!N:\quad\quad:\overset{-}{C}\!\equiv\!\overset{+}{O}:$$

We can imagine forming a molecule of carbon monoxide from a molecule of N_2 by removing a proton from one nitrogen nucleus and adding it to the other.

These molecules provide examples for the calculation of formal charge for molecules that have both lone-pair and bond-pair electrons. Whereas bond-pair electrons are divided equally in the formal charge procedure, the lone-pair electrons are assigned wholly to the atom to which they are "attached." Consequently, the carbon atom is assigned five electrons in CO, and because it has only four valence electrons in the atomic state $(2s^2 2p^2)$, its formal charge is -1. A similar calculation for the oxygen atom results in a charge of $+1$. Of course, this does not mean that these charges are fully present, but only that the bonding results in a nonuniform charge distribution. In fact, the experimental dipole moment of CO does indicate a distribution of charge corresponding to

$$\overset{\delta-}{:C}\!\equiv\!\overset{\delta+}{O}:$$

The presence of multiple bonds increases the bond order between atoms, with a consequent increase in bond dissociation energy and decrease in bond length. The homonuclear diatomic molecules N_2, O_2, and F_2 illustrate the bond order–dissociation energy–bond length relationship:

Molecule:	N_2	O_2	F_2
Bond order:	3	2	1
Bond energy (kcal mole^{-1}):	225	118	37
Bond length (Å):	1.10	1.21	1.42

Further examples are provided by the hydrocarbons acetylene, ethylene, and ethane:

Molecule:	C_2H_2	C_2H_4	C_2H_6
C—C bond order:	3	2	1
C—C bond energy (kcal mole^{-1}):	230	163	88
C—C bond length (Å):	1.21	1.35	1.54

An interesting problem arises in the Lewis structural formulation of the transient cyanogen molecule (CN). A closed-shell configuration cannot be constructed for CN because there is an odd number of valence electrons. The cyanogen molecule has nine valence electrons, four valence electrons associated with the carbon atom and five electrons with the nitrogen atom. Thus either C or N will "own" only seven electrons in the CN molecule. We choose C because it is less electronegative than N. Therefore the best Lewis structure for CN is ·C≡N: .

Lewis structures for molecules such as CN, which have an odd number of electrons, necessarily cannot have closed shells associated with each atom. At least one atom, carbon in the CN example, is left with an "open shell." As a consequence of this open-shell structure two molecules of CN combine to form the dimer $(CN)_2$, which is called *dicyanogen*. The driving force for this reaction is readily seen to be the formulation of a new C—C bond without any significant weakening of the C—N triple bond:

$$·C≡N: + ·C≡N: → :N≡C—C≡N:$$

The odd electron in CN is unpaired. Consequently we would predict the cyanogen molecule to be paramagnetic, a prediction that is in agreement with experimental data. A multiple-bonded molecule with particularly vexing magnetic properties (for Lewis's structural theory) is O_2, which is known to have two unpaired electrons in its ground state and to be paramagnetic. An unusual structure such as :Ö≡Ö: or :Ö—Ö: would be required to explain this magnetic behavior. However, the observed bond length and bond energy of O_2 are completely consistent with the simple double-bond structure :Ö=Ö: . We will see in Chapter 4 that the molecular-orbital theory provides a satisfactory explanation of both the paramagnetism and the bond properties of the oxygen molecule.

Students often have difficulty in writing appropriate Lewis structures.

Although a strict set of rules cannot be developed, the following summary may prove helpful:

1. Determine the total number of valence electrons in the molecule or ion.

2. Write the atomic symbols in the proper order and first "satisfy" the external or *terminal* atoms. To satisfy these atoms means that the *maximum* number of electrons surrounding any given terminal atom is two for hydrogen or helium and eight for all other atoms in the periodic table.

3. If the central atom has not achieved an octet and if lone electron pairs (or nonbonded electrons) are available on the terminal atoms, allow these lone electrons to bond with the central atom by forming multiple bonds.

4. Among several Lewis structures, those with low formal charges $(0, \pm 1)$ on each atom are preferred.

2-13 Bonding to Heavier Atoms

The octet rule has been extremely valuable as a guide in writing electronic formulas. For second-row nonmetallic elements (B, C, N, O, F), exceptions to the rule are very rare. It is easy to explain why this is so. Atoms of the second-row elements have stable $2s$ and $2p$ orbitals, and the "magic number" of eight corresponds to the closed valence-orbital configuration $2s^2 2p^6$. Adding a ninth, tenth, or larger number of electrons to such a configuration is impossible, because the next atomic orbital available to a second-row element is the highly energetic $3s$ orbital.

Beyond the second row in the periodic table the octet rule is not obeyed with such satisfying regularity. However, it remains a useful rule, as illustrated by molecules such as PH_3, PF_3, H_2S, and SF_2:

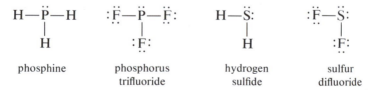

| phosphine | phosphorus trifluoride | hydrogen sulfide | sulfur difluoride |

Atoms of the heavier elements do more than obey the octet rule. Some of them show a surprising ability to bind more atoms (or associate with more electron pairs) than would be predicted from the octet rule. For example, phosphorus and sulfur form the compounds PF_5 and SF_6, respectively. Lewis structures for these compounds use all the valence electrons of the heavy element in bonding:

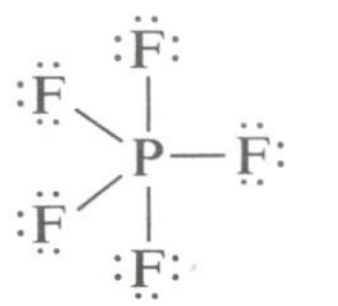

Phosphorus
pentafluoride

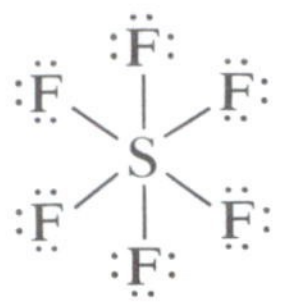

Sulfur
hexafluoride

That phosphorus shares 10 electrons and sulfur shares 12 electrons obviously violates the octet rule. The theory of atomic structure helps us see why the violation has occurred. The noble gas in the third row with phosphorus and sulfur is argon. The argon electronic structure fills the $3s$ and $3p$ orbitals, but leaves the five $3d$ orbitals vacant. If some of these $3d$ orbitals are used for electron-pair sharing, extra bonds are possible. The atomic theory thus provides an explanation of the enhanced bonding versatility of elements in the third row and beyond.

Perhaps the most important consequence of the use of d orbitals is the formation of an important series of oxyacids. The best-known examples are phosphoric acid (H_3PO_4), sulfuric acid (H_2SO_4), and perchloric acid ($HClO_4$). It is possible to write a Lewis structure for sulfuric acid that obeys the octet rule:

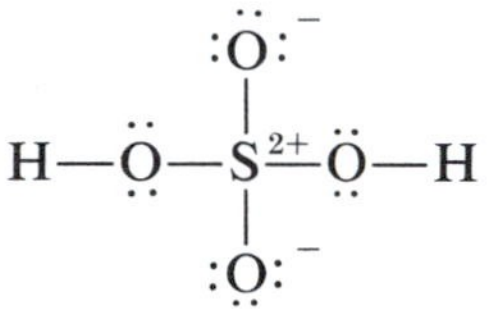

However, examination of this structure reveals that a formal charge of $+2$ is on the sulfur atom. Development of a large positive formal charge on an electronegative nonmetal atom is not very reasonable. The formal charge can be removed if we write two sulfur-oxygen double bonds, thereby allowing the sulfur atom to share 12 electrons:

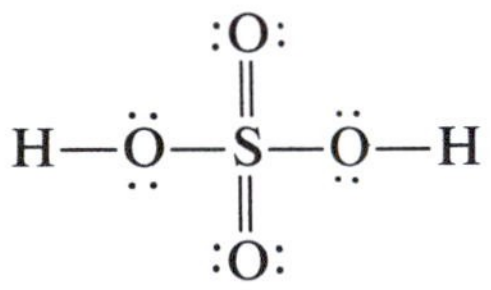

sulfuric acid

Similar Lewis formulas can be written for other oxyacids:

phosphoric acid   perchloric acid

2-14 Resonance

There are molecules and ions for which more than one satisfactory Lewis formula can be drawn. For example, the nitrite ion, NO_2^-, can be formulated as either

In either case the octet rule is satisfied. If either of these structures were the "correct" one, the ion would have two distinguishable nitrogen-oxygen bonds, one single and one double. Double bonds are shorter than single bonds, but structural studies of NO_2^- show that the two N—O bonds are indistinguishable.

Consideration of NO_2^- and many other molecules and ions shows that our simple scheme for counting electrons and assigning them to the valence shells of atoms as bonds or unshared pairs is not entirely satisfactory. Fortunately, this simple model may be altered fairly easily to fit many of the awkward cases. The problem with NO_2^- is that the ion is actually more symmetrical than either one of the Lewis electronic structures that we have written. However, if we took photographs of the two formulas shown previously and superimposed the pictures, we would obtain a new formula having the same symmetry as the molecule. The photographic superimposition method is the same as writing a formula such as

This formula would imply, "NO_2^- is a symmetrical ion, having partial double-bond character in each of the N—O bonds." For some purposes the formula is adequately informative. However, keeping track of the electrons in such a formula requires some rather special notation. What we actually do most of the time in such situations is to write two or more Lewis formulas and connect them with a symbol that means: "Superimpose these formulas to get a reasonable representation of the molecule." Applied to NO_2^- the formulas are

The double-headed arrow is the symbol for superimposition. It should not be confused with the symbol consisting of two arrows pointing in opposite directions, $\rightleftarrows$, which indicates that a reversible chemical reaction is occurring. The double-headed arrow conveys no implication of dynamic action.

The method of combining two or more structural formulas to represent a single chemical species is called the *resonance method*. The method is used not only for the construction of electronic formulas, but also as the basis of one method for making approximate quantum-mechanical analyses of molecular structures.

When we consider the benzene molecule, C_6H_6, which has six carbon atoms arranged in a ring, we can draw two formulas that are equally satisfactory:

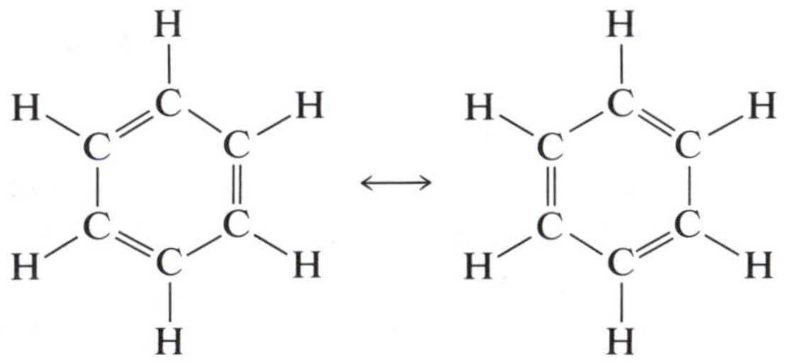

Both resonance structures show the ring to be composed of alternate single and double bonds. However, structural studies reveal that all of the carbon-carbon bond distances are equal. The full symmetry of the molecule can be indicated by the double-headed arrow between the two structures or by the single structure

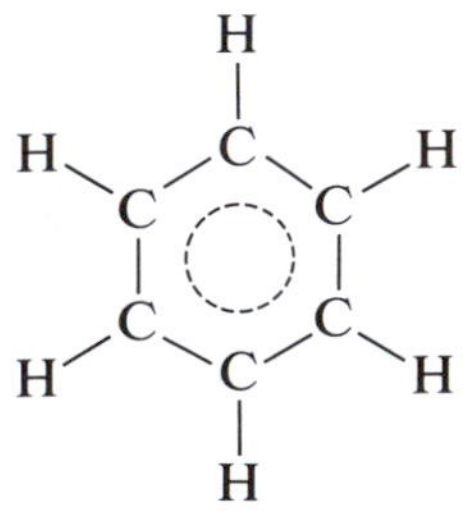

Resonance notation is required in many cases other than those in which it is demanded by symmetry. For example, compare two well-known anions, nitrate (NO_3^-) and nitroamide (^-O_2NNH). Nitrate has threefold symmetry so we can write a set of three equivalent resonance structures:

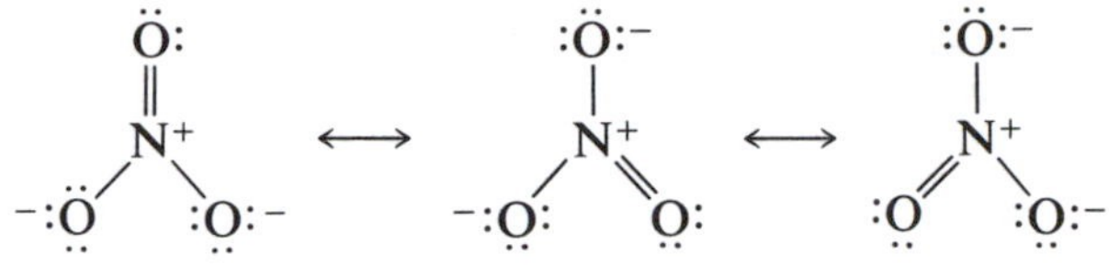

For the nitroamide ion we can write two equivalent structures, plus a third that is not equivalent to the other two:

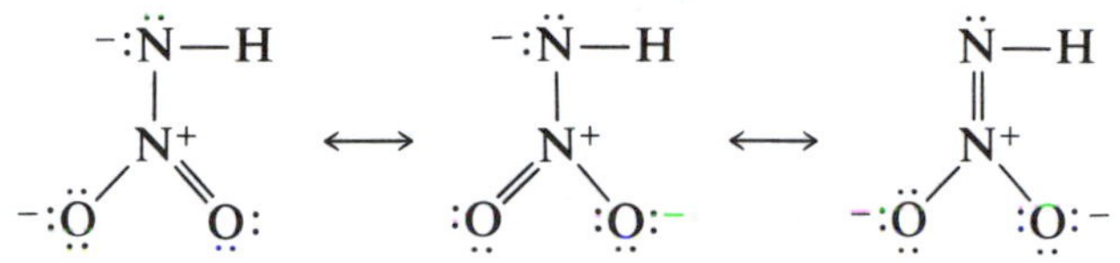

Common sense tells us that all three structures should contribute to our description of the ion. Since the structures are not equivalent, the resonance symbol no longer means: "Mix these structures equally in your thinking." It merely means: "Mix them." Therefore no quantitative implications are intended by the double-headed arrow. When we become semiquantitative in our description we state that structure III "contributes" more to the structure of the nitroamide ion than either of the equivalent structures I and II because III places both formal negative charges on the oxygen atoms.

The last of the polyatomic molecules we will discuss is the anion obtained by removing the two protons in sulfuric acid, the sulfate ion, SO_4^{2-}. As in the case of H_2SO_4, an octet-rule structure with only single bonds can be written by assigning three lone pairs to each oxygen atom:

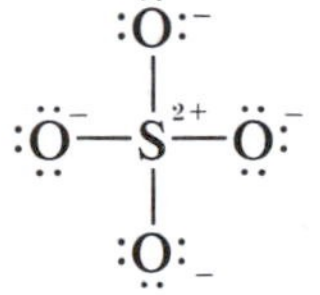

However, if we consider the large positive formal charge on the sulfur atom we conclude that this is not a particularly appropriate structure. A much better representation of the bonding in SO_4^{2-} removes the +2 formal charge on the central sulfur atom by forming two sulfur-oxygen double bonds.

There are six equivalent structures with two S=O bonds and two S—O bonds. Thus we represent the bonding in SO_4^{2-} as a resonance hybrid of the following six equivalent structures:

I   II   III

IV   V   VI

The resonance hybrid of the six equivalent structures (I–VI) of SO_4^{2-} would have an average S—O bond order of $1\frac{1}{2}$. In accord with this model of partial double-bond character is the fact that the observed S—O bond length in SO_4^{2-} (1.49 Å) is 0.27 Å shorter than the standard S—O single bond length of 1.76 Å, which is obtained by adding the atomic radii of sulfur (1.04 Å) and oxygen (0.72 Å); see Figure 2-1.

2-15 Molecular Geometry

Up to this point we have been concerned mainly with writing appropriate Lewis structures for atoms and molecules. Now we wish to extend our understanding of molecular bonding by investigating molecular geometry. Minimum energy is the criterion that determines the structure adopted by a molecule. The ground-state molecular geometry of $BeCl_2$ is linear, whereas that of OCl_2 is angular (bent); BF_3 is trigonal planar, NF_3 is trigonal pyramidal, and ClF_3 is T-shaped; SeF_4 is tetrahedral, SF_4 is irregular tetrahedral (seesaw), and ClF_4^- is square planar; PF_5 is trigonal bipyramidal, but ClF_5 is square pyramidal; finally, SF_6 is octahedral, but XeF_6 is irregular octahedral. In this section we wish to understand why the minimum energy structure corresponds to those observed.

There are four energy factors that determine the geometry of a molecule. These are: (1) electron-electron repulsions, (2) nuclear-nuclear repulsions, (3) electron-nuclear attractions, and (4) the kinetic energies of the electrons.

An exact understanding of molecular geometry would require that all four of these factors be considered in an exact calculation. What we wish to do in this section is to utilize a qualitative method which considers *only* electron-electron repulsions to predict molecular geometry. This method is appropriately called the *valence-shell electron-pair repulsion* (VSEPR) method. In Chapter 5 we will show that it also is possible to predict molecular geometry using the *molecular-orbital* method. The shapes of molecules containing transition-element atoms will be dealt with in Chapter 6.

The valence-shell electron-pair repulsion method and molecular geometry

The simplest approach to the prediction of molecular geometry was published by Nevil V. Sidgwick and Herbert M. Powell in 1940, and extended by Ronald J. Gillespie and Sir Ronald S. Nyholm in 1957. This approach, the valence-shell electron-pair repulsion method, is commonly called the VSEPR method and states that the bonding electron pairs and lone electron pairs of an atom will adopt a spatial arrangement that minimizes electron-pair repulsion around that atom.

Before applying the VSEPR method to molecular systems, let us consider the most favorable arrangement for *n* electron pairs circumscribed on a sphere. For values of *n* from 2 through 6 the predicted arrangement is shown in Figure 2-8. The predicted arrangement for 7 through 12 electron pairs will be discussed later in this section.

To apply the VSEPR method to molecules we simply count the number of lone electron pairs and the number of atoms around the central atom in a polyatomic molecule. We will call the total number of lone-pair electrons and attached atoms the *steric number*, SN. If there are no lone-pair electrons around the central atom (A) and the SN arises only from attached atoms (X), the observed molecular geometry agrees with that shown in Figure 2-8:

Molecule	Steric number	Predicted geometry		Example
AX_2	2	linear	$180°$	$BeCl_2$
AX_3	3	trigonal planar	$120°$	BF_3
AX_4	4	tetrahedral	$109.5°$	SiF_4
AX_5	5	trigonal bipyramidal	$90°$ $120°$	PF_5
AX_6	6	octahedral	$90°$ $90°$	SF_6

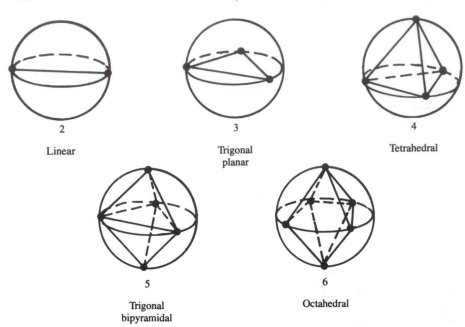

Figure 2-8 The arrangement of electron pairs on the surface of a sphere so that the distance between electron pairs is maximized.

In each case the predicted geometry corresponds to separating the bonding electron pairs as widely as possible, to minimize electron-pair repulsion. It should be emphasized that multiple bonds between atoms do not alter the prediction of molecular geometry. For example, beryllium dihydride and carbon dioxide each have SN = 2 and are predicted to be linear:

$$H\text{—}Be\text{—}H \qquad :\ddot{O}\text{=}C\text{=}\ddot{O}:$$

If one of the attached atoms is replaced by a lone electron pair, the molecular geometry changes. *By the molecular geometry we mean the positions of the atoms that we can determine experimentally, but not the placement of lone-pair electrons, which can only be inferred.* Thus for AX_3 and AX_2E where X represents an attached atom and E a lone pair of electrons, we obtain the following structures:

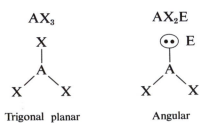

AX_3 — Trigonal planar

AX_2E — Angular

We can not only predict that an AX_2E molecule will be angular but we also can predict that the XAX angle will be *less* than 120°. This prediction results from the postulate that lone electron pairs occupy more space and therefore exert a larger repulsive effect than do bonding electron pairs.

We can apply this same postulate to predict the relative bond angles in the isoelectronic sequence CH_4, $:NH_3$, and $H_2\ddot{O}:$. Each of these molecules has SN = 4. However, due to the lone pair–bond pair repulsions in NH_3 and H_2O, we predict the bond angles in these two molecules to be less than the tetrahedral angle (109.5°). The experimental bond angles agree with these predictions:

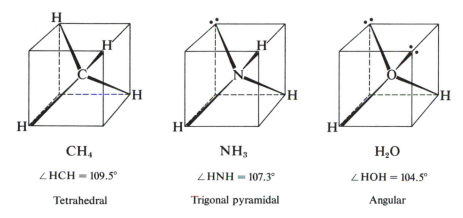

CH$_4$	NH$_3$	H$_2$O
∠HCH = 109.5°	∠HNH = 107.3°	∠HOH = 104.5°
Tetrahedral	Trigonal pyramidal	Angular

Examples of predicted molecular shapes from SN = 2 through SN = 6 are shown in Figure 2-9.

The structure of a molecule that contains different atoms attached to a central atom exhibits distortions from the idealized structures shown in Figure 2-9. Thus CH_3Cl has HCH angles of 110.5° and ClCH angles of 108.5°, compared to the perfect tetrahedron for which all bond angles are 109.5°. Further examples of these distortions from an idealized geometry are ethylene and formaldehyde. In both of these molecules the carbon atoms have SN = 3, for which the idealized bond angles are 120°. The observed structures are

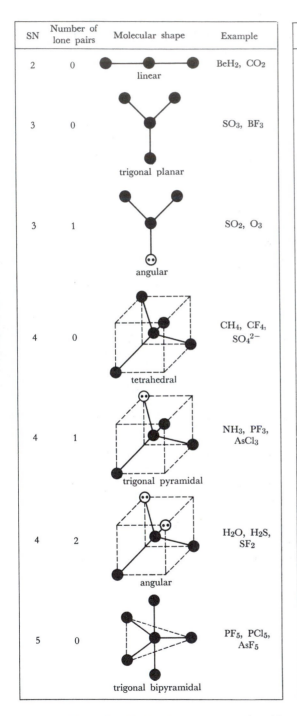

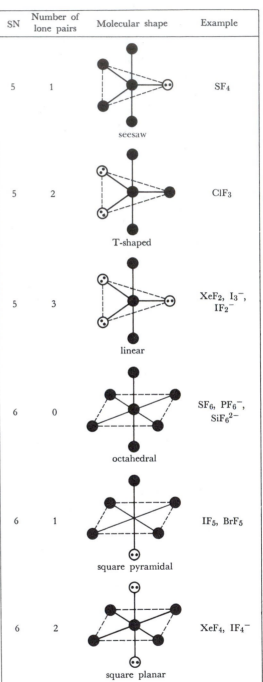

Figure 2-9 Molecular shapes predicted by the VSEPR method.

The placement of electron pairs for SN = 5 deserves special comment. The spatial arrangement for five electron pairs is a trigonal bipyramid, in which there are three equatorial and two axial positions:

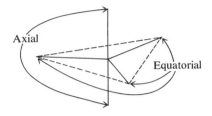

The shape of a molecule such as PF_5 must be trigonal bipyramidal because there are no lone pairs surrounding the central phosphorus atom. However, it is not necessary that the axial bond lengths be equal to the equatorial bond lengths. Since each axial bond experiences three 90° repulsions compared to only two 90° repulsions for the equatorial positions, we predict that the axial bonds will be longer than the equatorial bonds.* Experimentally, the axial P—F bond lengths are 1.557 Å compared to 1.534 Å for the equatorial bond lengths.

Next we consider the SF_4 molecule which has four bond pairs and one lone pair surrounding the sulfur atom. Where should we place the lone electron pair in SF_4? According to the VSEPR method, the most prohibitive repulsion is (lone pair)–(lone pair), followed in order by (lone pair)–(bond pair) and (bond pair)–(bond pair). Therefore in SF_4 the "worst" repulsion is (lone pair)–(bond pair) because there is only one lone pair. If we place the lone pair in one of the equatorial orbitals, it will repel only two bonded pairs at a 90° angle, whereas in an axial orbital it will repel three pairs at 90°:

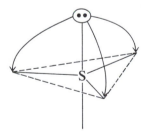

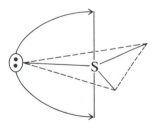

Three 90° interactions
(axial placement of lone pair)

Two 90° interactions
(equatorial placement of lone pair)

Therefore the VSEPR choice is equatorial placement of the lone electron pair, which results in a smaller number of 90° interactions. The resulting shape is sometimes referred to as the "seesaw" or "butterfly" shape. Accordingly, we also place the second (e.g., ClF_3) and third (e.g., I_3^-) lone pairs into equatorial orbitals, and predict the shapes shown in Figure 2-9.

The equatorial position of the lone pairs in SF_4 and ClF_3 should result in distortions of the idealized 90° and 120° angles between bonds. Indeed, the experimental geometries indicate the expected distortions:

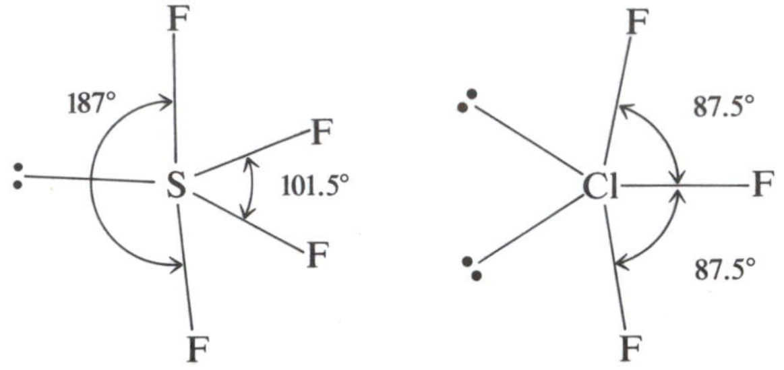

As our last example in the discussion of $SN = 5$, we consider those molecules in which the attached atoms are not all identical. Examples of such molecules are CH_3PF_4 and SOF_4. In each of these molecules, the less electronegative groups are found to occupy the equatorial positions and to cause distortions from the idealized 90° and 120° angles, similar to those caused by lone electron pairs. Thus we find the following structures:

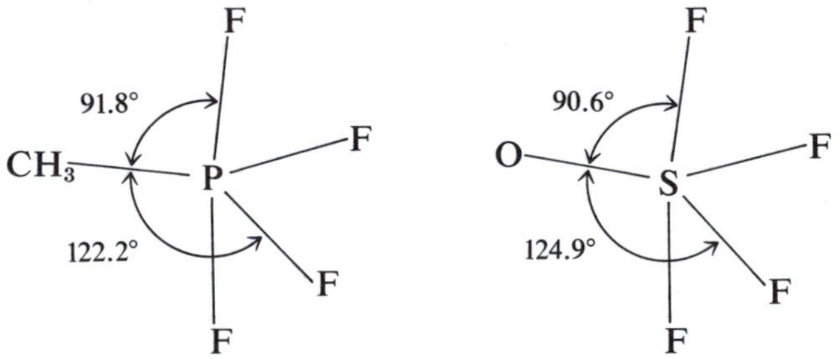

The VSEPR method is easy to use and gives the correct molecular shape for a remarkably large number of molecules. All of the predicted shapes in Figure 2-9 are in agreement with experimentally determined molecular structures.

The VSEPR method can be summarized by three simple rules:

1. The electron pairs around a central atom will adopt a spatial arrangement that minimizes electron-pair repulsion.

*The 90° interactions are much larger than 120° or 180° interactions, because electron-pair repulsions decrease very rapidly as the distance between pairs increases (see Section 7-3).

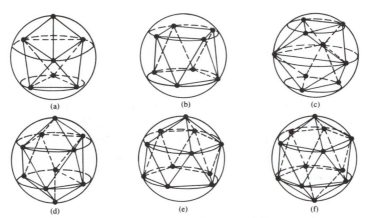

Figure 2-10 Arrangements of 7 through 12 electron pairs on a sphere to maximize distance between pairs. (a) 7 pairs: monocapped octahedron; (b) 8 pairs: square antiprism; (c) 9 pairs: tricapped trigonal prism; (d) 10 pairs: bicapped square antiprism; (e) 11 pairs: monocapped pentagonal antiprism; (f) 12 pairs: icosahedron.

2. The most prohibitive repulsion is (lone pair)–(lone pair), followed in order by (lone pair)–(bond pair) and (bond pair)–(bond pair).

3. Among several structures involving 90° interactions, the most favored structure is the one that results in a smaller number of 90° lone-pair interactions.

VSEPR Applied To Molecules With Steric Number Greater Than Six

The idealized geometry for 7 through 12 electron pairs is shown in Figure 2-10. For any real molecular system there are several different geometries that have nearly the same energy. Thus, rather than the monocapped-octahedral structure shown in Figure 2-10, the IF_7 molecule (SN $= 7$) has a pentagonal-bipyramidal structure:

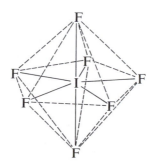

Another molecule with SN = 7 is XeF_6. Its molecular geometry is predicted to be *ir*regular octahedral due to the presence of the lone pair on the xenon atom. The experimental structure has been unsettled because the geometry is not static—an intramolecular rearrangement process rapidly interconverts fluorine positions. In agreement with the VSEPR prediction, the structure definitely is not regular octahedral.

Examples of species with SN equal to 8 and 9 are TaF_8^{3-} and ReH_9^{2-}. These systems obtain the square-antiprismatic and tricapped-trigonal-prismatic structures, respectively. Steric numbers higher than 9 are rare and will not be discussed here.

Exceptions to the VSEPR Rules

So far we have applied the VSEPR method to predict the molecular shapes of molecules that have all their electrons in pairs. If a molecule has one electron that is unpaired, we cannot predict molecular shape so simply. Consider the isoelectronic series BeH_3^{2-}, BH_3^-, CH_3, and NH_3^+, all of which have the Lewis electron dot structure

$$H—\dot{A}—H$$
$$|$$
$$H$$

Experimental evidence indicates that NH_3^+ and CH_3 are trigonal planar.

No experimental evidence exists for BeH_3^{2-} and BH_3^- but accurate quantum-mechanical calculations predict the degree of distortion from trigonal planar to trigonal pyramidal to follow the order $NH_3^+ < CH_3 < BH_3^- < BeH_3^{2-}$.

Another class of molecules that are not readily understood in terms of the VSEPR rules are the third-, fourth-, and subsequent-row compounds of Groups V and VI. The geometries of the hydrides of these compounds are compared to those of the second-row elements in Table 2-6. It is clear that only the second-row compounds exhibit bond angles near the expected tetrahedral angle. The subsequent-row compounds all exhibit bond angles near 90°. A possible rationalization of this phenomenon is that due to the larger size of the central atom, there is less bond pair–bond pair repulsion in these compounds and hence the bond angles are considerably smaller than the tetrahedral angle.

Other exceptions to the VSEPR rules are provided by some of the alkaline earth dihalide molecules. Most of these MX_2 molecules are found to be linear, X—M—X, as predicted. However, the molecules CaF_2, SrF_2, $SrCl_2$, BaF_2, $BaCl_2$, $BaBr_2$, and BaI_2 are found to be angular. Also, the ions $TeCl_6^{2-}$,

TABLE 2-6. BOND ANGLES IN
GROUP V AND VI HYDRIDES.

NH_3	107.3°	H_2O	104.5°
PH_3	93.3°	H_2S	92.2°
AsH_3	91.8°	H_2Se	91.0°

$TeBr_6^{2-}$, and $SbBr_6^{3-}$ are found to be regular octahedral, even though they are all of the AX_6E type and should be distorted just as predicted for XeF_6. This is rationalized after the fact by noting that the Br and Cl atoms are fairly bulky and hence the atom–atom repulsions perhaps dominate over the lone pair–bond pair repulsions.

These are not the only exceptions to the VSEPR rules, but they are sufficient to point out that although these empirical rules are capable of predicting molecular geometry for a wide range of molecules, they overlook features such as atom–atom repulsion which are critical in some instances.

2-16 The Use of Lewis Structures to Predict Molecular Topology

In this section we present and discuss some simple empirical rules that allow one to predict *molecular topology,* that is, the order or sequence in which the atoms of a molecule are connected. Suppose we are asked to predict the molecular structure for the molecule HCOF, which has 18 valence electrons. Will this molecule be linear HCOF, HOCF, HFCO, HFOC, HOFC, or HCFO? Or will it be nonlinear, and if so, will it be planar? Compounds with the same molecular formula, but different structures, are called *isomers*. We have seen that once we know the sequence in which the atoms are bonded, we can predict the structure by using the VSEPR method. We now wish to discuss cases in which we are given the molecular formula but not the sequence in which the atoms are connected. We are interested in predicting the topology of covalent molecules that do not involve electron deficient bonding. We are therefore restricted mainly to a discussion of molecules containing the hydrogen atom, and atoms from Groups IV, V, VI, and VII.

The following empirical "rules" seem to be generally valid for predicting the structure of the *most stable* isomer of a molecule. The priority of the rules should be taken in the order listed.

1. Hydrogen atoms are terminal. There are exceptions to this rule: for example, F—H—F⁻ and B_2H_6 (Figure 2-7).

2. Halogen atoms are terminal. There are also exceptions to this rule: for example, H—F—F⁺. Furthermore, the heavier halides (Cl, Br, I) are central if the possibility exists that they can bond to more than one oxygen or fluorine atom: for example, ClF_3 and HOClO.

3. Of the remaining atoms, if any two are from the same period (row) in the periodic table, the less electronegative atom will be surrounded by at least as many atoms as the more electronegative atom. A similar statement applies if any two atoms are from the same group (column) in the periodic table. Other things being equal, the less electronegative atom will be surrounded by more atoms than the more electronegative atom.*

   This rule can be summarized qualitatively by stating that highly electronegative atoms generally occupy terminal positions because they then need to share their electron density with only one other atom. By contrast, less electronegative atoms are more capable of sharing their electron density; consequently, they bond to more than one atom.

4. Discard structures with three-membered rings. These structures involve "strain" energy because the interbond angle is 60° whereas no two orbitals may make an angle of less than 90° with each other. For example, two isomers of C_3H_6 are cyclopropane and propylene; the latter is more stable by 7.5 kcal mole⁻¹:

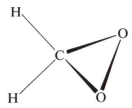

| Cyclopropane | Propylene |

There are exceptions to this rule. Dioxirane is isoelectronic with cyclopropane and it is known to have a three-membered ring structure:

5. Structures in which the terminal atoms obtain the closed-shell octet are preferred. Obviously, the closed-shell configuration for the hydrogen atom is only two electrons, not the octet. In general, the octet also will be impossible for terminal atoms that are from Groups IA, IIA, and IIIA in the periodic table (e.g., Li, Be, and B). However, bonding for these elements generally involves ionic or electron deficient bonding which we are excluding here.

---

*"Other things being equal" refers to equal satisfaction of the subsequent rules.

6. The Lewis electron dot structure for a given topology should obey the Pauling *electroneutrality principle*. That is, low formal charges on the atoms are preferred. If formal charges are required, positive charge should reside on the less electronegative atoms and negative charge on the more electronegative atoms, whenever possible. Discard structures that require adjacent atoms to have formal charges of the same signs.

7. Structures that form the maximum number of bonds are preferred.

Applying these rules to the molecule HCOF, we can write three isomers, each involving a different topology:

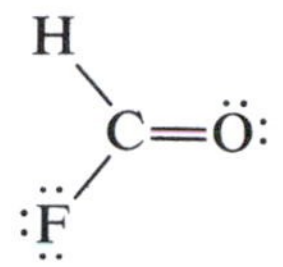

| Planar | Nonlinear | Nonlinear |
| I | II | III |

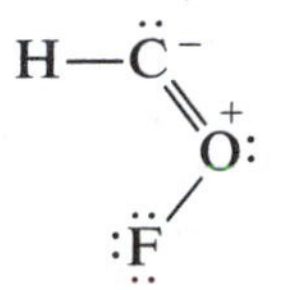

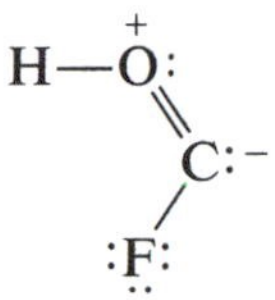

Structure I best obeys rules 3 and 6. Structures II and III explicitly disobey rule 6 in that the carbon atom has a negative formal charge while the more electronegative oxygen atom has a positive formal charge. Structure I corresponds to the most stable isomer for HCOF.

Now let us discuss another 18-valence-electron molecule: HNSO. Applying the rules to HNSO results in the following possible isomers:

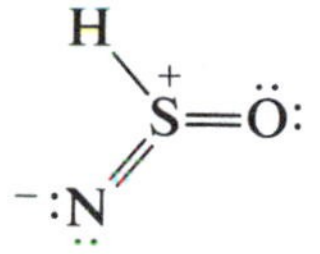

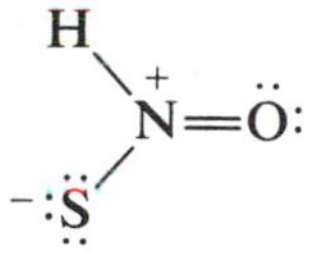

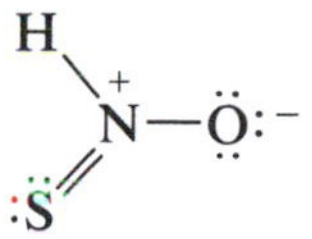

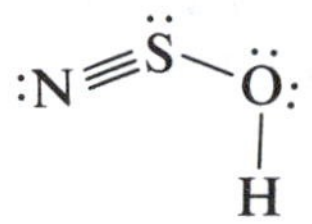

Recall that the closed-shell octet may be exceeded for the sulfur atom, since it is a third-row element. Other structures in which the oxygen atom is in a central position are excluded, because oxygen is more electronegative than sulfur (rule 3). Structures I, IV, and V have a total of five bonds each, but structures II, III, and VI have only four bonds each. Of the five-bond structures, IV and V are favored over I because structure I exhibits formal

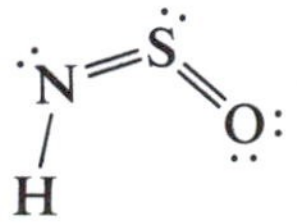

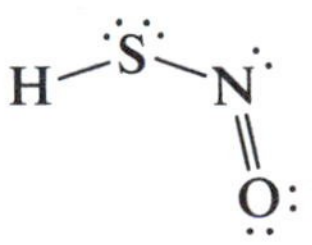

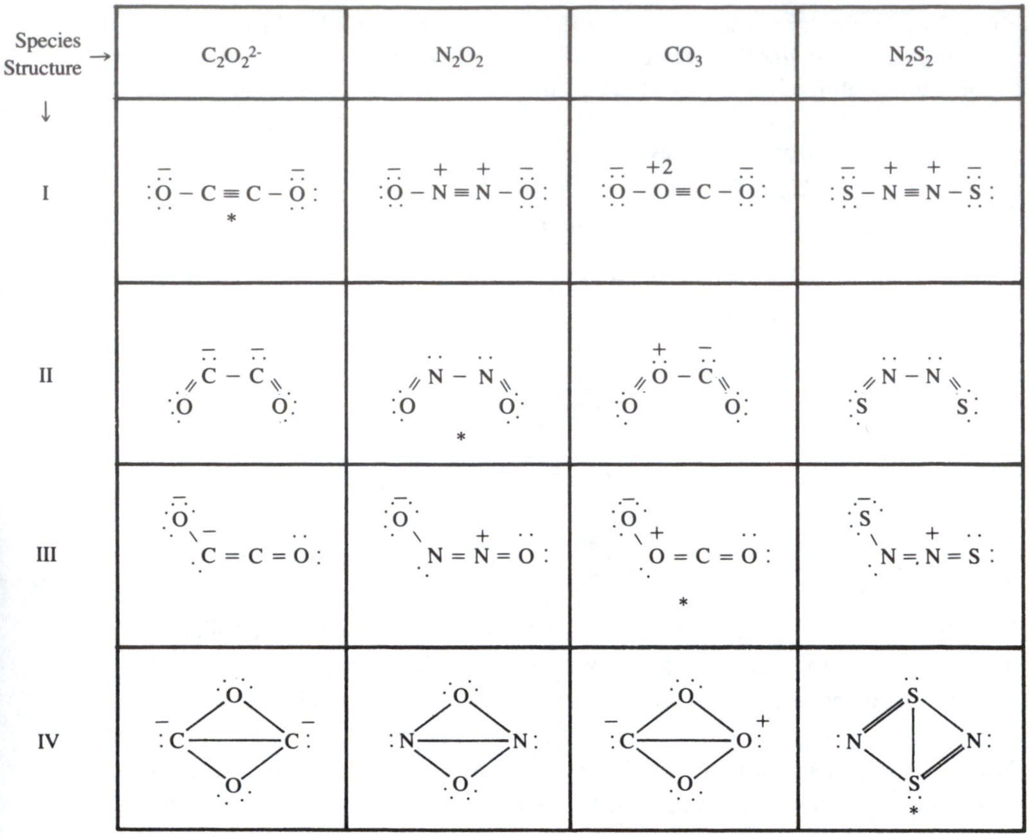

Species Structure →	$C_2O_2^{2-}$	N_2O_2	CO_3	N_2S_2
I	$:\overset{-}{O} - C \equiv C - \overset{-}{O}:$ *	$:\overset{-}{O} - \overset{+}{N} \equiv \overset{+}{N} - \overset{-}{O}:$	$:\overset{-}{O} - O \overset{+2}{\equiv} C - \overset{-}{O}:$	$:\overset{-}{S} - \overset{+}{N} \equiv \overset{+}{N} - \overset{-}{S}:$
II	C–C with O, O:	N–N with O, O: *	O–C with O, O:	N–N with S, S:
III	O, C=C=O:	O, N=N=O:	O, O=C=O: *	S, N=N=S:
IV	C—C ring with O, O	N—N ring with O, O	C—O ring with O, O	N—N ring with S, S *

Figure 2-11 Possible Lewis electron dot structures for the 22 valence electron species $C_2O_2^{2-}$, N_2O_2, CO_3, and N_2S_2. For each species, the experimentally observed structure is indicated with an asterisk (*).

charges. Between structures IV and V we predict structure V to be more stable because we expect the nitrogen atom to be surrounded by more atoms than the oxygen atom (rule 3). Therefore we predict structure V to be the most stable; it has been observed experimentally. Structures IV and VI also have been observed, but only at low temperature conditions.

We already have discussed two molecules containing 18 valence electrons, HCOF and HNSO. These molecules have different molecular structures; we have seen that the topology of the most stable isomer can be predicted on the basis of the seven simple rules presented previously.

Now we wish to consider a series of four species, each containing 22 valence electrons, in an attempt to predict the observed molecular structure. These species are $C_2O_2^{2-}$, N_2O_2, CO_3, and N_2S_2. In Figure 2-11 we present four possible molecular structures for each of these species. The structure designated by * is the one which has been observed experimentally for each species.

For $C_2O_2^{2-}$, structure I is the only one that places the excess negative charge on the more electronegative (oxygen) atoms. In the case of N_2O_2, each of the structures I through IV predicts a total of five bonds, but only structures II and IV have zero formal charges. The cyclic structure (IV) should be strained, and consequently structure II is favored. For the carbon trioxide molecule, CO_3, structure III comes closest to satisfying the electroneutrality principle. For N_2S_2, both structures II and IV have zero formal charge on each atom, but because sulfur can expand its valence beyond eight, structure IV contains seven bonds compared to only five for structure II. Structure II is rejected also because the more electronegative nitrogen atoms are central, whereas the less electronegative sulfur atoms are terminal.

Let us consider next the molecules HCN and HNC, for which we can write the following Lewis structures

$$H-C\equiv N: \qquad \text{and} \qquad H-\overset{+}{N}\equiv\overset{-}{C}:$$

Experimentally, HCN is a stable gas-phase molecule, whereas HNC has been detected only at low temperatures. The molecule HNC is disfavored on two counts: (1) It places the more electronegative nitrogen atom in a central position and (2) the HNC structure results in formal charges on the nitrogen and carbon atoms. Of these two points, the second seems to be more important. To illustrate this fact let us consider the molecule HCP, which is isoelectronic with HCN. Phosphorus has a lower electronegativity than carbon, yet experimentally the molecule exists as HCP rather than HPC. The HPC structure is rejected because the resulting Lewis structure has formal charges on the phosphorus and carbon atoms.

We emphasize that the foregoing rules are useful only in predicting molecular structure for molecules that are predominantly covalent. For mainly ionic molecules we find the most electronegative atom in the central position. One example is Li_2O, which by analogy with H_2O, we predict to be bent. In fact, Li_2O is observed to be linear (Li—O—Li). Other related examples are the linear molecules LiOH and CsOH. Consider also the molecule Al_2O, which if covalent would be predicted to be linear $:Al-Al=\ddot{O}:$. Experimentally, the oxygen atom is in the central position (Al—O—Al).

2-17 Molecular Symmetry

Webster's Dictionary defines *symmetry* as the "correspondence in size, shape, and relative position of parts that are on opposite sides of a dividing

TABLE 2-7. SYMMETRY ELEMENTS AND SYMMETRY OPERATIONS

Element	Symbol	Operation
Identity	E	Leaves each particle in its original position
n-fold proper axis	C_n	Rotation about the axis by $360°/n$ (or by a multiple)
Plane	σ	Reflection in the plane
Inversion center	i	Inversion through center
n-fold improper axis	S_n	Rotation by $360°/n$ (or by a multiple) followed by reflection in a plane perpendicular to the axis

line or median plane or that are distributed about a center or axis." This definition is equally applicable to the intricate beauty of a flower, to a work of art, or to a molecule. In this section we consider the application of symmetry in describing molecular structure. In subsequent chapters we will utilize our knowledge of molecular symmetry for taking proper combinations of orbitals in bonding theory and for the interpretation of molecular spectra.

A useful definition of symmetry for our discussion is the following: If a molecule has two or more orientations that are indistinguishable we say that the molecule possesses symmetry. We shall see that it is possible to quantify this definition by specifying the number and types of ways in which we can shift the molecule from one indistinguishable orientation to another. There are only a few distinct ways by which we can reorient a molecule to an indistinguishable configuration. These "ways" are called *symmetry operations*. A symmetry operation is some transformation of the molecule, such as a rotation or reflection, which leaves the molecule in a configuration in space that is indistinguishable from its initial configuration. A *symmetry element* is that point, line, or plane in the molecule about which the symmetry operation takes place. The possible symmetry elements and symmetry operations are tabulated in Table 2-7. We see that there are only five elements of symmetry needed to specify the symmetry of a molecule. These are the identity, n-fold proper axis, plane, center of inversion, and an n-fold improper axis. We shall find that a molecule need not possess all five elements to have its symmetry specified.

If a molecule possesses more than one symmetry element, all of the elements pass through a common *point* in space. Because all the symmetry operations for any molecule also obey the properties of a mathematical *group*, we refer to the *point group* symmetry of a molecule.

We may understand the symmetry elements and operations more readily from a few examples. Consider first the water molecule, Figure 2-12. There is a proper axis passing through the oxygen atom and bisecting the H—O—H

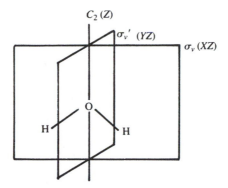

Figure 2-12 Symmetry elements of the water molecule. There are two mutually perpendicular σ_v planes and a C_2 axis that passes through the oxygen atom and bisects the H—O—H angle.

angle. Rotation about this axis by $360°/2 = 180°$ leaves the O unchanged and interchanges the positions of the H atoms. The resulting configuration is indistinguishable (though different) from the original one. If the molecule is rotated twice by 180° it is returned to the original configuration. This symmetry element consists of the Z axis and is labeled $C_2(Z)$. The element is called a *twofold* axis. The operations around it are distinguished as C_2 and C_2^2. As the last produces the original configuration, we may write $C_2^2 = E$.

In the H_2O molecule there are also two planes of symmetry. The first lies in the plane of the molecule. We will label this symmetry element $\sigma(XZ)$ if the molecule lies in the XZ plane. Reflection in the $\sigma(XZ)$ plane leaves all of the atoms unshifted. The second plane of symmetry is perpendicular to the plane of the molecule and contains the twofold axis; it is labeled $\sigma(YZ)$. The operation of reflection in the $\sigma(YZ)$ plane leaves the O atom unshifted but the two H atoms are interchanged.

As a second example, consider the trigonal bipyramidal PF_5 molecule shown in Figure 2-13. This molecule has a C_3 axis passing through the phosphorus and the axial fluorine atoms. Rotation about this axis by $360°/3 = 120°$ leaves the phosphorus and axial fluorine atoms unchanged but moves each equatorial fluorine atom into the position of the next. The molecule also may be rotated twice about this axis by 120° to produce another indistinguishable configuration, but if the molecule is rotated three times successively by 120° it is returned to the original configuration. Thus the symmetry element C_3 has operations distinguished as C_3, C_3^2, and $C_3^3 = E$.

In the PF_5 molecule there are also three C_2 axes, one along each $P—F_{eq}$ bond. A rotation of $360°/2$ about such an axis leaves the F_{eq} and P atoms

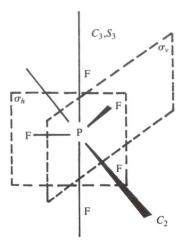

Figure 2-13 Phosphorus pentafluoride has the symmetry elements C_3 (through P and the axial F atoms), three σ_v (containing one P—F_{eq} bond and both P—F_{ax} bonds), a horizontal plane of symmetry σ_h (which contains the three P—F_{eq} bonds), an S_3 axis coincident with C_3, and three C_2 axes (one along each P—F bond). We show only one of the three vertical planes and one of the twofold axes.

that lie on the axis unchanged. The other two equatorial fluorine atoms are interchanged as are the two axial fluorine atoms.

If more than one symmetry axis is present, the one of highest order (largest value of n) is termed the *principal axis,* and by convention this is the one that is placed vertically. In the PF$_5$ molecule the C_3 axis is the principal axis and the $3C_2$ axes are called *subsidiary axes.* A further convention states that planes of symmetry that contain the principal or vertical axis are called *vertical planes* and are given the symbol σ_v. A reflection plane that is perpendicular to the vertical axis is called a *horizontal plane,* with the symbol σ_h.

The PF$_5$ molecule has three σ_v planes and one σ_h plane. The σ_h plane contains the phosphorus atom and the three equatorial fluorine atoms. The operations of reflection in the σ_h plane leaves the atoms that lie in the plane unshifted but interchanges the two axial fluorine atoms. Each of the σ_v planes contains the phosphorus atom, the two axial fluorine atoms, and one of the equatorial fluorine atoms. Reflection in one of these planes leaves the atoms contained in the plane unaltered and exchanges the other two equatorial fluorine atoms.

The final symmetry element for the PF$_5$ molecule is a threefold improper axis, S_3. The S_3 axis is coincident with the C_3 axis; its operation is to rotate by 120° and then reflect in the σ_h plane. In this operation, the axial fluorine atoms are interchanged and each equatorial fluorine atom is moved into the

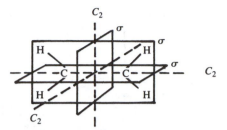

Figure 2-14 Symmetry elements of the ethylene molecule. There are three mutually perpendicular C_2 axes and three mutually perpendicular planes of symmetry. There is also a center of symmetry at the midpoint of the C—C bond.

position of the next. In summary, the symmetry elements of PF_5 are E, C_3, $3C_2$, $3\sigma_v$, σ_h, and S_3.

The equatorial and axial fluorine atoms in PF_5 are *nonequivalent*. In terms of molecular symmetry, nonequivalent atoms are those that cannot be interchanged by a symmetry operation of the molecule. Experimentally, the axial P—F bonds are longer (1.577 Å) than the equatorial P—F bonds (1.534 Å). However, even if all five bond lengths were equal the axial and equatorial fluorine atoms would still be nonequivalent because no symmetry operation exists to interchange the axial and equatorial positions.

The H_2O and PF_5 molecules have an atom (O and P, respectively) at the point at which all the symmetry elements intersect. Not all molecules have such an atom. Consider the ethylene molecule, Figure 2-14. The symmetry elements of ethylene are E, $3C_2$, 3σ, and an inversion center i. In this molecule, the point through which all of the symmetry elements pass is at the midpoint of the C—C bond.

We should note that any linear molecule has an infinite number of σ_v planes containing the molecular axis. In addition, there is a proper axis, C_n, which coincides with the molecular axis and exists for any value of n from one to infinity. That is, rotation by any angle, large or small, leaves the molecule in an indistinguishable position.

2-18 Polar and Nonpolar Polyatomic Molecules

In section 2-10 we discussed the electronic structure of HCl and pointed out that heteronuclear diatomic molecules are polar whereas homonuclear diatomic molecules are nonpolar. A nonpolar molecule has zero (or nearly zero) dipole moment. For polyatomic molecules there are numerous instances in which individual bonds are polar although the molecule as a whole is

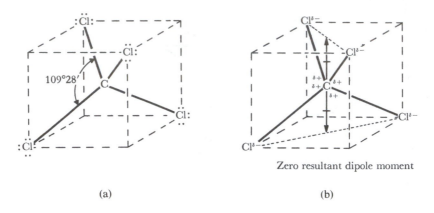

Zero resultant dipole moment

(a)   (b)

Figure 2-15 (a) Tetrahedral structure of CCl_4; (b) canceling bond dipoles in CCl_4.

nonpolar. An example is CCl_4. The molecular structure of CCl_4 is shown in Figure 2-15(a). Since chlorine is more electronegative than carbon, the bonding electron pairs are pulled toward the chlorine atoms. Thus each C—Cl unit has a small bond dipole. The bond dipoles can be resolved into equal and opposite CCl_2 dipoles, as shown in Figure 2-15(b). The symmetrical (tetrahedral) molecular shape of CCl_4 results in a zero dipole moment; thus CCl_4 is nonpolar.

An example of a polar polyatomic molecule is CH_3Cl. Since the electronegativities of carbon and hydrogen are nearly the same, the contribution of the three C—H bonds to the net dipole moment is negligible. The electronegativity difference between carbon and chlorine is large, however, and it is this highly polar bond and the lone-pair electrons on chlorine that account for most of the dipole moment of CH_3Cl ($\mu = 1.87$ D):

Another example of a polar polyatomic molecule is H_2O. Because oxygen is more electronegative than hydrogen, the electron pairs in the two O—H bonds are pulled more toward the oxygen atom. In addition, the oxygen atom has two lone pairs of electrons. The result is a separation of charge in the H_2O molecule, in which the oxygen atom is relatively negative and the hydrogen atoms are relatively positive:

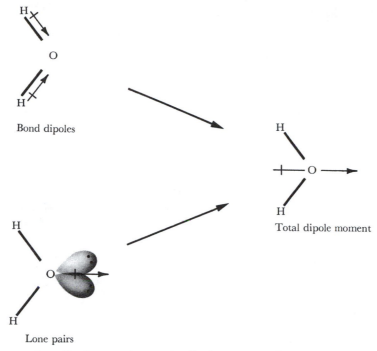

Figure 2-16 Contributions to the 1.85 D dipole moment of H_2O.

Because of the angular shape of H_2O, the H—O bonds and lone-pair contributions combine, as shown in Figure 2-16, to give the dipole moment of 1.85 D.

We can pursue the examination of polar and nonpolar compounds further by studying ethylene and the various isomers of dichloroethylene. As a result of their symmetrical structures, ethylene and *trans*-1,2-dichloroethylene are nonpolar molecules:

$\mu = 0.00$ D

Ethylene

$\mu = 0.00$ D

trans-1,2-dichloroethylene

However, the two other isomers of dichloroethylene are polar molecules:

$$\mu = 1.90 \text{ D} \qquad\qquad \mu = 1.34 \text{ D}$$

cis-1,2-dichloroethylene   1,1-dichloroethylene

We can analyze these dipole moments in terms of the hypothetical bond dipoles μ_{CH} and μ_{CCl} (Figure 2-17). Let us assume an idealized structure in which all of the bond angles are 120°. Then the dipole moment for the cis-isomer can be expressed as

$$\mu_{cis} = 2(\mu_{CCl} + \mu_{CH}) \cos 30°$$

and for the 1,1-isomer as

$$\mu_{1.1} = 2(\mu_{CCl} + \mu_{CH}) \cos 60°$$

We can calculate a value for $(\mu_{CCl} + \mu_{CH})$ from the experimental dipole moment of the cis molecule:

$$(\mu_{CCl} + \mu_{CH}) = \frac{\mu_{cis}}{2 \cos 30°}$$

$$= \frac{1.90 \text{ D}}{2(0.866)}$$

$$= 1.09 \text{ D}$$

Now we can substitute this into the expression for the 1,1-isomer to obtain

$$(\mu_{1.1})_{calc.} = 2(1.09 \text{ D}) \cos 60° = 1.09 \text{ D}$$

This calculated value is less than the experimental value of 1.34 D. The discrepancy between the calculated and experimental dipole moments is partly a result of using the idealized geometry and partly because individual bond dipoles are not strictly transferable from one molecule to another.

The relationship between dipole moment and molecular geometry can be further illustrated by a comparison of the three isomers of dichlorobenzene:

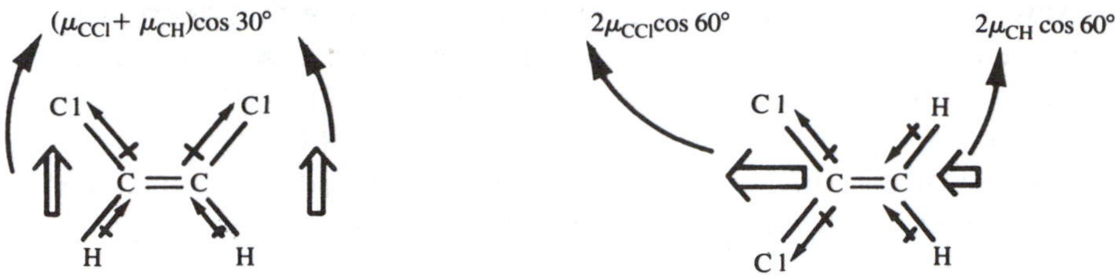

$$\mu_{cis} = 2(\mu_{CCl} + \mu_{CH})\cos 30°$$
$$= 1.90\,D$$

$$\mu_{1,1} = 2(\mu_{CCl} + \mu_{CH})\cos 60°$$
$$= 1.34\,D$$

$$(\mu_{1,1})_{calc} = 1.09\,D$$

Figure 2-17 Bond dipole analysis for *cis*-1,2-dichloroethylene and 1,1-dichloro-ethylene.

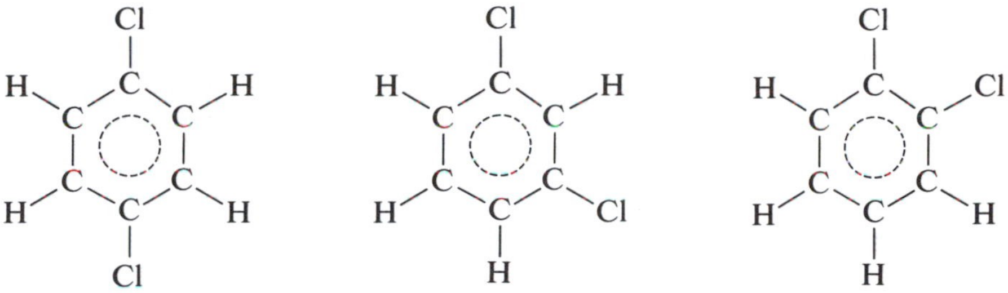

$\mu = 0.00$ D

para-dichlorobenzene

$\mu = 1.72$ D

meta-dichlorobenzene

$\mu = 2.50$ D

ortho-dichlorobenzene

It is clear that the *para*-isomer must have a zero dipole moment by reason of its symmetry. Consideration of the individual C—Cl bond dipoles also leads to the conclusion that the *ortho*-isomer should have a larger dipole moment than the *meta*-isomer, as observed.

Our interest in the relationship between dipole moments and molecular structure is not simply from the viewpoint of electronic structure. In Chapter 7 we will illustrate that a molecular dipole has an important influence upon the properties of a molecule. These charge separations provide strong intermolecular forces that influence the boiling point, melting point, and reactivity of the molecules.

In summary, polar molecules have bond dipoles that add to give a resultant nonzero dipole moment. Nonpolar molecules have either pure covalent bonds (equal sharing) or bond dipoles that cancel due to a "symmetrical" molecular shape. For molecules of the general form AX_n where A is the central atom, "symmetrical" refers to a molecule that is linear $(n = 2)$, trigonal planar $(n = 3)$, tetrahedral $(n = 4)$, trigonal bipyramidal $(n = 5)$, or octahedral $(n = 6)$. Table 2-8 gives the molecular shapes and the polarities of several representative polyatomic molecules.

TABLE 2-8. MOLECULAR SHAPES AND DIPOLE MOMENTS OF
SELECTED POLYATOMIC MOLECULES

Molecule	Shape	Dipole Moment (D)	Classification
CS_2	Linear	0	Nonpolar
HCN	Linear	2.98	Polar
COS	Linear	0.712	Polar
BF_3	Trigonal planar	~0	Nonpolar
SO_3	Trigonal planar	~0	Nonpolar
CF_4	Tetrahedral	~0	Nonpolar
CCl_4	Tetrahedral	~0	Nonpolar
NH_3	Trigonal pyramidal	1.47	Polar
PF_3	Trigonal pyramidal	1.03	Polar
AsF_3	Trigonal pyramidal	2.59	Polar
H_2O	Angular	1.85	Polar
H_2S	Angular	0.97	Polar
SO_2	Angular	1.63	Polar
NO_2	Angular	0.316	Polar
O_3	Angular	0.53	Polar

QUESTIONS AND PROBLEMS

1. Write orbital electronic configurations and then draw Lewis structures for atomic sodium, silicon, phosphorus, and sulfur. How many unpaired electrons are there in each atom?

2. The ionization energy of astatine (At) is not given in Table 2-1 because this element is not available in large quantities and no accurate *IE* measurement has been made. Using the information available in Table 2-1 estimate the values of *IE* for astatine.

3. Explain why the electron affinities of both silicon and sulfur are larger than that of phosphorus.

4. Make a plot of IE_2 values versus atomic number for the elements helium through calcium. Explain the differences between this plot and that given for the IE_1 values (Figure 2-2). Why are the maximum IE_2 values not those of the noble gases?

5. The ionization energy of atomic hydrogen is 13.598 eV, and its electron affinity is 0.754 eV. Convert these quantities to cm^{-1} and to kcal $mole^{-1}$.

6. Predict the relative effective radii of the species H^-, He, and Li^+. Explain your choices.

7. Write the electronic configuration ($1s^2 2s^2$ . . .) for the following: F^-, Na^+, Ne, O^{2-}, and N^{3-}. What would you predict about the relative sizes of these species?

8. Examine the 12 ionization energies of Mg found in Table 2-1 using the IE_m/m method of Section 2-3. Are the results consistent with the known electronic configuration of Mg: $1s^2 2s^2 2p^6 3s^2$?

9. Which atom in each of the following pairs would you expect to have the larger electron affinity (EA)? (a) Cu or Zn; (b) K or Ca; (c) S or Cl; (d) H or Li; (e) As or Ge.

10. Show that e^2 has a value of 332 kcal $mole^{-1}$ Å.

11. The bond distance in diatomic LiF is 1.52 Å. Assuming ionic bonding, calculate the energy required to dissociate LiF into Li^+ and F^-.

12. Calculate the dissociation energies of BeO and CaO to M^{2+} and O^{2-} ions using an ionic bonding model. Bond distances are given in Table 4-9. The *second* electron affinity of oxygen atoms is approximately -8 eV. That is, the reaction $O^{2-} \rightarrow O^- + e^-$ is exothermic by 8 eV.

13. Calculate the electronegativity of hydrogen (EN_H) assuming the value 3.98 for EN_F from Table 2-4. Do this calculation for both the Pauling and the Mulliken scales.

14. In some tabulations of electronegativity, the most electronegative atom is listed as neon. This is contrary to the usual statement that "the most electronegative atom is fluorine." Using the Mulliken definition of electronegativity, explain why neon has a higher electronegativity than fluorine.

15. Write Lewis structures for the molecules and ions listed below. Show resonance structures if appropriate. Also indicate in each case whether

there are formal charges on one or more atoms.

(a) FBr   (b) S_2 (c) Cl_2 (d) P_2
(e) NCO^- (f) CNO^- (g) $BeCl_2$ (h) CS_2
(i) BF_3 (j) SO_3 (k) CO_3^{2-} (l) CF_4
(m) $SiBr_4$ (n) BF_4^- (o) NCl_3 (p) PF_3
(q) CH_3^- (r) SF_2 (s) XeO_3 (t) SO_2
(u) SF_6 (v) Na_2O (w) ClO_2 (x) N_2F_2
(y) CsF   (z) SrO

16. Nitrogen forms a trifluoride, NF_3, but NF_5 exists only as ionic $NF_4^+F^-$. For phosphorus both PF_3 and PF_5 are known as molecules. Write Lewis structures for NF_3, PF_3, and PF_5. Present possible explanations for the fact that PF_5 is stable, whereas molecular NF_5 is not. In light of your explanations, which of the following molecules would you expect to be nonexistent: OF_2, OF_4, OF_6, SF_2, SF_4, SF_6? Write Lewis structures and appropriate comments to support your case.

17. Write Lewis structures for CO_2 and SO_2. Are the C—O bonds primarily ionic or covalent? The SO_2 molecule has a dipole moment, whereas CO_2 does not. What shape do you expect for each molecule?

18. The acetylene molecule, HCCH, is linear. Write a Lewis structure for acetylene. Do you expect the C—C bond to be longer in C_2H_2 than in C_2H_4? Compare the energies of the C—C bonds in C_2H_4 and C_2H_2. Is C_2H_2 polar or nonpolar?

19. Iodine forms several oxyions of the type IO_x^{n-}. Write Lewis structures for IO_3^-, IO_4^-, and IO_6^{5-}. Predict the relative I—O bond lengths in these oxyions.

20. Write a Lewis structure for S_2. Do you expect the molecule to be paramagnetic or diamagnetic?

21. Write Lewis structures for BF_3 and NO_3^-. Do these molecular species have anything in common? The dipole moment of BF_3 is zero. What geometrical structure do you expect for BF_3? What geometrical structure might NO_3^- have?

22. Write Lewis structures for CN^- and CO. The C—O bond length in CO is shorter than in CO_2. Explain.

23. Xenon forms a number of interesting molecules and ions with fluorine and oxygen. Write a Lewis structure for each of the following: XeO_4, XeO_3, XeF_8^{2-}, XeF_6, XeF_4, XeF_2, and XeF^+. Show the placement of formal charges in the Lewis structures. Avoid structures with formal

charge separation, if possible. Using expected trends in effective atomic radii, predict whether the Xe—F bond length in XeF_4 will be longer or shorter than the I—F bond length in the related ion IF_4^-.

24. Write Lewis structures for H_2O and HF. The water molecule is non-linear. Do H_2O and HF have dipole moments?

25. The electronic structure of the thiocyanate ion, NCS^-, can be represented as a hybrid of two resonance structures. Write these two structures and give the C—N and C—S bond orders for each structure.

26. For each of the following cases give the Lewis structure of a known chemical example:
    (a) A diatomic molecule with one unpaired electron
    (b) A triatomic molecule with two double bonds
    (c) A diatomic molecule with formal charge separation
    (d) A diatomic molecule with partial ionic character
    (e) An alkaline earth oxide
    (f) A molecule or ion with two equivalent resonance structures
    (g) A molecule or ion with three equivalent resonance structures

27. The atoms of the yet-to-be-discovered "hypotransition" elements, starting at $Z = 121$, will have electrons in the $5g$ orbitals.
    (a) How many elements will there be in the hypotransition metal series?
    (b) How many of the atoms will be diamagnetic?
    (c) Which electronic configurations in the series will have seven unpaired electrons?
    (d) What is the maximum number of unpaired electrons an atom can have in the series? Will this be a new record for atoms in the periodic table?
    (e) What is the *IE* of an electron in a $5g$ orbital of atomic hydrogen? Is this likely to be larger, or smaller, than the *IE* of a $5g$ electron in one of the hypotransition elements? Briefly explain your choice.
    (f) In atomic hydrogen the $5s$, $5p$, $5d$, $5f$, and $5g$ orbitals all have the same energy. Will this be true for the hypotransition elements? If not, what will the energy order be? Explain briefly.

28. Periodic acid, $(HO)_5IO$, is one compound in the class of oxyacids discussed in Section 2-13. Write a Lewis structure for periodic acid. Explain why the I—O bond length (1.78 Å) is shorter than the I—OH bond lengths (1.89 Å).

29. Predict the molecular shape and polarity for each of the following molecules:

    (a) CS_2 (b) SO_3 (c) ICl_3 (d) BF_3
    (e) CBr_4 (f) SiH_4 (g) SF_2 (h) SeF_6
    (i) PF_3 (j) ClO_2 (k) IF_5 (l) OF_2
    (m) H_2Te

30. For the molecules and ions CO_2, NO_2^+, NO_2, NO_2^-, and SO_2:
    (a) Predict their molecular shapes and indicate which neutral molecules are polar.
    (b) Predict the number of unpaired electrons for each molecule.
    (c) Predict the bond angles in NO_2^+ and NO_2^-.
    (d) Predict relative N—O bond lengths for NO_2^+, NO_2, and NO_2^-.

31. Use VSEPR theory to predict the molecular shapes of XeF_4, XeO_4, XeO_3, XeF_2, and $XeOF_2$.

32. Indicate the VSEPR theory geometrical structure for AsH_3, ClF_3, and $SeCN^-$.

33. Use VSEPR theory to predict the molecular geometry (e.g., bent, linear, pyramidal) of each of the following molecules and ions: (a) SO_3^{2-}, (b) SO_3, (c) BrF_5, (d) I_3^-, (e) CH_3^+, (f) CH_3^-, (g) PCl_3F_2.

34. Which of the neutral molecules in Problem 33 would you expect to have dipole moments?

35. Borazine has the formula $(BHNH)_3$. The combination of a boron atom and a nitrogen atom is isoelectronic with two carbon atoms. How would you formulate the electronic structure of borazine?

36. A compound is found to have the empirical formula $(NH_4)_2SbCl_6$. X-ray diffraction studies reveal that the anion consists of discrete units containing one Sb and six Cl's. The compound is diamagnetic. The electronic configuration of Sb is $[Kr]5s^25p^3$. Would you expect the compound to be only $(NH_4)_2SbCl_6$, or an equal mixture of $(NH_4)_3SbCl_6$ and $(NH_4)SbCl_6$? Why?

37. Three geometrical structures commonly are found for three-coordinate molecules of the main-group elements. What are these structures? In each case how many lone electron pairs does the central atom have, according to the VSEPR theory? Using chlorine atoms bonded to a central atom, give an example of each structure. Which molecules among these examples are polar?

38. Tellurium (Te) is a rather vile element, so only a few chemists work with it. To inspire the handful of tellurium chemists to greater achieve-

ment, predict the formulas of the tellurium-fluorine molecules (or ions) that would exhibit the following geometrical structures: (a) angular, (b) T-shaped, (c) trigonal pyramidal, (d) seesaw, (e) square planar, (f) square pyramidal, (g) trigonal bipyramidal, (h) octahedral.

39. Predict the shapes of the following ions, using VSEPR theory: (a) AlF_6^{3-}, (b) TlI_4^{3-}, (c) $GaBr_4^-$, (d) NO_3^-, (e) NCO^-, (f) CNO^-, (g) $SnCl_3^-$, (h) $SnCl_6^{2-}$.

40. Is XeF_5^+ a reasonable species? What geometry would you predict it to have?

41. Use the empirical rules given in Section 2-16 to predict the sequence of atoms and the molecular shape of the following molecules: (a) C_3O_2, (b) NOF, (c) NSF, (d) $ClClF^+$, (e) $H_2B_2O_3$, (f) B_2O_3, (g) HNCO, (h) HNO_3, (i) HNO_4.

42. Determine all of the symmetry elements for each of the following molecules: (a) PF_3, (b) SO_3, (c) H_2Te, (d) IF_5, (e) ClF_3, (f) XeF_4, (g) $(BHNH)_3$.

43. The dipole moment of H_2O is 1.85 D whereas that of F_2O is only 0.297 D, although the bond angles in the two compounds are nearly identical. Explain why the dipole moment of F_2O is much smaller than that of H_2O.

44. Use the bond dipole approach ($\mu_{CCl} + \mu_{CH}$) and the experimental dipole moment of *ortho*-dichlorobenzene ($\mu = 2.50$ D) to calculate the expected dipole moment for *meta*-dichlorobenzene. (For the structures of these molecules see Section 2-18). Compare the calculated dipole moment to the experimental value for *meta*-dichlorobenzene ($\mu = 1.72$ D).

Suggestions for Further Reading

Bent, H.A., "Isoelectronic Systems," *J. Chem. Educ.* 43: 170 (1966).

Brooks, D.W., Meyers, E.A., Silicio, F., and Nearing, J.C., "Electron Affinity: The Zeroth Ionization Potential," *J. Chem. Educ.* 50: 487 (1973).

Chen, E.M., and Wentworth, W.E., "The Experimental Values of Atomic Electron Affinities," *J. Chem. Educ.* 52: 486 (1975).

Companion, A., *Chemical Bonding,* 2nd ed., New York: McGraw-Hill, 1979.

Emeléus, H.J., and Sharpe, A.G., *Modern Aspects of Inorganic Chemistry,* 4th ed., London: Routledge and Kegan Paul, 1973.

Gillespie, R.J., *Molecular Geometry,* London: Van Nostrand Reinhold, 1972.

Gimarc, B.M., "The Shapes of Simple Polyatomic Molecules and Ions," *J. Amer. Chem. Soc.* 92: 266 (1970).

Gray, H.B., *Chemical Bonds,* Menlo Park, Calif.: Benjamin, 1973.

Hall, M.B., "Valence Shell Electron Pair Repulsions and the Pauli Exclusion Principle," *J. Amer. Chem. Soc.* 100: 6333 (1978).

Hall, M.B., "Stereochemical Activity of *s* Orbitals," *Inorg. Chem.* 17: 2261 (1978).

Liebman, J.F., "Comments and a Warning About Isoelectronic Systems," *J. Chem. Educ.* 48: 188 (1971).

Mackay, K.M., and Mackay, R.A., *Introduction to Modern Inorganic Chemistry,* 2nd ed., London: Intertext, 1972.

Palke, W.E., and Kirtman, B., "Valence Shell Electron Pair Interactions in H_2O and H_2S. A Test of the Valence Shell Electron Pair Repulsion Theory," *J. Amer. Chem. Soc.* 100: 5717 (1978).

Pauling, L., *The Chemical Bond,* Ithaca, N.Y.: Cornell Univ. Press, 1967.

Pimentel, G.C., and Spratley, R.D., *Understanding Chemistry,* San Francisco: Holden Day, 1971.

Sanderson, R.T., *Chemical Bonds and Bond Energy,* 2nd ed., New York: Academic Press, 1976.

Schnuelle, G.W., and Parr, R.G., "On the Shapes and Energetics of Polyatomic Molecules," *J. Amer. Chem. Soc.* 94: 8974 (1972).

Thompson, H.B., Wells, M., and Weaver, J.E., "A Simple Model of Bond Geometry and the Stereoactivity of Lone Pairs," *J. Amer. Chem. Soc.* 100: 7213 (1978).

3 The Valence Bond and Hybrid Orbital Descriptions of Chemical Bonding

The Lewis structure and the attendant VSEPR method allow us to predict the number of bond pairs and lone pairs, and the molecular shape, of any particular molecule. However, the *type* of electron (e.g., *s*, *p*, or *d*) involved in the bonding is not specified. In this chapter we examine the *valence bond* (VB) description of chemical bonding which does specify the nature of the electrons involved in the bonding. This description of bonding retains the electron-pair concept that is so useful in writing Lewis electron dot structures and thus it uses a *localized* orbital approach. It was first described by Walter Heitler and Fritz London in 1927, and later developed more fully by Linus Pauling who also introduced the idea of *orbital hybridization*.

In Chapters 4, 5, and 6, the *molecular-orbital* (MO) method will be used to describe chemical bonding. The molecular-orbital method utilizes *delocalized* orbitals. In Chapter 4 we will present a detailed comparison of the valence bond and molecular-orbital theories as applied to H_2. Suffice it to say here that neither of these theories is exact. The method to use depends largely on the molecular property of interest. If one is investigating a property of the ground-state molecule (e.g., molecular geometry or bond dissociation energy), it is generally simpler to employ the localized orbital model. However, the delocalized orbital model is generally preferred for describing spectroscopic properties of molecules (e.g., ionizations or excited states arising from electronic transitions). Mathematically, it is possible to transform from the localized (VB) orbitals to the delocalized (MO) orbitals. Therefore, it is impossible to state that one approach is inherently more accurate than the other.

3-1 Valence Bond Theory for the Hydrogen Molecule

We begin our discussion of valence bond theory with the simplest molecule, H_2. Imagine that the two hydrogen atoms are sufficiently separated so that

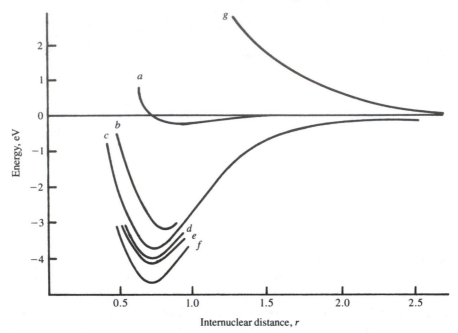

Figure 3-1 Calculated and experimental energy curves for H_2. The experimental energy is respresented by curve f.

the electrons on each hydrogen atom are attracted by only one nucleus. The appropriate product wave function describing this system is

$$\psi_1 = 1s_a(1)1s_b(2) \tag{3-1}$$

The numbers 1 and 2 designate electrons 1 and 2; the letters a and b refer to the two hydrogen nuclei. The potential energy curve calculated from ψ_{I} is presented in curve a of Figure 3-1 and is compared to the experimental energy (curve f in the figure). Now, it is clear that Equation 3-1 is not adequate to describe the bonding in molecular hydrogen. We have neglected the fact that at shorter internuclear distances we cannot distinguish electrons 1 and 2. A more appropriate wave function will result if we take linear combinations of these product wave functions, which will allow for the indistinguishability of the electrons,

$$\psi_{\mathrm{II}}^{(+)} = 1s_a(1)1s_b(2) + 1s_a(2)1s_b(1) \tag{3-2}$$

and

$$\psi_{\mathrm{II}}^{(-)} = 1s_a(1)1s_b(2) - 1s_a(2)1s_b(1)$$

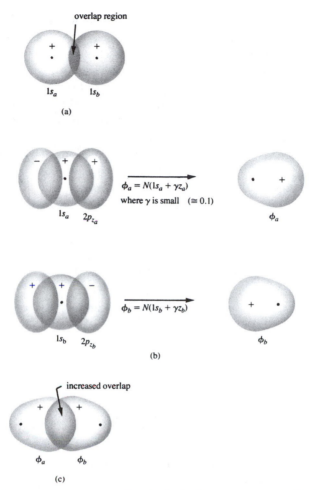

Figure 3-2 Representation of bonding in the hydrogen molecule: (a) overlap of the 1s orbitals; (b) mixing a small amount of 2p with 1s on each hydrogen atom to form polarized 1s orbitals; (c) overlap of the polarized orbitals.

Curve b of Figure 3-1 shows that the $\psi_{\text{II}}^{(+)}$ wave function represents a considerable improvement in describing the energy of the system, whereas $\psi_{\text{II}}^{(-)}$ describes a repulsive (nonbonding) state (curve g). Consideration of the electron spins associated with the functions $\psi_{\text{II}}^{(+)}$ and $\psi_{\text{II}}^{(-)}$ shows that the electrons are spin-paired in the $\psi_{\text{II}}^{(+)}$ orbital but unpaired in $\psi_{\text{II}}^{(-)}$ (Section 1-13).

Further improvement in the calculated energy results if the value of Z is allowed to vary in the atomic hydrogen wave function

$$\psi(1s) = \frac{1}{\sqrt{\pi}}\left(\frac{Z}{a_0}\right)^{3/2} e^{-Zr/a_0}$$

Because each electron is now attracted to two nuclei rather than just one nucleus, the "effective" nuclear charge increases from 1.0 in the hydrogen atom to 1.17 in the H_2 molecule. The value of $Z = 1.17$ provides the lowest possible energy for Equation 3-2. As seen in curve c of Figure 3-1, the total energy is lowered relative to that achieved for $Z = 1.0$ (curve b, Figure 3-1). Even with this correction, however, we are still unable to reproduce accurately the experimental bond energy although the computed bond length is in good agreement with the experimental value.

A closer approximation to the true wave function can be obtained by considering that each hydrogen atom *polarizes* or distorts the electron cloud on the other hydrogen atom. That is, each hydrogen atom no longer has a spherically symmetric electron cloud but the electron density is distorted towards the bonding region in the molecule. Since the 1s atomic orbital is spherically symmetric it does not represent this distortion very well, as can be seen in Figure 3-2(a). However, the 2p orbital is not spherically symmetric (Figure 1-15), so that a small admixture of 2p character with the 1s orbital distorts the spherical electron density, as in Figure 3-2(b). The wave function that describes this polarized electron cloud is given by ϕ_a on atom a and ϕ_b on atom b:*

$$\phi_a = N\ (1s_a + \gamma z_a) \tag{3-3}$$

$$\phi_b = N\ (1s_b + \gamma z_b) \tag{3-4}$$

The internuclear axis is taken as the Z axis, N is a normalization constant, and γ represents the amount of $2p_z$ orbital that is *mixed* with the 1s orbital on each atom.

The electron density distribution for the mixed atomic orbital ϕ_a is given by ϕ_a^2. Since we require that ϕ_a be normalized, we can calculate the normalization constant N:

$$\int \phi_a^2\ dv = N^2 \int\ (1s_a + \gamma z_a)(1s_a + \gamma z_a)\ dv$$
$$= N^2 \Big[\int\ (1s_a)^2\ dv + \gamma \int\ (1s_a)(z_a)dv$$
$$+ \gamma \int\ (z_a)(1s_a)\ dv + \gamma^2 \int\ (z_a)^2\ dv \Big] \tag{3-5}$$

$$\int \phi_a^2\ dv = N^2\ [1 + 0 + 0 + \gamma^2] = 1 \tag{3-6}$$

Notice that of the four integrals in Equation 3-5, the first and last are equal to *one* because the 2s and $2p_z$ atomic orbitals are normalized. The second and third integrals are zero because of the orthogonality of the atomic orbitals (Section 1-12). From Equation 3-6 we find

*We will abbreviate $2p_{z_a}$ as z_a and $2p_{z_b}$ as z_b.

$$N^2 = \frac{1}{1 + \gamma^2}$$

and

$$N = \pm \frac{1}{\sqrt{1 + \gamma^2}}$$

Since the sign of the wave function is arbitrary (the square of the wave function determines the electron density), we can arbitrarily choose the positive root for N. Therefore

$$N = \frac{1}{\sqrt{1 + \gamma^2}}$$

From the normalization and orthogonality conditions we see that the contribution of each atomic orbital to ϕ_a is in proportion to the square of its coefficient. Since

$$\int \phi_a^2 \, dv = \frac{1}{1 + \gamma^2} \left[\int (1s_a)^2 \, dv + \gamma^2 \int (z_a)^2 \, dv \right] = 1$$

the contribution of the $1s_a$ atomic orbital to the electron density ϕ_a^2 is $1/(1 + \gamma^2)$ and the $2p_{z_a}$ contribution is $\gamma^2/(1 + \gamma^2)$.

The value of γ that provides the lowest energy for H_2 has been found by calculation to be about 0.1. This results in $1/1.01 = 0.99$ $1s_a$ orbital contribution and $0.01/1.01 = 0.01$ $2p_{z_a}$ orbital contribution. That is, the ϕ_a orbital is 99 percent $1s_a$ and 1 percent $2p_{z_a}$. An analogous conclusion can be drawn for the ϕ_b orbital. However, even though the $2p$ orbitals make only a 1 percent contribution to the wave function, this results in a 5 percent improvement in the calculated energy (curve d, Figure 3-1). The reason for this significant lowering of energy is that the polarized orbitals are able to enhance the amount of electron density in the bonding region of the molecule. This can be seen by comparing Figure 3-2(a) with Figure 3-2(c).

The wave function used to obtain curve d of Figure 3-1 is determined by utilizing the Heitler-London wave function, Equation 3-2, and replacing $1s_a$ by ϕ_a and $1s_b$ by ϕ_b. The appropriate wave function is then

$$\psi_{\text{III}} = \phi_a(1)\phi_b(2) + \phi_a(2)\phi_b(1) \tag{3-7}$$

At this point we might inquire further as to how the $2p$ orbital can be involved in the bonding in H_2 when it is not needed to describe the ground state of the hydrogen atom. One way to imagine this is to hypothesize that in order for the $2p$ orbital to contribute to the bonding in H_2, the hydrogen atom electron must be *promoted* from the $1s$ to the $2p$ orbital. Then instead of allowing only the orbital overlap $1s_a$–$1s_b$ [Figure 3-2(a)], we also have the overlaps $1s_a$–$2p_b$, $2p_a$–$1s_b$, and $2p_a$–$2p_b$ (Figure 3-3). These four possibilities together provide the increased bonding electron density shown in Figure 3-2(c). Notice that there are two hypothetical processes involved in utilizing the $2p$ orbitals. These are: (1) promotion of the $1s$ electron to the $2p$ orbital, and (2) mixing of the $1s$ orbital with the $2p$ orbital in the bonding process. This mixing is represented by Equations 3-3 and 3-4 and is referred to as

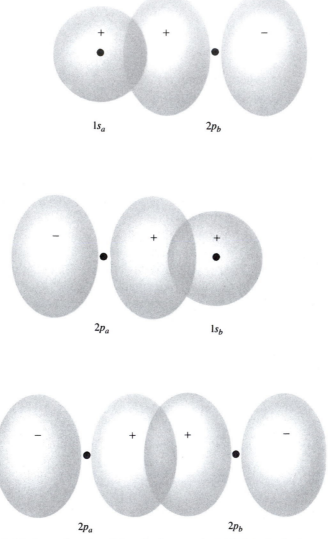

Figure 3-3 Orbital overlaps involving the $2p$ atomic orbital in the hydrogen molecule.

hybridization. The increase in bonding that results from hybridization more than compensates for the promotion energy.

Examination of Figure 3-1 shows that curve *d* has still failed to correctly reproduce the experimental bond energy. A further improvement to the wave function can be obtained by considering both electrons to be placed on one of the hydrogen atoms: $\phi_a(1)\phi_a(2)$ or $\phi_b(1)\phi_b(2)$. We can add these *ionic* forms to the ψ_{III} wave function, Equation 3-7, with the weighting factor λ:

$$\psi_{IV} = \phi_a(1)\phi_b(2) + \phi_a(2)\phi_b(1) + \lambda\phi_a(1)\phi_a(2) + \lambda\phi_b(1)\phi_b(2) \qquad (3-8)$$

Because H_2 is a covalent molecule we expect the value of λ to be much less than one. The best value of λ has been found to be about $\frac{1}{4}$, which was determined by varying λ until the lowest total energy was achieved for Equation 3-8. The value of λ^2 is about 0.06, indicating that the contribution of the ionic forms is about 6 percent. The resulting improvement in the calculated energy is only 2–3 percent (curve *e*, Figure 3-1).

There is a close connection between the wave function given in Equation 3-8 and the Lewis electron-dot resonance forms:

H_a—H_b	$:H_a^-H_b^+$	$H_a^+H_b^-:$
Covalent	Ionic	Ionic

The covalent resonance structure is represented by the first two terms in the wave function, and each of the ionic forms corresponds to one of the last two terms (Equation 3-8).

The applications of VB theory to the hydrogen molecule illustrates an important procedure that arises in all quantum-mechanical problems: a trial wave function is chosen and improvements are made on it to provide a better fit with the experimental results. For the hydrogen molecule, a 50-term wave function reproduces the experimental bond energy to within 0.0001 eV.

3-2 Valence Bond Theory for Hydrogen Fluoride

The Lewis electron dot structure for hydrogen fluoride is H—$\ddot{\underset{..}{F}}$: . Since the atomic electron configurations are $1s^1$ for the hydrogen atom and $1s^2\,2s^2\,2p_x^2\,2p_y^2\,2p_z^1$ for the fluorine atom, the simplest VB wave function for the HF molecule is

$$\psi_I = 1s_F^2\,2s_F^2\,2p_{xF}^2\,2p_{yF}^2[1s_H(1)\,2p_{zF}(2) + 1s_H(2)\,2p_{zF}(1)]$$

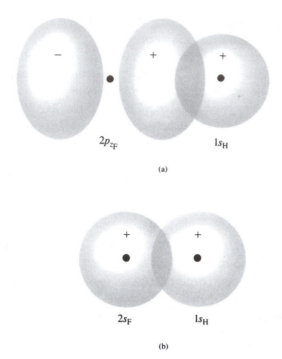

Figure 3-4 Valence bond description of the bonding in HF: (a) overlap of the $2p_{z_F}$ orbital with the $1s_H$ orbital; (b) overlap of the $2s_F$ orbital with the $1s_H$ orbital.

We can write a simplified form of this wave function by neglecting the non-bonding electrons on the fluorine atom to obtain simply

$$\psi_{II} = 1s_H(1)\ 2p_{z_F}(2) + 1s_H(2)\ 2p_{z_F}(1) \tag{3-9}$$

The electron-pair bond represented by ψ_{II} is shown in Figure 3-4(a). However, the ψ_{II} wave function does not adequately describe the bond energy of HF. Further improvement in the wave function results if the $2s$ orbital on the fluorine atom is allowed to bond. This can be accomplished by hybridizing the fluorine $2p_z$ and $2s$ atomic orbitals. In Equation 3-9 we can replace $2p_{z_F}$ by

$$\phi_F = N\ (2p_{z_F} + \gamma\ 2s_F) \tag{3-10}$$

where

$$N = \frac{1}{\sqrt{1 + \gamma^2}}$$

Ionic terms also can be introduced into the wave function. Based on electronegativity considerations, we expect the ionic structure H^+F^- to contribute much more than H^-F^+. Keeping this in mind we can write an improved wave function:

$$\psi_{III} = 1s_H(1)\phi_F(2) + 1s_H(2)\phi_F(1) + \lambda\phi_F(1)\phi_F(2)$$

Although the exact amount of ionic character is not known for certain, there is no doubt that it contributes more than the 6 percent found for H_2. Pauling has estimated that the ionic component comprises almost 50 percent of the wave function for HF.

We can now inquire about the mechanism of $2s$–$2p$ hybridization represented by Equation 3-10. Since the $2s$ orbital is completely filled in the fluorine atom, it might appear to be unavailable for bonding to the hydrogen atom. However, it can become available for bonding by considering the hypothetical promotion of a $2s$ electron to the $2p$ orbital:

$$1s^22s^22p_x^22p_y^22p_z^1 \xrightarrow{\text{promotion}} 1s^22s^12p_x^22p_y^22p_z^2$$

Now that the $2s$ orbital has only one electron it can overlap and pair with the $1s$ electron on the hydrogen atom, as shown in Figure 3-4(b). The net effect of involving both the $2p_{z_F}$ and $2s_F$ orbitals in the bonding is to hybridize these orbitals as represented by Equation 3-10. For HF, the mixing parameter γ is less than $\frac{1}{2}$. Therefore, the contribution of the $2s$ orbital to the bonding in HF is less than 25 percent $[(\frac{1}{2})^2 \times 100]$. This relatively small contribution results because of the large promotion energy required to excite the $2s$ electron to the $2p$ orbital. For the fluorine atom the promotion energy amounts to almost 500 kcal mole^{-1}!

If this much promotion energy is required, what advantage is there in hybridization? There are two distinct advantages that arise from hybridization. First, the bonding electron density is concentrated in the region between the nuclei, leading to an enhancement of bonding character. Let us suppose that the value of γ found to minimize the energy for HF is 0.2. The effect of mixing 0.2 $2s$ character with the $2p$ orbital is shown in Figure 3-5(a). Clearly this will result in an increase in electron density in the bonding region; see Figure 3-5(b). The second advantage that arises from hybridization is a decrease in electron–electron repulsion. Since part of the $2s$ orbital was mixed with the $2p$ orbital in the hybridization process, the lone-pair orbital is no longer simply $2s^2$. In order to be orthogonal to the bonding hybrid orbital, the lone-pair orbital must have the form

$$\phi_{lp} = N(2s - \gamma 2p)$$

The shape of this orbital is represented in Figure 3-5(c). It is seen that whereas the bonding hybrid orbital is concentrated on the side of the fluorine nucleus toward the hydrogen atom, the lone-pair hybrid orbital is concentrated toward the opposite side [Figure 3-5(d)]. Consequently, the electron–electron repulsion will be reduced between these two pairs of electrons.

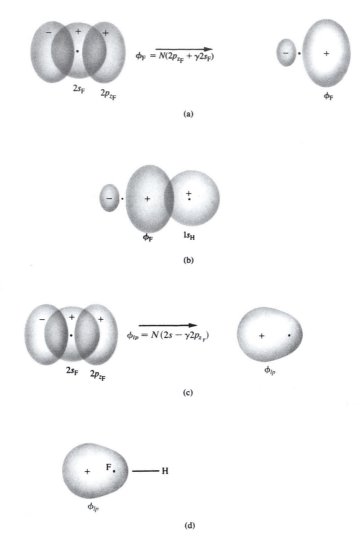

Figure 3-5 Hybrid-orbital description of the bonding in HF. (a) Formation of the bonding hybrid orbital, ϕ_F. (b) Overlap of the bonding hybrid orbital, ϕ_F, with the hydrogen $1s$ atomic orbital. (c) Formation of the lone-pair hybrid orbital, ϕ_{lp}. (d) Electron density of the lone-pair hybrid orbital in HF. Notice that the lone-pair electron density is oriented away from the HF bond.

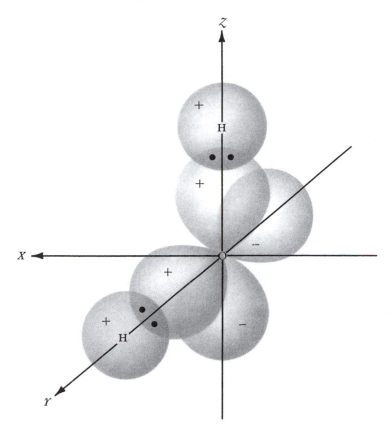

Figure 3-6 Simple picture of the bonding in H_2O. Only the oxygen $2p$ orbitals are used, thus an H—O—H bond angle of 90° is predicted.

3-3 Valence Bond Theory for the Water Molecule

The Lewis structure of the water molecule exhibits two bond electron pairs and two lone electron pairs, H_2O: . We now consider the application of VB theory to the water molecule. The oxygen atom has the electronic configuration $1s^2 2s^2 2p_x^2 2p_y^1 2p_z^1$. The $2p_y$ and $2p_z$ orbitals are available to form electron-pair bonds because they each contain only one electron. These unpaired electrons can pair with the two hydrogen atom $1s$ electrons to form two bonds at right angles, as shown in Figure 3-6. If we neglect the $1s^2 2s^2 2p_x^2$ electrons on the oxygen atom, the valence bond wave function for this bonding model is

$$\psi = [1s_{H_a}(1)2p_{yO}(2) + 1s_{H_a}(2)2p_{yO}(1)]$$

$$\cdot \, [1s_{H_b}(3)2p_{zO}(4) + 1s_{H_b}(4)2p_{zO}(3)]$$

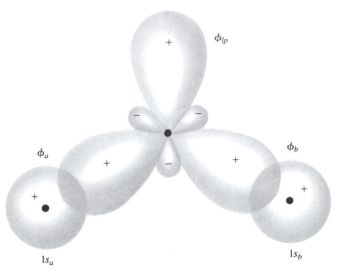

Figure 3-7 Formation of the bonding hybrid orbitals ϕ_a and ϕ_b and the lone-pair hybrid orbital ϕ_{lp} in the water molecule. The bonding hybrid orbitals are formulated to maximize the overlap with the hydrogen atom $1s$ orbitals at the experimental bond angle of 104.5°.

The deficiency of this model is that it predicts the water molecule to have a 90° bond angle (Figure 3-6), although the observed bond angle is 104.5°. If we mix the $2s$ orbital with the $2p_y$ and $2p_z$ orbitals we can form two hybrid orbitals that point directly toward the two hydrogen atoms and therefore provide maximum overlap with the $1s$ atomic orbitals of the hydrogen atoms (Figure 3-7). A third lone-pair hybrid orbital that points away from the hydrogen atoms also is shown in Figure 3-7.

Let us examine carefully how the two bonding hybrid orbitals (ϕ_a, ϕ_b) and the lone-pair hybrid orbital (ϕ_{lp}) can be formed. If we use the coordinate system shown in Figure 3-8, we can write

$$\phi_a = N[\cos (\theta/2) \, 2p_z + \sin (\theta/2) \, 2p_y + \gamma 2s]$$
$$\phi_b = N[\cos (\theta/2) \, 2p_z - \sin (\theta/2) \, 2p_y + \gamma 2s]$$

We can evaluate the mixing parameter γ by requiring that these two hybrid orbitals be orthogonal:

$$\int \phi_a \phi_b \, dv = \int [\cos (\theta/2)2p_z + \sin (\theta/2)2p_y + \gamma 2s]$$
$$\cdot [\cos (\theta/2)2p_z - \sin (\theta/2)2p_y + \gamma 2s) \, dv$$
$$= \cos^2 (\theta/2) \int 2p_z 2p_z \, dv - \sin^2 (\theta/2) \int 2p_y 2p_y \, dv$$
$$+ \gamma^2 \int 2s2s \, dv = 0$$

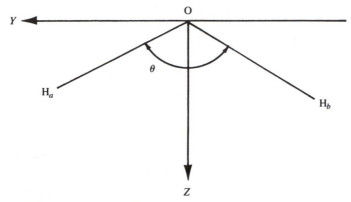

Figure 3-8 Coordinate system for the water molecule.

Because each of the atomic orbitals is orthogonal to the other atomic orbitals on the same nucleus, we have neglected all of the cross terms (e.g., $\int 2p_z 2p_y\ dv = 0$). Furthermore, each of the atomic orbitals is normalized, so we obtain

$$\int \phi_a \phi_b\ dv = \cos^2\ (\theta/2) - \sin^2\ (\theta/2) + \gamma^2 = 0$$

Employing some simple trignometric relationships, this reduces to $\cos\theta + \gamma^2 = 0$, or

$$\cos\theta = -\gamma^2 \tag{3-11}$$

Since $\cos\ (104.5°) = -0.25 = -\gamma^2$, we obtain $\gamma = 0.50$.

If we also evaluate $\cos\ (\theta/2)$ and $\sin\ (\theta/2)$, we obtain the wave functions

$$\phi_a = N\ (0.61\ 2p_z + 0.79\ 2p_y + 0.50\ 2s)$$
$$\phi_b = N\ (0.61\ 2p_z - 0.79\ 2p_y + 0.50\ 2s)$$

Before proceeding to determine the form of the lone-pair hybrid orbital, let us normalize ϕ_a and ϕ_b hybrid orbitals:

$$\int \phi_a{}^2\ dv = N^2[(0.61)^2 \int (2p_z)^2\ dv + (0.79)^2 \int (2p_y)^2\ dv$$
$$+ (0.50)^2 \int (2s)^2\ dv]$$
$$= N^2[(0.61)^2 + (0.79)^2 + (0.50)^2]$$
$$= N^2(1.25) = 1.0$$

Therefore $N = 1/\sqrt{1.25}$ and the normalized hybrid orbitals are

$$\phi_a = 0.55 \; 2p_z + 0.71 \; 2p_y + 0.45 \; 2s$$
$$\phi_b = 0.55 \; 2p_z - 0.71 \; 2p_y + 0.45 \; 2s$$

Now we can determine the form of the lone-pair hybrid orbital. It can be written in general as

$$\phi_{lp} = c_1 2p_z + c_2 2p_y + c_3 2s$$

We can easily evaluate the three coefficients by using the constraints of normalization, orthogonality, and the *unit-orbital* contribution (see Section 3-7). The unit-orbital contribution simply states that the contribution of any atomic orbital to all three hybrid orbitals must add up to 1.0. Applying this to the $2s$ atomic orbital we have

$$(0.45)^2 + (0.45)^2 + c_3{}^2 = 1.0$$

Therefore $c_3{}^2 = 0.60$ and $c_3 = \pm 0.77$. Since the sign of the wave function is arbitrary, we can specify that $c_3 = +0.77$. We also notice that the coefficient c_2 must be zero since the $2p_y$ orbital is completely utilized in the ϕ_a and ϕ_b hybrid orbitals $[(0.71)^2 + (0.71)^2 = 1.0]$.

The coefficient c_1 can be determined from the normalization of ϕ_{lp}:

$$\int \phi_{lp}{}^2 dv = c_1{}^2 \int (2p_z)^2 dv + (0.77)^2 \int (2s)^2 dv = 1.0$$

$$c_1{}^2 + (0.77)^2 = 1.0$$

$$c_1{}^2 = 0.40$$

$$c_1 = \pm 0.64$$

It is clear from Figures 3-7 and 3-8 that the direction of the $2p_z$ orbital is opposite to that of ϕ_{lp} and consequently c_1 must be negative. We can also deduce that c_1 must be negative by requiring the orthogonality conditions

$$\int \phi_a \phi_{lp} dv = 0 \qquad \text{and} \qquad \int \phi_b \phi_{lp} dv = 0$$

In summary, the bonding hybrid orbitals in H_2O contain 80 percent $2p$ character $\{[(0.55)^2 + (0.71)^2] \times 100\}$ and 20 percent $2s$ character $[(0.45^2 \times 100]$. The lone-pair hybrid orbital contains 40 percent $2p$ character $[(0.64)^2 \times 100]$ and 60 percent $2s$ character $[(0.77)^2 \times 100]$. The total contribution for the $2p$ orbitals add up to 200 percent because both the $2p_y$ and the $2p_z$ orbitals are involved. The appropriate wave functions are then

$$\phi_a = \sqrt{0.80} \; 2p_a + \sqrt{0.20} \; 2s$$
$$\phi_b = \sqrt{0.80} \; 2p_b + \sqrt{0.20} \; 2s$$

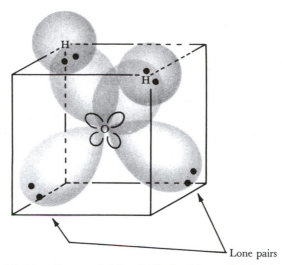

Figure 3-9 Hyrbid-orbital bonding model for H_2O. In this model the two bonding hybrid orbitals are equivalent and the two lone-pair hybrid orbitals are equivalent.

$$\phi_{lp} = \sqrt{0.40}\, 2p_{lp} + \sqrt{0.60}\, 2s$$

In these equations, $2p_a$, $2p_b$ and $2p_{lp}$ refer to the contributions of the $2p_y$ and $2p_z$ atomic orbitals to the appropriate hybrid orbitals. The actual contribution of any particular atomic orbital ($2p_y$, $2p_z$) will depend upon the coordinate system. However, the combined contribution of $2p_y$ and $2p_z$ must always total 80 percent to each bonding hybrid orbital and 40 percent to the lone-pair hybrid orbital.

In this hybrid-orbital model of chemical bonding for the water molecule, one of the lone pairs on the oxygen atom maintains pure $2p_x$ character and the other lone pair consists of 40 percent $2p$ and 60 percent $2s$ character. We could proceed further with the hybrid-orbital model by requiring that the two lone-pair orbitals also be equivalent just as the two bonding orbitals are equivalent. Since we have a total of 140 percent $2p$ character in the lone-pair orbitals (100 percent $2p_x$ and 40 percent $2p_z$), each equivalent lone-pair orbital must contain 70 percent $2p$ character. Therefore, each lone-pair orbital contains 30 percent $2s$ character. This model results in the electron distribution shown in Figure 3-9. These orbitals can be generated easily from the previous lone-pair orbitals (ϕ_{lp} and $2p_x$) by simply taking linear combinations of $2p_x$ and ϕ_{lp}:

$$\phi_{lp_1} = \frac{1}{\sqrt{2}}\, (2p_x + \phi_{lp}) \tag{3-12}$$

$$\phi_{lp_2} = \frac{1}{\sqrt{2}} (\phi_{lp} - 2p_x) \tag{3-13}$$

The formation of ϕ_{lp_1} and ϕ_{lp_2} from $2p_x$ and ϕ_{lp} is shown in Figure 3-10.

At this point it is worthwhile to inquire as to which representation of the lone-pair orbitals is preferred, the $(2p_x, \phi_{lp})$ representation shown in Figure 3-7 or the $(\phi_{lp_1}, \phi_{lp_2})$ representation shown in Figure 3-9. The answer is that since the latter pair of orbitals is related to the former by the linear combinations shown in Equations 3-12 and 3-13, it is impossible for us to state that one representation is better than the other. In other words, these are simply two alternative, but equivalent, schemes of representing the total electron density distribution in the water molecule. The equivalence of these two representations can be seen clearly by comparing the total electron density for both representations. The total electron density for the $(\phi_{lp}, 2p_x)$ model is the integral over all space of the function

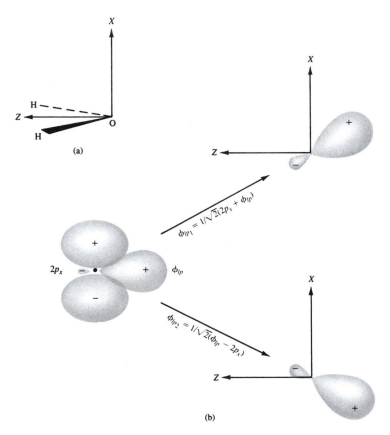

Figure 3-10 (a) Coordinate system for the water molecule; (b) formation of two equivalent lone-pair hybrid orbitals on the oxygen atom.

$$\phi_{lp}^2 + 2p_x^2$$

For the $(\phi_{lp_1}, \phi_{lp_2})$ model, this function works out to be

$$\phi_{lp_1}^2 + \phi_{lp_2}^2 = \left[\frac{1}{\sqrt{2}}(2p_x + \phi_{lp})\right]^2 + \left[\frac{1}{\sqrt{2}}(\phi_{lp} - 2p_x)\right]^2 = \phi_{lp}^2 + 2p_x^2$$

Therefore, the electron density distribution in both models is the same!

3-4 Valence Bond Theory for the Ammonia Molecule

The Lewis electron dot structure for ammonia exhibits only one lone pair, :NH_3. Since the nitrogen atom has the electronic configuration $1s^2 2s^2 2p_x^1$ $2p_y^1 2p_z^1$, there are three unpaired electrons that can pair with the three hydrogen atom $1s$ electrons. In this bonding model, three bonds are at right angles (Figure 3-11); the observed bond angle is 107.3°. Just as for the water

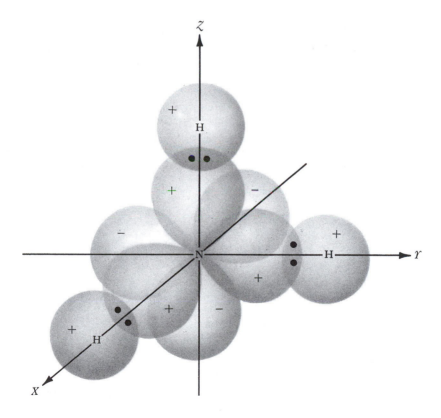

Figure 3-11 Simple picture of the bonding in NH_3, using only the nitrogen $2p$ orbitals. In this model the NHN bond angle is predicted to be 90°.

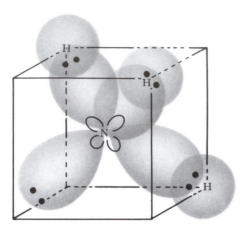

Figure 3-12 Hyrbid orbital bonding model for NH_3.

molecule, we can form orbitals that overlap directly with the hydrogen atom
$1s$ atomic orbitals by hybridizing the nitrogen $2s$ orbital with the nitrogen
$2p$ orbitals. We can write the hybrid orbitals in the general form

$$\phi = \frac{1}{\sqrt{1 + \gamma^2}}(2p + \gamma 2s)$$

where $2p$ refers to some linear combination of $2p_x$, $2p_y$ and $2p_z$ atomic
orbitals. The value of γ can then be obtained by using Equation 3-11:

$$\cos 107.3° = -0.30 = -\gamma^2 \qquad \text{or} \qquad \gamma = 0.55$$

Therefore, each of the hybrid orbitals has the general form

$$\phi = 0.88 \ 2p + 0.48 \ 2s$$

Each of the bonding hybrid orbitals contains 77 percent $2p$ character
$[(0.88)^2 \times 100]$ and 23 percent $2s$ character $[(0.48)^2 \times 100]$. The lone-pair
hybrid orbital contains the remaining $2p$ and $2s$ character. This amounts
to 69 percent $2p$ $[3.00 - 3(0.77)]$ and 31 percent $2s$ character $[1.00 - 3(0.23)]$.
These hybrid orbitals are shown in Figure 3-12.

In summary, we see that the hybrid orbital model predicts that the lone-
pair orbital on the nitrogen atom contains more $2p$ character than $2s$ character. The nonhybrid orbital model predicted that the lone-pair orbital was
100 percent $2s$.

3-5 Valence Bond Theory for Molecules Containing No Lone–Pair Electrons

We now discuss the VB theory for molecules in which all of the electrons surrounding the central atom are involved in bonding to the terminal atoms. We will use BeH_2, BH_3, CH_4, PH_5, and SH_6 as prototype molecules. Of these prototype molecules, only CH_4 is stable under normal conditions; BH_3 has been detected as a transient species, but molecular BeH_2, PH_5, and SH_6 never have been detected experimentally.

VB Theory for BeH_2

Let us first examine the chemical bonding in the linear BeH_2 molecule. The electronic configuration of the beryllium atom is $1s^2 2s^2 2p^0$. According to simple VB theory, therefore, beryllium is not capable of forming any electron-pair bonds since there are no unpaired electrons available to pair with the hydrogen $1s$ electrons. However, two electrons become available to form electron-pair bonds if we hypothetically promote one of the $2s$ electrons to the $2p_z$ orbital:

$$1s^2 2s^2 2p_z^0 \xrightarrow{\text{promotion}} 1s^2 2s^1 2p_z^1$$

We can now consider one of the hydrogen $1s$ electrons to bond with the beryllium $2s$ electron and the other hydrogen $1s$ electron to bond to the beryllium $2p_z$ electron. This model would lead to nonequivalent bonds, however, since the $2p_z$ electron distribution is different from that of the $2s$ orbital.

It is possible to form two *equivalent* bonding orbitals by hybridizing the $2s$ and the $2p_z$ atomic orbitals. Following the method outlined for H_2O and NH_3 we obtain

$$\phi_a = \frac{1}{\sqrt{1+\gamma^2}}(\gamma 2s + 2p_z)$$

and

$$\phi_b = \frac{1}{\sqrt{1+\gamma^2}}(2s - \gamma 2p_z)$$

The mixing parameter γ can be determined by utilizing Equation 3-11:

$$\cos 180° = -1 = -\gamma^2 \qquad \text{or} \qquad \gamma = 1$$

We also could have determined that $\gamma = 1$ by noting the requirement that ϕ_a and ϕ_b must be equivalent orbitals. Therefore, the $2p_z$ orbital contribution must be the same in both ϕ_a and ϕ_b, as must also be true for the $2s$ contribution. The result is that the two hybrid orbitals are

$$\phi_a = \frac{1}{\sqrt{2}}(2s + 2p_z)$$

$$\phi_b = \frac{1}{\sqrt{2}}(2s - 2p_z)$$

The formation of these hybrid orbitals is illustrated in Figure 3-13. It is seen that the $2s$ and $2p$ orbitals hybridize to give two equivalent *linear* sp hybrid orbitals. One hybrid orbital, ϕ_a, is directed at H_a and strongly overlaps the $1s_a$ orbital. The other hybrid orbital, ϕ_b, is directed at H_b and strongly overlaps the $1s_b$ orbital. The two localized valence orbitals are shown in Figure 3-14. The four valence electrons participate in two localized electron-pair

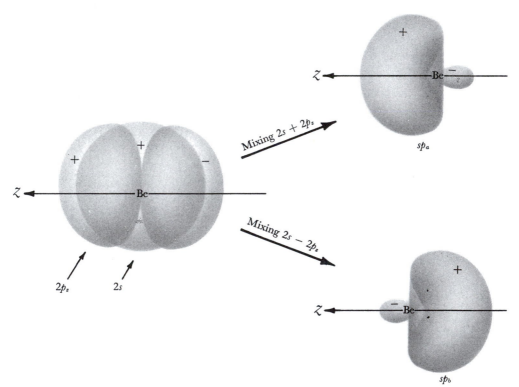

Figure 3-13 Formation of the two sp orbitals by linear combination of $2s$ and $2p_z$ orbitals.

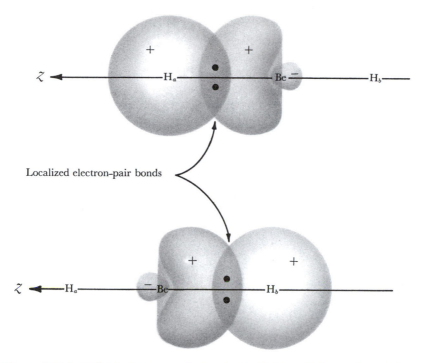

Figure 3-14 Localized electron-pair bonds for BeH_2 which are formed from two equivalent sp hybrid orbitals centered at the Be nucleus. Each Be sp orbital forms a localized bonding molecular orbital with a hydrogen $1s$ orbital.

bonds, analogous to the Lewis structure for BeH_2. Each of the linear sp hybrid orbitals has half s and half p character, and the two sp hybrid orbitals are sufficient to attach two hydrogen atoms to the central beryllium atom in BeH_2.

VB Theory for BH_3

Now we consider VB theory for the trigonal planar BH_3 molecule. In order to form three equivalent bonds between the boron and the hydrogen atoms we must carry out the hypothetical promotion

$$1s^2 2s^2 2p_x{}^1 2p_y{}^0 2p_z{}^0 \xrightarrow{\text{promotion}} 1s^2 2s^1 2p_x{}^1 2p_y{}^1 2p_z{}^0$$

and orbital hybridization of the $2s$, $2p_x$, and $2p_y$ atomic orbitals.

We can now determine the form of each of the three equivalent hybrid orbitals,

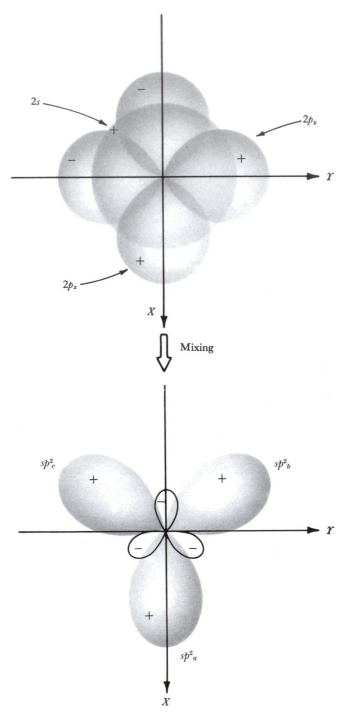

Figure 3-15 Formation of three equivalent sp^2 hybrid orbitals by linear combination of the $2s$ and $2p$ orbitals. The sp^2 hybrid orbitals are trigonal planar.

$$\phi = \frac{1}{\sqrt{1+\gamma^2}}(\gamma 2s + 2p)$$

The term $2p$ refers to some linear combination of $2p_x$ and $2p_y$. The exact form of these hybrid orbitals will be determined in Section 3-7. For the moment, let us simply concentrate on the percent contribution of the $2s$ and $2p$ orbitals to each hybrid orbital. Obviously, if all three orbitals are to be equivalent, each must contain 33 percent $2s$ and 67 percent $2p$ character. This can also be determined by Equation 3-11:

$$\cos 120° = -0.50 = -\gamma^2 \qquad \text{or} \qquad \gamma^2 = 0.50$$

The $2s$ contribution to each hybrid orbital is $\gamma^2/(1+\gamma^2) = 0.50/1.50 = 0.33$, so that its contribution is indeed 33 percent. The $2p$ contribution to each hybrid orbital is $1/(1+\gamma^2) = 1/1.50 = 0.67$, or 67 percent.

Since these hybrid orbitals result from combining one $2s$ and two $2p$ orbitals, they are called sp^2 hybrid orbitals. Figure 3-15 illustrates the formation of these hybrid orbitals from the constituent atomic orbitals. Since any two p orbitals lie in the same plane and the s orbital is nondirectional, the sp^2 hybrid orbitals lie in a plane. The three sp^2 hybrid orbitals form three equivalent, localized bonding orbitals with the three hydrogen $1s$ orbitals. Each of the sp^2–$1s$ bonding orbitals is occupied by an electron pair, as illustrated in Figure 3-16.

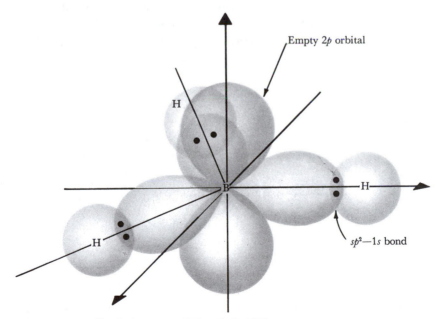

Figure 3-16 Localized electron-pair bonds in BH_3.

VB Theory for CH$_4$

Methane, CH_4, is an example of a molecule with four equivalent atoms attached to a central atom. All the carbon valence orbitals are needed to attach the four hydrogen atoms. In order to form four equivalent bonds between the carbon and hydrogen atoms we must carry out the hypothetical promotion

$$1s^2 2s^2 2p_x{}^1 2p_y{}^1 2p_z{}^0 \xrightarrow{\text{promotion}} 1s^2 2s^1 2p_x{}^1 2p_y{}^1 2p_z{}^1$$

and orbital hybridization of the $2s$, $2p_x$, $2p_y$, and $2p_z$ atomic orbitals. By requiring that each of the four hybrid orbitals be equivalent, we can immediately see that each hybrid orbital must contain 25 percent contribution from each of the $2s$, $2p_x$, $2p_y$, and $2p_z$ atomic orbitals. This result also follows from Equation 3-11. If each of the hybrid orbitals is written as

$$\phi = \frac{1}{\sqrt{1+\gamma^2}}(\gamma 2s + 2p),$$

then

$$\cos 109.5\,° = -0.33 = -\gamma^2 \qquad \text{or} \qquad \gamma^2 = 0.33$$

The $2s$ contribution to each hybrid orbital is $\gamma^2/(1+\gamma^2) = 0.33/1.33 = 0.25$, so that its contribution is 25 percent. The total $2p$ contribution to each hybrid orbital is $1/(1+\gamma^2) = 1/1.33 = 0.75$, or 75 percent. Since three $2p$ orbitals are involved, each one must contribute 25 percent. The exact form of each of these hybrid orbitals will be determined in Section 3-7.

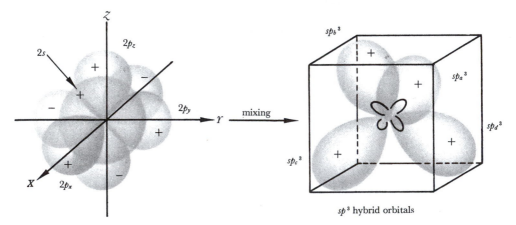

Figure 3-17 Formation of four equivalent, tetrahedral sp^3 hybrid orbitals.

Since these hybrid orbitals result from combining one $2s$ and three $2p$ orbitals, they are called sp^3 hybrid orbitals. Figure 3-17 illustrates the formation of these tetrahedral hybrid orbitals from the constituent atomic orbitals. Four localized bonding orbitals can be made by combining each hydrogen $1s$ orbital with the sp^3 hybrid orbital, as indicated in Figure 3-18. There are eight valence electrons (four from the carbon atom and one from each of the four hydrogen atoms) to distribute in the four localized bonding orbitals. These eight electrons account for the four equivalent, localized electron-pair bonds shown in Figure 3-18.

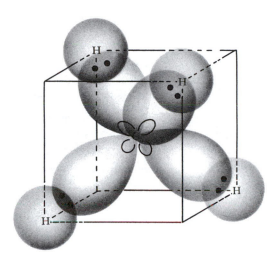

Figure 3-18 Localized electron-pair structure for CH_4.

VB Theory for PH_5 and SH_6

Next we consider the hypothetical PH_5 and SH_6 molecules. We shall not treat these molecules in the same detail as given to BeH_2, BH_3, and CH_4.

For PH_5, we must form five hybrid orbitals, one for each attached hydrogen atom. In order to form these hybrid orbitals, we must first unpair one of the phosphorus atom electrons by carrying out a hypothetical promotion:

$$[Ne]3s^2 3p_x{}^1 3p_y{}^1 3p_z{}^1 \xrightarrow{\text{promotion}} [Ne]3s^1 3p_x{}^1 3p_y{}^1 3p_z{}^1 3d_{z^2}{}^1$$

We have arbitrarily indicated that the $3s$ electron is promoted to the $3d_{z^2}$ orbital. We can now form five dsp^3 hybrid orbitals. These are depicted in

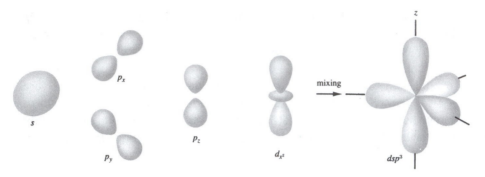

Figure 3-19 Formation of dsp^3 hybrid orbitals, which are trigonal bipyramidal.

Figure 3-19; the dsp^3 hybrid orbitals are directed toward the corners of a trigonal bipyramid. Five localized bonding orbitals can be made by combining each hydrogen $1s$ orbital with a dsp^3 hybrid orbital. There are ten valence electrons (five from the phosphorus atom and one from each of the five hydrogen atoms) to distribute in the five localized bonding orbitals. These ten electrons account for the five localized electron-pair bonds shown in Figure 3-20. The PH_5 molecule is not known, but the structure of the analogous PF_5 molecule is trigonal bipyramidal, so that the dsp^3 hybrid orbital model provides a good description of the bonding in these compounds.

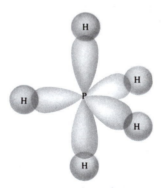

Figure 3-20 Localized electron-pair structure for the hypothetical PH_5 molecule.

As our final example of the hybrid orbital description in this section, we treat the hypothetical molecule SH_6. In this case we must promote two of the sulfur electrons to the $3d$ orbitals in order to obtain six d^2sp^3 hybrid orbitals.

$$[Ne]3s^23p_x^23p_y^13p_z^1 \xrightarrow{\text{promotion}} [Ne]3s^13p_x^13p_y^13p_z^13d_{z^2}^13d_{x^2-y^2}^1$$

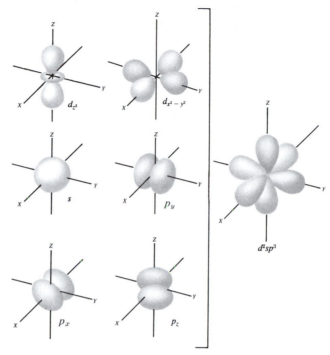

Figure 3-21 Formation of d^2sp^3 hybrid orbitals, which are octahedral.

The six d^2sp^3 hybrid orbitals are directed toward the corners of an octa-hedron, as shown in Figure 3-21. By combining each hydrogen $1s$ orbital with a d^2sp^3 hybrid orbital, we can make a total of six localized bonding orbitals. Since the sulfur atom has 6 valence electrons and each hydrogen atom has 1, there are a total of 12 electrons to distribute in the six localized bonding orbitals. The six localized electron-pair bonds shown in Figure 3-22 account for these 12 electrons. Although SH_6 is unknown, the analogous SF_6 molecule is octahedral and therefore the d^2sp^3 hybrid orbitals provide an appropriate description of the chemical bonding.

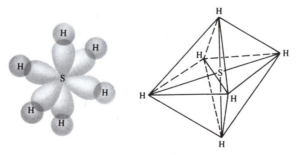

Figure 3-22 Localized electron-pair structure for the hypothetical SH_6 molecule.

Hybrid orbital descriptions for BeH_2, BH_3, CH_4, PH_5, and SH_6 are summarized in Table 3-1.

TABLE 3-1. HYBRID-ORBITAL DESCRIPTION AND MOLECULAR GEOMETRY

Example Molecule	Groups Attached to Central Atom	Hybrid Orbitals Appropriate for Central Atom	Molecular Geometry
BeH_2	2	sp	Linear [angle (H—Be—H) = 180°]
BH_3	3	sp^2	Trigonal planar [angle (H—B—H) = 120°]
CH_4	4	sp^3	Tetrahedral [angle (H—C—H) = 109°28′]
PH_5	5	dsp^3	Trigonal bipyramidal [angle (H_e—P—H_e) = 120°]* [angle (H_e—P—H_a) = 90°]
SH_6	6	d^2sp^3	Octahedral [angle (H—S—H) = 90°]

*H_e refers to an equatorial hydrogen atom and H_a to an axial hydrogen atom.

3-6 Hybrid-Orbital Description of Single and Multiple Bonds in Carbon Compounds

Carbon atoms have a remarkable ability to form bonds with hydrogen atoms and other carbon atoms. If one hydrogen atom in CH_4 is replaced with a CH_3 group, the C_2H_6 (ethane) molecule is obtained. The C_2H_6 molecule contains one C—C bond, and the structure around each carbon atom is tetrahedral. The actual HCH bond angle in ethane is 107.4°, slightly less than the ideal bond angle of 109.5° predicted for a regular tetrahedral structure around each carbon atom. Nonetheless, this bond angle is sufficiently close to tetrahedral so that we can say that each carbon atom in ethane uses approximately sp^3 hybrid orbitals, as shown in Figure 3-23.

By continually replacing hydrogen atoms with CH_3 groups, we obtain the many hydrocarbons with the full sp^3 bonding structure at each carbon atom. Such hydrocarbons are said to be *saturated* because each carbon atom uses all its valence orbitals to bond the maximum number (four) of atoms. The bonds in CH_4 and C_2H_6 are called σ (sigma) bonds, because the orbitals involved are directed along the bond axis between the nuclei.* By

*Clearly, all of the bonds we discussed for H_2, HF, H_2O, and NH_3 are also sigma bonds.

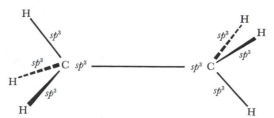

Figure 3-23 Localized bond structure for C_2H_6.

comparison, in *unsaturated* organic molecules a carbon atom uses only two or three of its four valence orbitals for σ bonding, which leaves one or two $2p$ orbitals available for π (pi) bonding. A π orbital is characterized by a concentration of electron density above and below the plane of the molecule. For example, in ethylene, C_2H_4, there are three atoms around each carbon atom. One way to describe localized bond orbitals for C_2H_4 is first to use three sp^2 hybrid orbitals for each carbon atom to form five σ bonds, four C—H bonds and one C—C bond, as in Figure 3-24. The σ bonds account for 10 of the 12 valence electrons in C_2H_4. In this scheme each carbon atom has one valence p orbital not involved in σ bonding that is perpendicular to the set of sp^2 hybrid orbitals. Thus the two $2p$ carbon orbitals in ethylene can overlap to form a π bonding orbital. The $2p$–$2p$ overlap is largest when the two sp^2 orbital sets are oriented in the same plane, as shown in Figure 3-25. The remaining valence electron pair in C_2H_4 occupies the π bonding orbital. From the electronic structural model shown in Figure 3-25, we predict that the C_2H_4 molecule will have a planar structure, with H—C—H and H—C—C bond angles of 120°. From experimental data we know that the molecule is planar, and that the H—C—H bond angle is 116.6° and the

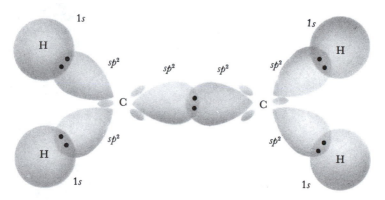

5 σ-bonding pairs = 10 electrons

Figure 3-24 The σ-bond structure for C_2H_4.

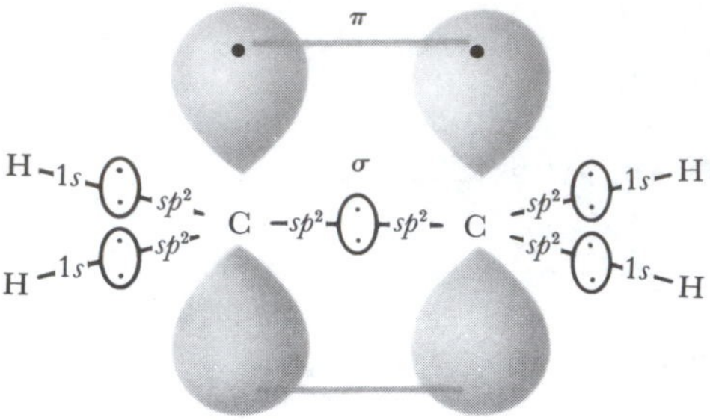

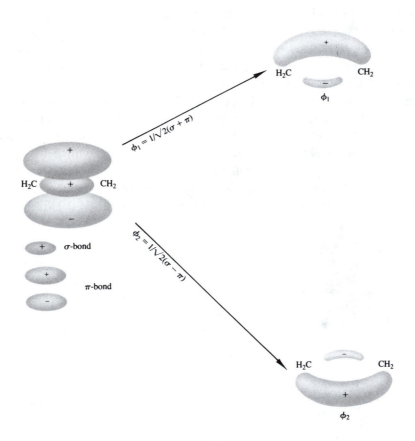

Figure 3-25 Representation of the bonding in C_2H_4, using localized orbitals with $2p$–$2p$ overlap to form a π bond.

Figure 3-26 Formation of the two bent bonds in ethylene from the σ and π bonds.

H—C—C bond angle is 121.7°. Thus the molecular structure of C_2H_4 is in close agreement with the predictions from the electronic structural model. Comparison of the hybrid-orbital description of bonding in C_2H_4 with the Lewis structure shows that the carbon–carbon double bond is composed of one σ bond and one π bond.

There is another bonding model that equally well describes the electron density distribution in ethylene. Instead of the σ-π bonding model presented in the previous paragraph, we can describe the ethylene carbon–carbon bond in terms of two equivalent *bent* bonds: $H_2C\frown CH_2$. Because of their shape, these bonds are sometimes referred to as "banana" bonds. We can form banana bonds very easily by simply taking linear combinations of the σ and π bonds. As shown in Figure 3-26, the following linear combinations are appropriate:

$$\phi_1 = \frac{1}{\sqrt{2}}(\sigma + \pi)$$

$$\phi_2 = \frac{1}{\sqrt{2}}(\sigma - \pi)$$

The ϕ_1 and ϕ_2 orbitals represent the two banana bonds.

Saturated and unsaturated carbon compounds differ considerably in the ease with which the molecules rotate around the carbon–carbon bonds. Rotation around the single bond in ethane requires very little energy, but rotation around the double bond in ethane does not occur at an appreciable rate at temperatures below about 400° C. An explanation of this restriction is the fact that the rotation of the CH_2 groups with respect to each other would twist the atomic $2p(\pi)$ orbitals out of alignment, thus breaking the π bond (see Figure 3-25). In terms of the bent-bond description, it also is clear that carbon–carbon bond rotation would break the bent bonds (Figure 3-26).

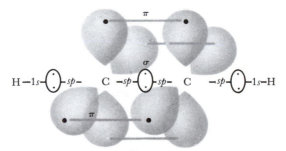

Figure 3-27 Localized orbital representation of the bonding in C_2H_2 that shows the overlap of the two $2p$ orbitals on each carbon to form two π bonds.

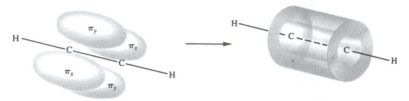

Figure 3-28 The two π-bonds form a cylindrically shaped electron cloud around the carbon-carbon bond in acetylene.

Acetylene

In acetylene, C_2H_2, only one carbon atom and one hydrogen atom are attached to either carbon atom. Thus we use two *sp* hybrid orbitals from each carbon atom to form σ bonds. For C_2H_2 there are three σ bonds, one C—C bond, and two C—H bonds. In addition, there are two *2p* orbitals on each carbon atom that are available to form two π bonds. The resulting electronic structure of C_2H_2 is shown in Figure 3-27. The observed molecular structure of C_2H_2 is linear, which is consistent with the electronic structural description. The triple bond in C_2H_2 consists of one σ and two π bonds. The two π bonds actually result in a *cylindrical* distribution of electron density about the molecular axis, as shown in Figure 3-28.

Benzene

The formula of benzene is C_6H_6, and the molecule has the planar structure shown in Figure 3-29. The planar hexagon of carbon atoms is an important molecular framework in structural chemistry, because it occurs in countless organic molecules.

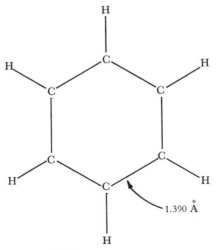

Figure 3-29 Planar structure of C_6H_6.

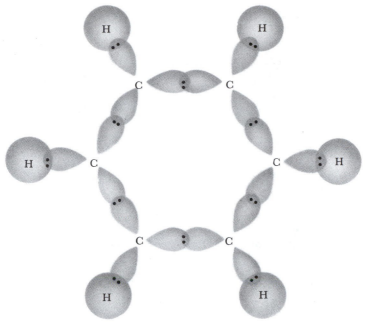

Figure 3-30 The σ-bond network in the benzene molecule.

Each carbon atom in the benzene ring is attached to two other carbon atoms and one hydrogen atom. Thus it is convenient to use sp^2 hybrid orbitals for each carbon atom to form the σ-bonding network shown in Figure 3-30. Since each carbon atom furnishes four valence electrons and each hydrogen atom furnishes one, there is one electron left for each carbon atom in a $2p$ orbital perpendicular to the plane of the molecule. In a localized π-bonding scheme three localized electron-pair bonds can be formed in the ways shown in Figure 3-31, which depicts the localized π-bond structures.

Simplified representations of the bonding in benzene are shown in Figure 3-32. Benzene actually is more stable than might be expected for a system of six carbon–carbon single bonds and three carbon–carbon π bonds. The added stability is due to the resonance or delocalization of the electrons in the three π bonds over all six carbon atoms, as is evident from the resonance structures shown in Figure 3-32. If we did not allow the delocalization of electrons in C_6H_6, we would have a system of three isolated double bonds (only one of the Kekulé structures shown in Figure 3-32).

The extra stability of benzene is called the experimental *resonance energy.* It is calculated by adding the bond energies of the C—C, C=C, and C—H bonds and comparing the total with the experimentally known value for the heat of formation of benzene. The difference indicates that benzene is about

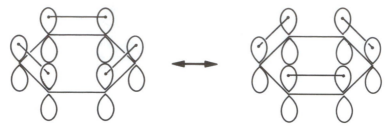

The two Kekulé structures

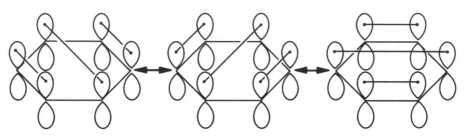

The three Dewar structures

Figure 3-31 The π-bond resonance structures for the benzene molecule.

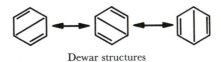

Kekulé structures

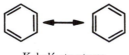

Dewar structures

Simple molecular-orbital picture

Figure 3-32 Simplified representations of bonding in a benzene molecule.

40 kcal mole^{-1} more stable than the sum of the bond energies for a system of six C—H, three C—C, and three isolated C=C bonds would suggest.

3-7 Mathematical Formulation of Hybrid Orbitals

In section 3-5 we outlined the hybrid-orbital description of bonding for molecules containing no lone-pair electrons on the central atom. The discussion in that section was mainly qualitative; in this section we will present a more quantitative description of hybrid orbitals.

Let us again consider the BeH_2 molecule discussed in Section 3-5. It is conceptually useful to think of orbital hybridization as occurring in two steps. In the first step, the Be valence electrons are promoted from the ground-state electronic configuration, $1s^2 2s^2$, to the excited state, $1s^2 2s^1 2p^1$. In the excited state there are two unpaired electrons; consequently, it is possible for the beryllium atom to form two electron-pair bonds as in BeH_2. In the second step, the valence orbitals are hybridized so that each of the valence electrons resides in an equivalent sp hybrid orbital, rather than in the individual $2s$ and $2p$ atomic orbitals.

To formulate hybrid orbitals mathematically, one must first choose a coordinate system: for BeH_2 we will choose the Z axis to be the molecular axis (Figure 3-13). Thus, for beryllium, we wish to hybridize the $2s$ and $2p_z$ atomic orbitals. If we start with two atomic orbitals we can form two hybrid orbitals; $\psi(sp_a)$, which is directed at hydrogen atom a, and $\psi(sp_b)$, which is directed at hydrogen atom b (Figure 3-13). These hybrid orbitals are linear combinations of the $2s$ and $2p_z$ atomic orbitals:

$$\psi(sp_a) = c_1 2s + c_2 2p_z$$
$$\psi(sp_b) = c_3 2s + c_4 2p_z$$

To determine the coefficients c_1, c_2, c_3, and c_4 we specify four requirements that the hybrid orbitals must satisfy.

1. Because the $2s$ orbital has spherical symmetry each equivalent hybrid orbital must contain an equal contribution from the $2s$ orbital. We will always choose a positive coefficient for the $2s$ orbital. (Recall that the sign of any wave function ψ is arbitrary because the electron density depends upon $|\psi|^2$.)

2. The hybrid orbitals must be *normalized*. For example,

$$\int |\psi(sp_a)|^2 \, dv = c_1^2 \int |2s|^2 \, dv + c_2^2 \int |2p_z|^2 \, dv$$
$$= c_1^2 + c_2^2 = 1$$

We have assumed that the $2s$ and $2p_z$ atomic orbitals are individually normalized.

3. The hybrid orbitals must be *orthogonal*. Two wave functions $\psi(sp_a)$ and $\psi(sp_b)$ are said to be orthogonal if $\int \psi(sp_a)\psi(sp_b)dv = 0$. That is,

$$\int \psi(sp_a)\psi(sp_b)\ dv = c_1c_3 \int |2s|^2\ dv + c_1c_4 \int (2s)(2p_z)\ dv$$
$$+ c_2c_3 \int (2s)(2p_z)\ dv + c_2c_4 \int |2p_z|^2\ dv$$
$$= c_1c_3 + c_2c_4 = 0$$

Here we have assumed that the $2s$ and $2p_z$ atomic orbitals already are orthogonal; that is, $\int (2s)(2p_z)dv = 0$. The $2s$ and $2p_z$ atomic orbitals are orthogonal because they have zero net overlap, as shown in Figure 1-24.

4. The square of the coefficients of any one component atomic wave function summed over all the hybrid orbitals in which it participates must be equal to unity. That is,

$$2s:\ c_1{}^2 + c_3{}^2 = 1$$
$$2p_z:\ c_2{}^2 + c_4{}^2 = 1$$

This is called the *unit-orbital contribution*.

To generate the two hybrid orbitals for BeH_2 we can determine the coefficient of the $2s$ orbital by combining rule 1 with rule 4:

Rule 1: $c_1 = c_3$

Rule 4: $c_1{}^2 + c_3{}^2 = 1$

Therefore, we obtain

$$c_1 = c_3 = \frac{1}{\sqrt{2}}$$

Next we wish to determine the coefficients of the $2p_z$ orbital (c_2 and c_4) in the two hybrid orbitals. From rule 2 we have

$$c_1{}^2 + c_2{}^2 = 1$$

$$\frac{1}{2} + c_2{}^2 = 1$$

$$c_2 = \pm \frac{1}{\sqrt{2}}$$

If we choose $c_2 = +1/\sqrt{2}$ we can obtain c_4 by application of rule 3:

$$c_1 c_3 + c_2 c_4 = 0$$

$$\frac{1}{2} + \frac{1}{\sqrt{2}} c_4 = 0$$

$$c_4 = -\frac{1}{\sqrt{2}}$$

In summary, the two hybrid orbitals for BeH_2 are

$$\psi(sp_a) = \frac{1}{\sqrt{2}} 2s + \frac{1}{\sqrt{2}} 2p_z$$

$$\psi(sp_b) = \frac{1}{\sqrt{2}} 2s - \frac{1}{\sqrt{2}} 2p_z$$

The shapes of these two hybrid orbitals have already been shown in Figure 3-13.

As our second example let us construct the wave functions for the three equivalent sp^2 hybrid orbitals. It is convenient to use the coordinate system for BH_3 in Figure 3-15, which shows the three sp^2 hybrid orbitals directed toward hydrogen atoms a, b, and c. The boron $2s$, $2p_x$, and $2p_y$ atomic orbitals are used to form the sp^2 hybrid orbitals.

We can write the three sp^2 hybrid orbitals as

$$\psi(sp^2_a) = c_1 2s + c_2 2p_x + c_3 2p_y$$

$$\psi(sp^2_b) = c_4 2s + c_5 2p_x + c_6 2p_y$$

$$\psi(sp^2_c) = c_7 2s + c_8 2p_x + c_9 2p_y$$

Now let us determine the coefficient of the $2s$ atomic orbital in each hybrid orbital.

Rule 1: $c_1 = c_4 = c_7$

Rule 4: $c_1^2 + c_4^2 + c_7^2 = 1$

Therefore,

$$c_1 = c_4 = c_7 = \frac{1}{\sqrt{3}}$$

The remainder of the coefficients could likewise be determined by strict application of rules 1 through 4. However, we can simplify our work considerably by taking account of the symmetry of the BH_3 molecule. Of the two p orbitals, only $2p_x$ is used to bond with atom a ($2p_y$ has zero overlap with a). That is, the coefficient c_3 must equal zero. For the first hybrid orbital we then obtain

$$\psi(sp^2{}_a) = \frac{1}{\sqrt{3}} 2s + c_2 2p_x$$

By applying the principle of normalization (rule 2) we obtain

$$\left(\frac{1}{\sqrt{3}}\right)^2 + c_2{}^2 = 1$$

$$c_2 = \pm\sqrt{\frac{2}{3}}$$

Arbitrarily, we will choose the positive root so that the large positive lobe of $\psi(sp^2{}_a)$ will be directed toward hydrogen atom a.

The complete hybrid orbital is

$$\psi(sp^2{}_a) = \frac{1}{\sqrt{3}} 2s + \sqrt{\frac{2}{3}} 2p_x$$

Notice that this wave function corresponds to one-third $2s$ character and two-thirds $2p_x$ character. The remaining one-third of the $2p_x$ orbital is divided equally between $\psi(sp^2{}_b)$ and $\psi(sp^2{}_c)$; that is, $c_5{}^2 = c_8{}^2$.

Figure 3-15 shows that the other two sp^2 hybrid orbitals, $\psi(sp^2{}_b)$ and $\psi(sp^2{}_c)$, are equivalent with respect to both the X and Y axes. Accordingly, we may calculate the coefficients of the $2p_x$ and $2p_y$ orbitals from the unit-orbital contribution (rule 4):

$$c_2{}^2 + c_5{}^2 + c_8{}^2 = 1$$

in which $c_5{}^2 = c_8{}^2$ and $c_2{}^2 = (\sqrt{2/3})^2$. Hence $c_5 = c_8 = \pm\sqrt{1/6}$. Figure 3-15 shows that both $\psi(sp^2{}_b)$ and $\psi(sp^2{}_c)$ point in a direction such that the $2p_x$ contribution is along the $-X$ axis. Hence c_5 and c_8 both have a value of

$$-\sqrt{\frac{1}{6}}.$$

To determine the coefficients c_6 and c_9 of the $2p_y$ orbital we again apply the unit-orbital contribution and the fact that the $2p_y$ orbital will contribute equally to $\psi(sp^2{}_b)$ and $\psi(sp^2{}_c)$. Thus

$$c_6{}^2 + c_9{}^2 = 1$$

in which $c_6{}^2 = c_9{}^2$ and therefore $c_6 = \pm\sqrt{1/2}$ and $c_9 = \pm\sqrt{1/2}$. Choosing the algebraic signs in the functions so that positive lobes are directed at hydrogen atoms b and c, we obtain $c_6 = +\sqrt{1/2}$ and $c_9 = -\sqrt{1/2}$. The complete set of sp^2 hybrid orbitals is therefore

$$\psi(sp^2{}_a) = \sqrt{\frac{1}{3}}2s + \sqrt{\frac{2}{3}}2p_x$$

$$\psi(sp^2{}_b) = \sqrt{\frac{1}{3}}2s - \sqrt{\frac{1}{6}}2p_x + \sqrt{\frac{1}{2}}2p_y$$

$$\psi(sp^2{}_c) = \sqrt{\frac{1}{3}}2s - \sqrt{\frac{1}{6}}2p_x - \sqrt{\frac{1}{2}}2p_y$$

The sp^3 hybrid orbitals may be derived in a way similar to that used for sp and sp^2. The results are presented here (coordinate system as shown in Figure 3-17).

$$\psi(sp^3{}_a) = \tfrac{1}{2}(2s + 2p_x + 2p_y + 2p_z)$$

$$\psi(sp^3{}_b) = \tfrac{1}{2}(2s - 2p_x - 2p_y + 2p_z)$$

$$\psi(sp^3{}_c) = \tfrac{1}{2}(2s + 2p_x - 2p_y - 2p_z)$$

$$\psi(sp^3{}_d) = \tfrac{1}{2}(2s - 2p_x + 2p_y - 2p_z)$$

In summary, we see that the four simple criteria presented in this section can be utilized to determine the form of the sp, sp^2, and sp^3 hybrid orbitals. Notice that each of the sp^2 hybrid orbitals has one-third $2s$ character and a total of two-thirds $2p$ character. Each of the sp^3 hybrid orbitals has one-fourth $2s$ character and a total of three-fourths $2p$ character. These relative contributions are in agreement with the symmetry requirements imposed on the equivalent hybrid orbitals.

3-8 Structure and Bonding in Boranes

The structures of several boron hydrides and borane anions are shown in Figure 3-33. Several of the higher boron hydrides, B_4H_{10}, B_5H_9, and $B_{10}H_{14}$, can be prepared by the thermal decomposition of B_2H_6. As shown in Figure 3-33, the boron-atom frameworks often are fragments of regular polyhedra. For example, the framework of B_5H_9 is that of an octahedron (compare it with $B_6H_6{}^{2-}$) which lacks one vertex, and the framework of $B_{10}H_{14}$ is that of an icosahedron (compare it with $B_{12}H_{12}{}^{2-}$) which lacks two adjacent vertices.

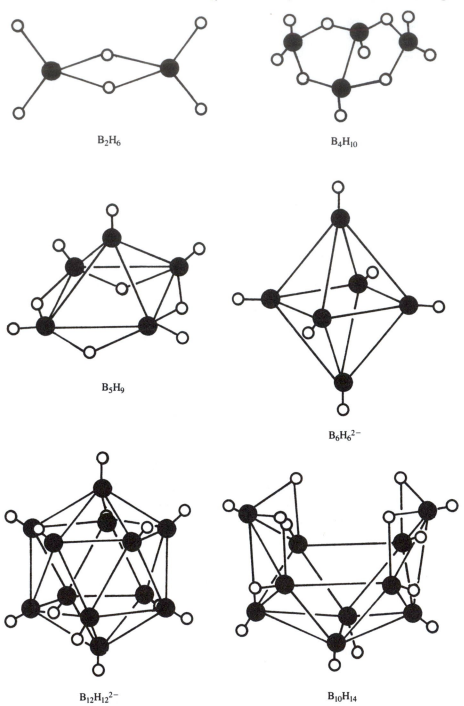

B_2H_6

B_4H_{10}

B_5H_9

$B_6H_6^{2-}$

$B_{12}H_{12}^{2-}$

$B_{10}H_{14}$

Figure 3-33 Structures of some boron hydrides and borane anions.

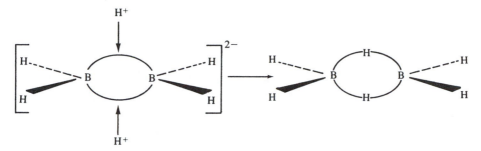

Figure 3-34 Hypothetical formation of B_2H_6. The two protons attack the two bent bonds in $B_2H_4^{2-}$. Therefore, the B—H—B bridge bonds are three-center, two-electron bonds.

In section 2-11 we pointed out that an ordinary Lewis electron-dot structure is not adequate to describe the bonding in the simplest boron hydride, B_2H_6. As shown in Figure 3-33, there are a total of eight bonds in B_2H_6: four B—H_t bonds and four B—H_b bonds (H_t refers to a terminal hydrogen atom and H_b to a bridging hydrogen atom). Diborane has only 12 valence electrons (3 from each boron atom and 1 from each hydrogen) and is therefore *electron deficient*. That is, if we constructed localized electron-pair bonds, we would need 16 electrons before all the atoms could be connected. Electron deficiencies of this type are characteristic of the boron hydrides, and we can see that an electronic structural model more complex than two-center, electron-pair bonds is necessary.

One way to formulate the electronic structure of B_2H_6 arises from the realization that B_2H_6 is isoelectronic with C_2H_4. We can imagine forming B_2H_6 in the following hypothetical reaction:

$$B_2H_4^{2-} + 2H^+ \rightarrow B_2H_6$$

Now, because $B_2H_4^{2-}$ is isoelectronic with C_2H_4, we can describe its electronic structure in terms of the bent-bond model (Figure 3-26). We then predict that one proton would attack the bent bond below the $B_2H_4^{2-}$ plane, and the other would attack the bent bond above the $B_2H_4^{2-}$ plane. This produces the observed molecular structure for B_2H_6 shown in Figure 3-34. In this model, each boron atom uses two sp^2 hybrid orbitals to bond the two terminal hydrogen atoms. Notice that the bridging hydrogen atoms are involved in *three-center* two-electron bonds. In contrast, the terminal hydrogen atoms are bonded by ordinary two-center two-electron bonds.

By including the three-center bond concept in our bonding description, we can write electron-dot structures for the boron hydrides and borane anions. The B—H—B bridge usually is represented by the structure

The only other type of three-center bond encountered in boron hydrides is the B—B—B bond, which can be represented by the structure

In addition to these two types of three-center bonds, we also can have ordinary two-center B—B and B—H bonds.

A satisfactory description of the bonding in these electron-deficient boron hydrides must fulfill the following three criteria:

1. It must show that each bond — both two-center and three-center — contains two electrons. In reality, this is a "conservation of electrons" criterion.

2. It must show that in the bonding each boron atom uses all four of its valence orbitals and each hydrogen atom its $1s$ valence orbital. The boron and hydrogen atoms must contribute a total of two valence orbitals to each two-center bond and three valence orbitals to each three-center bond. This is essentially a "conservation of orbitals" criterion.

3. The bonding it predicts must be consistent with the observed structure of the molecule. Generally, the bonds to each boron atom are in an approximate tetrahedral configuration.

It is useful to express the first two of these rules in the form of *topological equations*. Consider a formula, $B_b H_h{}^q$, in which q is the net charge. Let α represent the number of two-center bonds (B—B or B—H) and β the number of three-center bonds (B—H—B or B—B—B). Then criterion 1 may be stated as

$$2(\alpha + \beta) = 3b + h - q \tag{3-14}$$

In this equation the left side indicates that two electrons must reside in each bond and the right side indicates the total number of valence electrons in the species. Criterion 2 equates the total number of valence orbitals that must participate in the two-center and three-center bonds to the number of valence orbitals that are available from the boron and hydrogen atoms:

$$2\alpha + 3\beta = 4b + h \tag{3-15}$$

Solving Equations 3-14 and 3-15 for α and β, we obtain

$$\alpha = \tfrac{1}{2}(b + h - 3q) \tag{3-16}$$

$$\beta = b + q \tag{3-17}$$

Now we apply Equations 3-16 and 3-17 to describe the bonding for several of the boron compounds shown in Figure 3-33.

Exercise 3-1: Describe the bonding in B_2H_6.

Solution: For this molecule we have $b = 2$, $h = 6$, and $q = 0$. From Equations 3-16 and 3-17 we obtain $\alpha = 4$ and $\beta = 2$, corresponding to the bonding

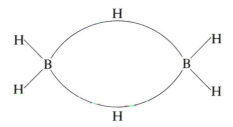

To satisfy criterion 3 (tetrahedral configuration of bonds to each boron atom) we require that the bridging hydrogen atoms lie in a plane perpendicular to the plane of the other atoms, which is consistent with the experimental structure shown in Figure 3-33.

Exercise 3-2: Describe the bonding in B_4H_{10}.

Solution: For this molecule we have $b = 4$, $h = 10$, and $q = 0$. From Equations 3-16 and 3-17 we obtain $\alpha = 7$ and $\beta = 4$. For this molecule there are two possible structures for these values.

Structure I
7 B—H bonds
3 B—H—B bonds
1 B—B—B bond
0 B—B bonds

Such a structure might be:

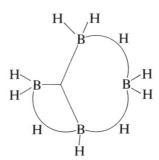

Structure II

6 B—H bonds
4 B—H—B bonds
1 B—B bond
0 B—B—B bonds

The structure shown here is the one identified by x-ray diffraction analyses and appears in Figure 3-33:

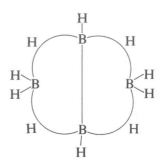

Exercise 3-3: Describe the bonding in B_5H_9.

Solution: For this molecule we have $b = 5$, $h = 9$, and $q = 0$. From Equations 3-16 and 3-17 we calculate $\alpha = 7$ and $\beta = 5$. A bonding pattern that is consistent with seven two-center and five three-center bonds and that is also consistent with the structure shown in Figure 3-33 is

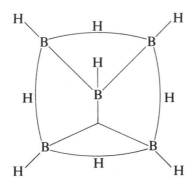

The boron framework of this molecule has the shape of a square pyramid (Figure 3-33). Because the four basal boron atoms are equivalent, the bonding must be described as a resonance hybrid of four bonding structures corresponding to permutation of the two B—B bonds and the B—B—B bond. B_5H_9 is the simplest boron hydride that uses all the possible types of two- and three-center bonds.

QUESTIONS AND PROBLEMS

1. Indicate an appropriate hybridization for the central-atom valence orbitals for each of the following molecules:

(a) CS_2 (b) SO_3 (c) BF_3 (d) CBr_4
(e) SiH_4 (f) SeF_6 (g) SiF_5^- (h) AlF_6^{3-}
(i) PF_4^+ (j) IF_6^+ (k) NO_2^+ (l) NO_3^-

2. Write a Lewis electron-dot structure for N_3. Predict its molecular shape using the VSEPR method. Discuss the electronic structure of N_3 in terms of hybridization at the central nitrogen atom and the number of σ and π electrons. How many unpaired electrons are there? Is the molecule polar? Predict the relative N—N bond lengths in N_2 and N_3.

3. Draw hybrid-orbital representations including π bonds for NO_3^-, NCO^-, and CNO^-.

4. The allyl cation has the formula $C_3H_5^+$. What structure does it have? Formulate the hybrid orbital model of the electronic structure of $C_3H_5^+$.

5. Consider the following two hybrid orbitals:

$$\phi_a = N(2p_z + \gamma 2s)$$
$$\phi_b = N(2s - \gamma 2p_z)$$

where $2p$ and $2s$ are atomic orbitals on one atom.
(a) Determine the normalization constant N, expressed in terms of the constant γ.
(b) Show that ϕ_a and ϕ_b are orthogonal.
(c) Sketch the shape of these hybrid orbitals for $\gamma \simeq 0.2$.
(d) Repeat part (c) for $\gamma = 1.0$.

6. The structure of ethylene indicates that the HCH and HCC angles do not correspond exactly to the sp^2 hybrid orbital angle of 120°:

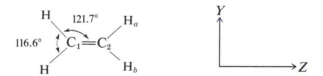

Use the coordinate system shown so that the molecule lies in the YZ plane:
(a) Calculate the coefficients of the three hybrid orbitals around carbon atom 2. Label the hybrid orbitals ϕ_a, ϕ_b, and ϕ_C where these hybrid orbitals point toward H_a, H_b, and C_1, respectively. Each of the hybrid orbitals will be of the general form $\phi = c_1 2p_z + c_2 2p_y + c_3 2s$. Be sure that each hybrid orbital is normalized and orthogonal to the other two hybrid orbitals.
(b) Calculate the percent contribution of $2p_z$, $2p_y$, and $2s$ in each of the hybrid orbitals.

7.   Formulate the hybrid-orbital description for pyridine, C_6H_5N. Notice that pyridine is isoelectronic with benzene.

8.   The ozone molecule, O_3, has a bond angle of 116.8°. Formulate the hybrid-orbital description for the central oxygen atom in ozone.
     (a) Use the general form $\phi = N(2p + \gamma 2s)$ and $\cos\theta = -\gamma^2$ to determine the form of the *bonding* hybrid orbitals.
     (b) Calculate the percent contribution of the $2s$ and $2p$ atomic orbitals to each of the bonding hybrid orbitals.
     (c) By using the unit-orbital contribution and the results obtained in (a) and (b), calculate the percent contribution of the $2s$ and $2p$ atomic orbitals in the lone-pair hybrid orbital.

9.   The cyclic form of ozone with the OOO angle of 60°, is only slightly less stable than the angular form (OOO angle of 116.8°).
     (a) Is it possible to formulate bonding hybrid orbitals with an inter-orbital angle of 60°? (*Hint:* Use $\cos\theta = -\gamma^2$; what happens to the value of γ for $\theta < 90°$?)
     (b) Formulate qualitatively the electronic structure of cyclic ozone by using the bent- or banana-bond approach, assuming the oxygen atoms to be sp^3 hybridized.

10. (a) Sketch the predicted shape of the two hybrid orbitals

$$\phi_\pm = \frac{1}{\sqrt{2}}(4s \pm 3d_{xz})$$

     (b) Would you predict the molecule AH_2 to be linear or bent if the atom A employed the hybrid orbitals $\phi_\pm$ to bond the two hydrogen atoms?

11. Solve Equations 3-16 and 3-17 for the unknown $B_3H_6{}^+$ and the known $B_6H_{11}{}^+$ ions and write a probable structure for each of these cations.

12. Solve Equations 3-16 and 3-17 for the $B_3H_8{}^-$ ion. Draw two possible structures and indicate which structure you think is more plausible.

13. In Section 3-7 we derived the following three sp^2 hybrid orbitals:

$$\psi(sp^2{}_a) = \sqrt{\tfrac{1}{3}}2s + \sqrt{\tfrac{2}{3}}2p_x$$

$$\psi(sp^2{}_b) = \sqrt{\tfrac{1}{3}}2s - \sqrt{\tfrac{1}{6}}2p_x + \sqrt{\tfrac{1}{2}}2p_y$$

$$\psi(sp^2{}_c) = \sqrt{\tfrac{1}{3}}2s - \sqrt{\tfrac{1}{6}}2p_x - \sqrt{\tfrac{1}{2}}2p_y$$

(a) Show that each of these hybrid orbitals is normalized.

(b) Show that $\psi(sp^2{}_a)$ is orthogonal to $\psi(sp^2{}_b)$.

(c) Show that the unit orbital contribution is obeyed for the $2p_x$ orbital.

(d) Calculate the total percent contribution of the $2p$ orbitals to each hybrid orbital. The total $2p$ contribution should be calculated by summing the $2p_x$ and $2p_y$ contributions.

14. Consider the sp, sp^2, and sp^3 hybrid orbitals. Use the general orbital form $\phi = N(2p + \gamma 2s)$ and $\cos \theta = -\gamma^2$. Calculate the value of γ for each of these types of hybrid orbitals. What is the percent contribution of the $2s$ atomic orbital to each of these types of hybrid orbitals?

15. Derive a pair of equations corresponding to Equations 3-14 and 3-15 for hydrocarbon molecules $C_cH_h{}^q$, where the hydrocarbon molecule is bonded only by two-center (α) and three-center (β) electron-pair bonds. Solve these equations for α and β corresponding to Equations 3-16 and 3-17.

16. Use the equations derived in problem 3-15 to obtain the number of two-center and three center bonds in $C_3H_5{}^+$, C_6H_6, $C_5H_5{}^+$, $C_3H_3{}^+$, $CH_5{}^+$, and $C_4H_4{}^{2+}$. Sketch Lewis electron dot structures for each of these species, in accord with the appropriate number of two-center and three-center bonds.

Suggestions for Further Reading

Coppens, P., "Experimental Electron Densities and Chemical Bonding," *Angew. Chem. Int. Ed.* 16:32 (1977).

Coulson, C.A., *The Shape and Structure of Molecules,* Oxford: Clarendon Press, 1973.

Gray, H.B., *Chemical Bonds,* Menlo Park, Calif.: Benjamin, 1973.

Gray, H.B., *Electrons and Chemical Bonding,* Menlo Park, Calif.: Benjamin, 1965.

Harvey, K.B., and Porter, G.B., *Introduction to Physical Inorganic Chemistry,* Reading, Mass.: Addison-Wesley, 1963.

Hsu, C.-Y., and Orchin, M., "A Simple Method for Generating Sets of Orthonormal Hybrid Atomic Orbitals," *J. Chem. Educ.* 50: 114 (1973).

Jaffé, H.H., and Orchin, M., *Symmetry in Chemistry*, Huntington, N.Y.: R.E. Krieger, 1976.

Jolly, W.L., *The Principles of Inorganic Chemistry*, New York: McGraw-Hill, 1976.

Lipscomb, W.N., "The Boranes and Their Relatives," *Science* 196: 1047 (1977).

McWeeny, R., *Coulson's Valence*, New York: Oxford, 1979.

Murrell, J.N., Kettle, S.F.A., and Tedder, J.M., *The Chemical Bond*, New York: Wiley, 1978.

Pauling, L., *The Chemical Bond*, Ithaca, N.Y.: Cornell Univ. Press, 1967.

Purcell, K.F., and Kotz, J.C., *Inorganic Chemistry*, Philadelphia: Saunders, 1977.

4 The Molecular–Orbital Theory of Electronic Structure and the Spectroscopic Properties of Diatomic Molecules

In Chapter 3 we introduced the valence bond and hybrid-orbital descriptions of chemical bonding. We saw the close connection between these bonding descriptions and the Lewis electron-dot structures. It was pointed out that these localized orbital descriptions are very appropriate in discussing ground-state properties of molecules, such as molecular geometry and bond dissociation energies. However, excited state properties of molecules, which are studied by the various types of spectroscopy are not readily treated by the valence bond description. In this chapter we introduce the molecular-orbital (MO) theory and apply it to diatomic molecules. As we shall see, molecular-orbital theory can be used to interpret ground-state and excited-state properties. Examples of excited-state properties are electronic transitions and ionizations, which were discussed in Chapter 1 for the hydrogen atom. In this chapter, we discuss these same properties for diatomic molecules.

We begin by illustrating the MO theory for the simplest molecular system, H_2^+. The Schrödinger equation can be solved exactly for this one-electron molecule. After outlining the theory for H_2^+, we will describe the MO theory for the two-electron hydrogen molecule, H_2. We discussed the VB theory for H_2 in Chapter 3. The VB and MO theories differ in the way in which they regard electron *correlation* or electron-electron interaction. In the VB theory (Section 3-1), the wave function is such that each of the two electrons of the electron-pair bond between two atoms tends to remain on its "own" atom. In contrast, the MO wave function is so formulated that the pair of electrons is in an electron orbital that extends over two or more atomic nuclei. Consequently, we say that the VB theory utilizes a localized bond approach whereas the MO theory utilizes a delocalized bond approach. In Section 4-2 we will carefully compare these two theories for the H_2 molecule.

4-1 Bonding Theory for H_2^+

When a hydrogen atom and a proton are separated by relatively large distances (10 Å or more), the atomic electron cloud of the hydrogen atom is not influenced significantly by the presence of the proton, as depicted in Figure 4-1(a). However, as the two nuclei approach each other the orbitals overlap and electron density increases between the nuclei, as shown in Figure 4-1(b). At shorter internuclear distances (1–2 Å) the electron density is distributed equally between the nuclei. This is represented in Figure 4-1(c).

The closer the nuclei come, the more electron density is attracted between them, the lower the energy falls, and the more stable the molecule-ion

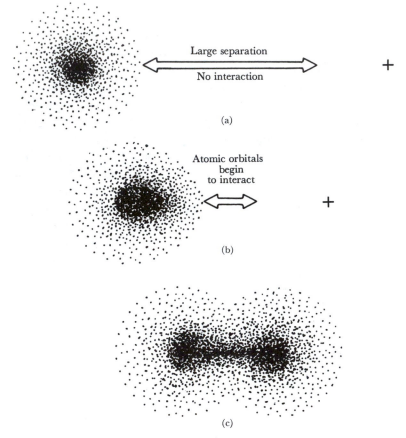

Figure 4-1 Representation of the formation of H_2^+ from a proton and a hydrogen atom. (a) At large internuclear separation the electron cloud of the hydrogen atom is unperturbed by the proton. (b) At shorter internuclear distances the electron cloud of the hydrogen atom becomes distorted by the proton, and electron density increases in the region between the nuclei. (c) At the equilibrium internuclear separation, electron density is equally distributed between the nuclei.

becomes. This behavior, however, persists only to a limit, after which, if the nuclei come closer, the repulsion between them begins to dominate. Beyond this limit, the increase in nuclear–nuclear repulsion is greater than the increase in nucleus-electron cloud attraction.

The distance at which these forces are balanced is called the *equilibrium internuclear distance*. Pull the atoms apart, and attractive forces pull them toward each other again. Push them together, and repulsive forces push back. The two atoms act very much as if they were connected by a spring. This condition of balance, or equilibrium distance, is what we normally mean when we speak of the *bond length* [Figure 4-1(c)].

The energy of a vibrating H_2^+ molecule is shown in Figure 4-2. Part of its energy is kinetic, E_K, the energy of motion of the atoms. The other part of

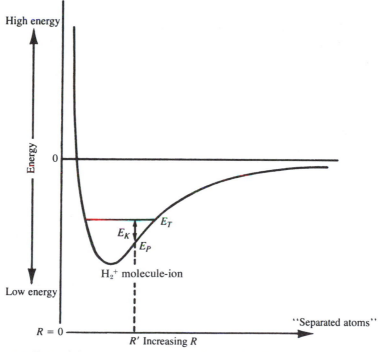

Firgure 4-2 Potential energy curve for the H_2^+ molecule. As the distance between nuclei decreases, the potential energy falls because of electron cloud-nucleus attraction, and then rises again because of nucleus-nucleus repulsion. The horizontal line marked E_T is the total energy of a vibrating molecule. At the extremes of vibration, where E_T touches the potential energy curve, kinetic energy is zero and all energy is the potential energy of the extended or compressed bond. In the center of the vibration, the kinetic energy of motion, E_K, is at a maximum; potential energy E_P, is at a minimum. At any other point in the vibration, the sum of kinetic and potential energies is constant: $E_T = E_K + E_P$. The "equilibrium bond length" is the bond length at the minimum E_P.

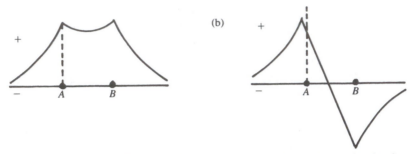

Figure 4-3 Wave functions for H_2^+: (a) plot of $1s_a + 1s_b$ along the nuclear axis; (b) plot of $1s_a - 1s_b$.

the energy is potential, E_P, the energy that motionless molecules with a given separation have because of attractive and repulsive forces. At the extreme limits of stretch or compression, the atoms are motionless at the instant of turnaround, but the forces on them are the greatest. In the middle of one vibration, the potential energy is least, but the atoms are moving most rapidly and the kinetic energy is greatest. So long as the molecule is not disturbed, the total energy, E_T, is constant. The point representing the molecule in the energy diagram of Figure 4-2 moves back and forth from one end of the horizontal E_T line to the other, but at all points $E_T = E_K + E_P$.

It is reasonable to assume that in the hydrogen molecule-ion, the electron is described by the appropriate atomic wave function when it is near one particular nucleus, and consequently largely influenced by that nucleus. Thus the simplest mathematical function that approximately describes the electronic wave function for the molecule is a linear combination of the atomic $1s$ wave functions. This approach usually is abbreviated LCAO–MO, which stands for *linear combination of atomic orbitals–molecular orbitals*.

For the hydrogen molecule-ion, there are only two possible combinations, $1s_a + 1s_b$ and $1s_a - 1s_b$. In Figure 4-3 we plot the value of $1s_a + 1s_b$ and $1s_a - 1s_b$ for a fixed internuclear distance. Actually, for any discussion of chemical bonding it is the *square* of the wave function that interests us, because this correponds to the probability of finding an electron at a given point. Schematic plots of $[1s_a + 1s_b]^2$ and $[1s_a - 1s_b]^2$ are presented in Figure 4-4 and compared to the square of the individual atomic wave functions.

The $1s_a + 1s_b$ wave function concentrates electron density between the nuclei, whereas the electron density for the $1s_a - 1s_b$ wave function is concentrated outside the internuclear region, and in fact has a *node* at the midpoint of the H—H bond. Obviously the molecule will be more stable if the electron is in the $1s_a + 1s_b$ orbital, because this will increase the amount of

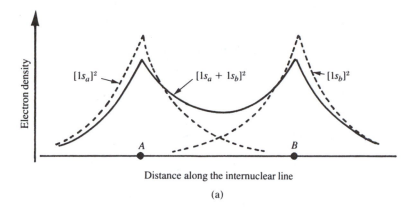

Distance along the internuclear line

(a)

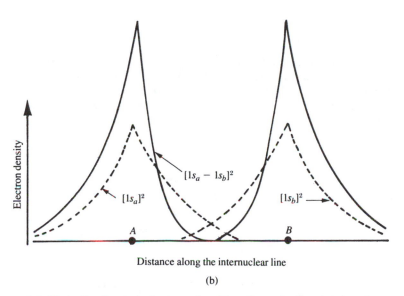

Distance along the internuclear line

(b)

Figure 4-4 Plot of estimated electron densities along the line joining the center of the two nuclei in H_2^+: (a) a comparison of the electron density of $[1s_a + 1s_b]^2$ with the atomic functions $[1s_a]^2$ and $[1s_b]^2$. The $[1s_a]^2$ and $[1s_b]^2$ lines represent one electron distributed between two nonbonded atoms. The electron density in the $1s_a + 1s_b$ orbital is concentrated between the nuclei. (b) The corresponding plot of the electron density for the $[1s_a - 1s_b]^2$ wave function. Notice that there is a node at the midpoint of the H—H bond. The electron density is concentrated in the outer regions of the molecule, not in the internuclear region.

electron-nuclear attraction and decrease the nuclear–nuclear repulsion. The calculated energy curves for H_2^+ are shown in Figure 4-5; it is clear that a curve similar to the experimental curve (Figure 4-2) results if the electron

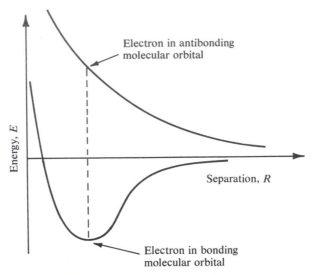

Figure 4-5 The energy of the H_2^+ molecule-ion with the electron in the $1s_a + 1s_b$ bonding orbital falls to a minimum at the observed interatomic distance. The energy of H_2^+ with the electron in the $1s_a - 1s_b$ antibonding orbital is greater than the energy of the completely separated proton and hydrogen atom. This represents a repulsive energy state.

is in the $1s_a + 1s_b$ orbital, whereas a repulsive curve results if the electron is in the $1s_a - 1s_b$ orbital. A molecular orbital that concentrates electron density in the region between the nuclei, thereby lowering the energy of the system, is called a *bonding molecular orbital* (for example, $1s_a + 1s_b$). An *antibonding molecular orbital* (for example, $1s_a - 1s_b$) is one which achieves the opposite effect.

The spatial-boundary picture of a molecular orbital, like that of an atomic orbital, outlines the volume that encloses most of the electron density, and consequently has a shape given by the molecular wave function. In Figure 4-6 we show schematically the "formation" of the bonding molecular orbital in H_2^+ by "adding" two $1s$ orbitals. The attraction that makes the molecule stable is the attraction of the nuclei for electron density concentrated between them. We can think of this concentration as an *overlap* of the $1s$ atomic orbitals, as shown in Figure 4-6.

Molecular orbitals are classified according to their shapes or angular properties, as atomic orbitals are classified as s, p, d, and so on. For homonuclear diatomic molecules there are two symmetry properties that are useful for classification. The first is *rotational* symmetry: the molecular orbital shown in Figure 4-6 is called sigma (σ) because it is symmetric when rotated about a line joining the nuclei. The second is *inversion* symmetry: the molecular

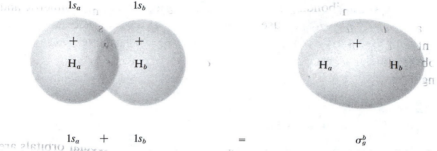

$$1s_a \quad + \quad 1s_b \qquad\qquad = \qquad\qquad \sigma_g^b$$

Figure 4-6 Schematic spatial boundary representation of the formation of the sigma gerade bonding (σ_g^b) molecular orbital of H_2^+. In the combination of the two $1s$ valence orbitals, electron density increases in the region between the two nuclei. Henceforth, the spatial boundary representation will be used instead of the electron-cloud picture shown in Figure 4-1(c). The σ_g^b molecular orbital is symmetric with respect to rotation around the internuclear axis and to inversion through the mid-point of the bond.

orbital shown in Figure 4-6 is called *gerade* (g) because inversion of any point in space through the center of symmetry (midpoint of the H—H bond) results in no change in sign of the wave function. Because the orbital also is bonding, its complete shorthand designation is σ_g^b, which stands for *sigma gerade bonding molecular orbital.*

The antibonding orbital shown in Figure 4-7 also is classified as a σ molecular orbital. That is, if the orbital is rotated by any arbitrary angle around the internuclear axis the orbital still looks the same. With respect to inversion

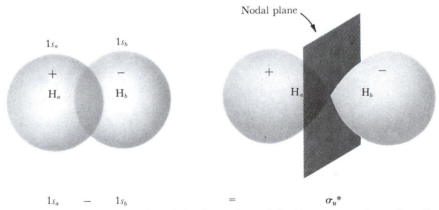

$$1s_a \quad - \quad 1s_b \qquad\qquad = \qquad\qquad \sigma_u^*$$

Figure 4-7 Schematic drawing of the formation of the sigma ungerade antibonding (σ_u^*) molecular orbital of H_2^+. In this subtractive combination of the two $1s$ valence orbitals, electron density decreases in the region between the nuclei. In the nodal plane there is zero probability of finding an electron. The σ_u^* molecular orbital is symmetric with respect to rotation around the internuclear axis but anti-symmetric with respect to inversion through the midpoint of the bond.

symmetry, the antibonding orbital shown in Figure 4-7 is antisymmetric and is called *ungerade* (*u*) because inversion of any point in space through the center of symmetry results in a change in sign of the wave function. We abbreviate this *sigma ungerade antibonding* orbital as σ_u^* (read sigma ungerade star).

Molecular-Orbital Energy Levels

The approximate wave functions for the σ_g^b and σ_u^* molecular orbitals are

$$\psi(\sigma_g^b) = N^b(1s_a + 1s_b) \tag{4-1}$$

$$\psi(\sigma_u^*) = N^*(1s_a - 1s_b) \tag{4-2}$$

These equations are simply the analytical expressions for the molecular orbitals shown in Figures 4-6 and 4-7, respectively. The values of the constants N^b and N^* in Equations 4-1 and 4-2 are fixed by the normalization condition

$$\int |\psi|^2 \, dx \, dy \, dz = \int |\psi|^2 \, dv = 1 \tag{4-3}$$

Let us evaluate N^b. First we substitute $\psi(\sigma_g^b)$ into Equation 4-3:

$$\int [\psi(\sigma_g^b)]^2 dv = 1 = \int [N^b(1s_a + 1s_b)]^2 dv$$

$$= (N^b)^2 \left[\int (1s_a)^2 \, dv + \int (1s_b)^2 \, dv + 2 \int (1s_a)(1s_b) \, dv \right] \tag{4-4}$$

Provided that the atomic orbitals $1s_a$ and $1s_b$ are already normalized,

$$\int (1s_a)(1s_a) \, dv = \int (1s_b)(1s_b) \, dv = 1$$

The integral involving both $1s_a$ and $1s_b$ is called the *overlap integral* and is denoted by the letter S:

$$S = \text{overlap integral} = \int (1s_a)(1s_b) \, dv$$

Thus Equation 4-4 reduces to

$$(N^b)^2[2 + 2S] = 1$$

and

$$N^b = \pm \sqrt{\frac{1}{2(1+S)}} \tag{4-5}$$

In our approximate scheme we shall neglect the overlap integral in determining the normalization constant.* Therefore, arbitrarily picking the positive sign in Equation 4-5, we obtain

$$N^b = \sqrt{\tfrac{1}{2}} \qquad (4\text{-}6)$$

The value of N^* is obtained in the same fashion, by substituting Equation 4-2 into Equation 4-3 and solving for N^*. The result is

$$N^* = \pm \sqrt{\frac{1}{2(1-S)}}$$

or, with the $S = 0$ approximation,

$$N^* = \sqrt{\tfrac{1}{2}}$$

The approximate molecular orbitals for H_2^+ therefore are

$$\psi(\sigma_g^b) = \frac{1}{\sqrt{2}} (1s_a + 1s_b) \qquad (4\text{-}7)$$

and

$$\psi(\sigma_u^*) = \frac{1}{\sqrt{2}} (1s_a - 1s_b)$$

The energies of the molecular orbitals are obtained from the Schrödinger equation,

$$\mathcal{H}\psi = E\psi \qquad (4\text{-}8)$$

Multiplying both sides of Equation 4-8 by ψ and then integrating, we have

$$\int \psi \mathcal{H} \psi \, dv = E \int \psi^2 \, dv. \qquad (4\text{-}9)$$

Because $\int \psi^2 \, dv = 1$, Equation 4-9 reduces to

$$E = \int \psi \mathcal{H} \psi \, dv \qquad (4\text{-}10)$$

*This approximation involves a fairly substantial error in the case of H_2^+. The overlap of $1s_a$ and $1s_b$ in H_2^+ is 0.585. Thus we calculate $N^b = 0.562$, as compared to $N^b = 0.707$ for the $S = 0$ approximation. In most other cases, however, the overlaps are smaller (usually between 0.2 and 0.3) and the approximation involves only a small error.

Substituting Equation 4-7 into Equation 4-10, we obtain

$$E[\psi(\sigma_g^b)] = \int [\psi(\sigma_g^b)]\mathcal{H}[\psi(\sigma_g^b)] \, dv$$

$$= \frac{1}{2} \int (1s_a + 1s_b)\mathcal{H}(1s_a + 1s_b) \, dv$$

$$= \frac{1}{2} \int (1s_a)\mathcal{H}(1s_a) \, dv + \frac{1}{2} \int (1s_b)\mathcal{H}(1s_b) \, dv$$

$$+ \frac{1}{2} \int (1s_a)\mathcal{H}(1s_b) \, dv + \frac{1}{2} \int (1s_b)\mathcal{H}(1s_a) \, dv \tag{4-11}$$

We shall not attempt to evaluate the various integrals in Equation 4-11, but instead shall replace them with the following shorthand:

$$q_a = \int (1s_a)\mathcal{H}(1s_a) \, dv$$

$$q_b = \int (1s_b)\mathcal{H}(1s_b) \, dv$$

$$\beta = \int (1s_a)\mathcal{H}(1s_b) \, dv = \int (1s_b)\mathcal{H}(1s_a) \, dv \tag{4-12}$$

In this case, since $1s_a$ and $1s_b$ are equivalent atomic orbitals,

$$q_a = q_b = q$$

We call q_a and q_b *coulomb integrals*. The coulomb integral represents the energy required to remove an electron from the valence orbital in question, in the field of the nuclei and other electrons in the molecule. Thus it is sometimes referred to as a *valence ionization potential*.

In this text we call β the *exchange integral*. In other sources, however, you may find β referred to as a *resonance* or *covalent integral*. We have seen that an electron in the σ_g^b molecular orbital has a large probability density in the overlap region common to both nuclei. Thus the electron is stabilized in this favorable position. The exchange integral β simply represents this added covalent-bonding stability.

Simplifying Equation 4-11, we finally have

$$E[\psi(\sigma_g^b)] = q + \beta$$

The energy of the σ_u^* molecular orbital is found in the same manner; that is, substitution into Equation 4-10 gives

$$E[\psi(\sigma_u^*)] = \tfrac{1}{2} \int (1s_a - 1s_b)\mathcal{H}(1s_a - 1s_b) \, dv = q - \beta$$

This result shows that the antibonding molecular orbital is less stable than the bonding molecular orbital by an amount equal to -2β. An electron in the σ_u^* molecular orbital has only a small probability of being found in the energetically favored overlap region. Instead, it is confined to the extreme ends of the molecule, which are positions of high energy relative to the internuclear region.

It is convenient to show the relative molecular-orbital energies in a diagram. Such a diagram for H_2^+ is shown in Figure 4-8. The valence orbitals of the combining atoms are represented in the outside columns and are ordered in terms of their coulomb energy. The most stable valence orbitals are placed lowest in the diagram. Because $1s_a$ and $1s_b$ have the same coulomb energy, these levels are placed directly opposite one another.

The molecular-orbital energies are indicated in the middle column. The σ_g^b orbital is more stable than the combining $1s$ valence orbitals, and the σ_u^* orbital is less stable.

The electron in the ground state of H_2^+ occupies the more stable molecular orbital; that is,

Ground state of $H_2^+ = (\sigma_g^b)^1$

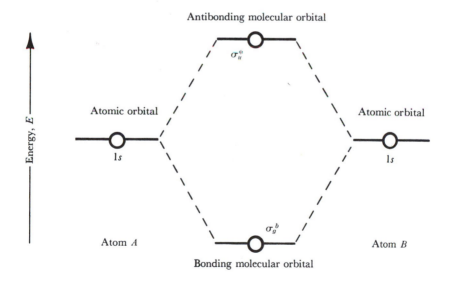

Figure 4-8 Relative energies of the two molecular orbitals for the hydrogen molecule-ion, and the two atomic $1s$ orbitals from which they come. The bonding MO is lower in energy, and the antibonding higher, than the energy of the original atomic orbitals.

Refinements in the Molecular-Orbital Treatment of H_2^+

At this point we should admit that the wave function $\psi(\sigma_g{}^b) = 1/\sqrt{2}$ $\cdot (1s_a + 1s_b)$ does not give a completely adequate description of the bonding in H_2^+. The bond dissociation energy calculated from Equation 4-16 is only 41 kcal mole^{-1}, whereas the experimental bond energy is 64 kcal mole^{-1}. Furthermore, the calculated equilibrium bond length is 1.32 Å compared to

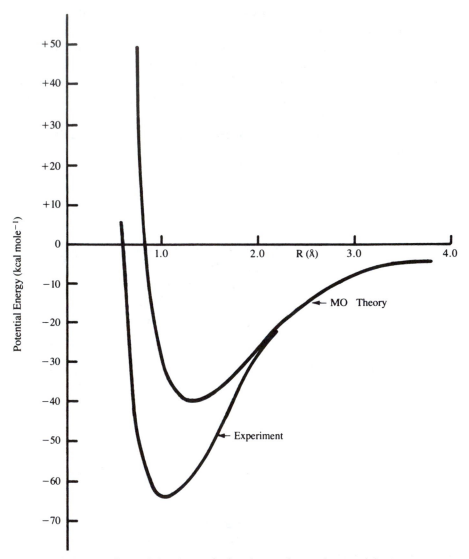

Figure 4-9 Comparison of the theoretical and experimental potential energy curves for H_2^+. The "MO theory" curve was calculated using the wave function of Equation 4-1.

the experimental value of 1.06 Å. Thus, although the calculated energy curve mimics the experimental curve, as seen by comparing Figures 4-2 and 4-5, the agreement is not satisfactory. This can be seen by accurately plotting the calculated and experimental curves on the same graph, as in Figure 4-9.

Two refinements can be used to improve the calculated energy for H$_2^+$. First, allow Z to vary in the $1s_a$ and $1s_b$ wave functions. Recall that

$$\psi(1s) = \frac{1}{\sqrt{\pi}} \left(\frac{Z}{a_0}\right)^{3/2} e^{-Zr/a_0}$$

By allowing Z to vary so as to obtain minimum energy in Equation 4-10, the calculated energy is lowered to 54 kcal mole^{-1} and the calculated equilibrium bond length is found to agree exactly with the experimental value (1.06 Å). The optimized value of Z is found to be 1.24, rather than 1.0 as in the H atom. Second, the wave function can be improved by allowing the $2p_z$ orbital to mix with the $1s$ orbital, just as was done in the valence-bond calculations for H$_2$ (Section 3-1). That is, we can replace $1s_a$ by ϕ_a where

$$\phi_a = N(1s_a + \gamma 2p_{z_a}) \tag{4-13}$$

A similar expression can be written for ϕ_b to replace $1s_b$. The value of γ that provides minimum energy is found to be 0.145. The resulting calculated energy is 63 kcal mole^{-1}, in excellent agreement with the experimental value of 64 kcal mole^{-1}. Recall from our discussion of the VB treatment on H$_2$ that the effect of mixing $2p$ character with the $1s$ orbital is to *polarize* the electron cloud, thereby allowing a build-up of more electron density in the internuclear region (Figure 3-2).

In summary, we see that by allowing a variation in Z and the $1s$–$2p$ mixing parameter γ, very good agreement can be obtained between experiment and theory for H$_2^+$.

4-2 Molecular-Orbital Theory and Valence Bond Theory for H$_2$

In constructing the MO wave function for molecular hydrogen, we start with the simple one-electron wave functions developed for H$_2^+$. To describe the bonding state for H$_2$ we simply place the two electrons in the $\psi(\sigma_g^b)$ molecular orbital (Equation 4-1):

$$\psi[\sigma_g^b(1)] = 1/\sqrt{2}[1s_a(1) + 1s_b(1)]$$
$$\psi[\sigma_g^b(2)] = 1/\sqrt{2}[1s_a(2) + 1s_b(2)]$$

The wave function that describes the hydrogen molecule is then the product of these two one-electron functions:

$$\psi_{MO} = \psi[\sigma_g{}^b(1)] \, \psi[\sigma_g{}^b(2)] \tag{4-14}$$

In terms of the molecular-orbital energy-level diagram developed for $H_2{}^+$, Equation 4-14 describes a system in which two electrons with antiparallel spins are placed in the bonding molecular orbital, to satisfy the Pauli exclusion principle.

The bond energy calculated from Equation 4-14 is only 62 kcal mole^{-1} compared to the experimental bond energy of 109.5 kcal mole^{-1}. The calculated energy can be improved considerably by allowing Z to vary. For the optimum Z (1.20), the bond energy is calculated to be 80 kcal mole^{-1}. Further improvement can be achieved by replacing the $1s_a$ and $1s_b$ functions by ϕ_a and ϕ_b, just as was done for $H_2{}^+$ (Equation 4-13).

It is worthwhile to compare the VB and MO wave functions for molecular hydrogen. In order to simplify the comparison, we consider the VB and MO wave functions in the absence of s–p mixing ($\phi = 1s$). The VB wave function (Equation 3-8) then becomes

$$\psi_{VB} = 1s_a(1)1s_b(2) + 1s_a(2)1s_b(1) + \lambda[1s_a(1)1s_a(2) + 1s_b(1)1s_b(2)]$$

We may compare this with the simple MO wave function:*

$$\psi_{MO} = \psi[\sigma_g{}^b(1)] \, \psi[\sigma_g{}^b(2)] = [1s_a(1) + 1s_b(1)][1s_a(2) + 1s_b(2)]$$

$$= 1s_a(1)1s_b(2) + 1s_a(2)1s_b(1) + 1s_a(1)1s_a(2) + 1s_b(1)1s_b(2)$$

It is readily apparent that the MO wave function corresponds to the VB wave function if $\lambda = 1$. Obviously the MO wave function gives too much weight to the ionic terms. (Recall that $\lambda \simeq \frac{1}{4}$ in VB theory.) This is because both electrons in the MO theory were placed in the same orbital, $\psi(\sigma_g{}^b)$, without any special attention being paid to the fact that these electrons would attempt to avoid one another. In other words, the simplest MO wave function neglects electron correlation, while the simplest VB wave function (with $\lambda = 0$) overestimates electron correlation (Equation 3-2). This proves to be a general difference between the two methods, and consequently we find that the VB theory is better at large internuclear distances for which the *isolated* atoms provide an extreme case, whereas the MO theory is better at

---

*Notice that we have neglected the normalization constant in ψ_{MO}: $(1/\sqrt{2})(1/\sqrt{2}) = \frac{1}{2}$. Actually, the ψ_{VB} wave function, Equation 3-8, should also have incorporated this normalization constant. For purposes of comparison, we can neglect this constant for both wave functions.

small internuclear distances where the *combined* atoms provide the extreme case. Both methods use atomic orbitals as their starting point, and relate bond strength to the degree of overlap of the atomic wave functions.

The overestimation of the ionic terms in ψ_{MO} can be removed by inclusion of the electronic configuration in which both electrons are in the σ_u^* orbital. This so-called configuration interaction results in a considerable improvement in the bond energy calculated from the ψ_{MO} wave function. Mathematically, we can express this improved wave function as

$$\psi_{MO}' = c_1\psi[\sigma_g^b(1)]\psi[\sigma_g^b(2)] - c_2\psi[\sigma_u^*(1)]\psi[\sigma_u^*(2)] \tag{4-15}$$

Intuitively, we expect $c_1 > c_2$, and this is found to be the case $(c_1/c_2 = 1.7)$. The calculated bond energy is 93 kcal mole^{-1} compared to only 62 kcal mole^{-1} in the absence of configuration interaction $(c_1 = 1, c_2 = 0$; Equation 4-15 reduces to Equation 4-14).

The way in which Equation 4-15 removes the overemphasis on the ionic terms can be shown readily. Expansion of the equation gives

$$\begin{aligned}
\psi_{MO}' &= (c_1/2)[1s_a(1) + 1s_b(1)][1s_a(2) + 1s_b(2)] \\
&\quad - (c_2/2)[1s_a(1) - 1s_b(1)][1s_a(2) - 1s_b(2)] \\
&= (c_1/2)[1s_a(1)1s_a(2) + 1s_a(1)1s_b(2) + 1s_b(1)1s_a(2) + 1s_b(1)1s_b(2)] \\
&\quad - (c_2/2)[1s_a(1)1s_a(2) - 1s_a(1)1s_b(2) - 1s_b(1)1s_a(2) + 1s_b(1)1s_b(2)]
\end{aligned}$$

Collecting the terms, we find

$$\begin{aligned}
\psi_{MO}' &= (c_1 + c_2)[1s_a(1)1s_b(2) + 1s_b(1)1s_a(2)] \\
&\quad + (c_1 - c_2)[1s_a(1)1s_a(2) + 1s_b(1)1s_b(2)]
\end{aligned}$$

Notice that the contribution of the ionic terms is reduced by the factor $(c_1 - c_2)$.

We can compare this configuration interaction MO wave function to the valence-bond wave function given earlier:

$$\begin{aligned}
\psi_{VB} &= 1s_a(1)1s_b(2) + 1s_a(2)1s_b(1) \\
&\quad + \lambda[1s_a(1)1s_a(2) + 1s_b(1)1s_b(2)]
\end{aligned}$$

In order to compare the two wave functions, let us examine the ratio of ionic to covalent terms. Since $\lambda \simeq \frac{1}{4}$ for the VB wave function, this ratio is about 1/4. For the MO wave function, the ratio of ionic to covalent terms is given by

$$\frac{c_1 - c_2}{c_1 + c_2} = \frac{c_1/c_2 - 1}{c_1/c_2 + 1} = \frac{1.7 - 1}{1.7 + 1} \simeq \frac{1}{4}$$

In other words, the VB and MO wave functions contain the same *proportions* of ionic to covalent terms. We have shown that the VB wave function that includes ionic terms is identical to the MO wave function that includes configuration interaction! Since the wave functions are identical, the calculated bond energies must also be identical, as is found to be the case (93 kcal mole^{-1}).

4-3 Net Bonding in Molecules with 1s Valence Atomic Orbitals

The net number of electron-pair bonds in a molecular system is equal to the total number of electron pairs that can be placed in bonding orbitals minus the total number that are forced (because of the Pauli principle) into antibonding orbitals. An electron in a bonding orbital gives the system stability, which is canceled by an electron in an antibonding orbital. We divide the net number of bonding electrons by two to preserve the idea of electron-pair bonds; that is, two net bonding electrons equal one "bond line" in a Lewis structure.

Let us consider in detail the MO theory for the simple molecules H_2^+, H_2, He_2^+, and He_2. Electronic structures are described for molecules in a way completely analogous to the process for individual atoms. That is, the electronic structures of the molecules are described by putting electrons into the lowest-energy molecular orbitals first and by observing the Pauli principle. The ground state is obtained by following Hund's rule. The energy-level scheme shown in Figure 4-8 is appropriate for diatomic molecules that have 1s valence orbitals.

The hydrogen molecule-ion, H_2^+, has the electronic structure $(\sigma_g^b)^1$; that is, it has one electron in a σ_g bonding orbital. Because there is one unpaired electron, H_2^+ is paramagnetic. Notice that H_2^+ has one-half of an electron-pair bond, or one-half of a σ_g bond. The hydrogen molecule, H_2, has two electrons in the bonding orbital; thus it has the electronic structure $(\sigma_g^b)^2$. Therefore H_2 has one net bond and is diamagnetic. The helium molecule-ion, He_2^+, which has been detected during electric discharges through helium gas, has three electrons in the molecular-orbital system. Two electrons occupy the bonding orbital and one electron occupies the high-energy antibonding orbital; thus the structure is $(\sigma_g^b)^2(\sigma_u^*)^1$. From this structure we predict one half a bond for He_2^+. For He_2 the electronic configuration is $(\sigma_g^b)^2(\sigma_u^*)^2$ from which we predict no net bonding for diatomic helium. In general, if the $n = 1$ shells are filled for both atoms of a diatomic molecule, there is no net bonding from the four $1s$ electrons. Thus, bonding (or the lack of bonding) depends on the electrons in higher-energy orbitals.

TABLE 4-1. COMPARISON OF SOME MOLECULAR-ORBITAL STRUCTURES, NET BONDING ELECTRONS, BOND LENGTHS, AND BOND ENERGIES

Molecule	Bonding Electrons	Anti-Bonding Electrons	Net Bonding Electrons	Bond Length, (Å)	Experimental Bond Energy (kcal mole⁻¹)
He_2	2	2	0	*	*
H_2^+	1	0	1	1.06	61
He_2^+	2	1	1	1.08	55
H_2	2	0	2	0.74	103

*Molecule is unstable.

A comparison of the molecular-orbital structures with bond energies and bond lengths is given in Table 4-1. The hydrogen molecule-ion, H_2^+, has a bond length of 1.06 Å, and H_2 has a bond length of 0.74 Å. The helium molecule-ion, He_2^+, exists and has a bond length of 1.08 Å. The helium molecule, He_2, does not exist, which is consistent with MO theory, from which we predict no net bonds. The bond energies are 61 kcal mole⁻¹ for the hydrogen molecule-ion, 103 kcal mole⁻¹ for the hydrogen molecule, and 55 kcal mole⁻¹ for the helium molecule-ion.* We see from these examples that the molecular-orbital electronic configuration is useful for interpreting the bonding in these molecules.

If the two electrons in the σ_g^b orbital were completely independent, one might expect the bond dissociation energy of H_2 to be twice that of H_2^+, because H_2 has two net bonding electrons compared to only one for H_2^+. The data in Table 4-1 show that the bond energy of H_2 (103 kcal mole⁻¹) is less than twice that of H_2^+ (61 kcal mole⁻¹). This is an indication of electron-electron repulsion, which arises because the two electrons in the σ_g^b orbital of H_2 occupy the same region of space. This repulsion results in the observed bond energy of H_2 being less than twice that of H_2^+.

Thus far, molecular-orbital theory explains the data well. The following is a summary of the method that we shall use to explain diatomic molecules of heavier atoms and more complicated molecules:

---

*The reader might notice that the bond energies quoted for H_2^+ and H_2 in Sections 4-1 and 4-2 differ slightly from those given here. The difference is due to the *zero-point* energy. This matter will be discussed in the following section and need not concern us here.

1. We shall first combine atomic orbitals in a suitable way to obtain a set of molecular orbitals. The total number of molecular orbitals we obtain will always be equal to the number of atomic orbitals with which be begin.

2. We shall then decide the order of energies of these molecular orbitals.

3. Next, we shall feed into these molecular orbitals all of the electrons in the molecule. We shall start from the lowest-energy orbital and work up, placing no more than two electrons in any one orbital.

4. Finally, we shall examine the filled bonding and antibonding orbitals to determine the net strength of the bonds. Two net bonding electrons correspond to what we have called a single bond in the Lewis model.

4-4 Molecular Spectroscopy

In the previous section we discussed the experimental bond lengths and bond dissociation energies for several diatomic molecules and ions. Experimental *vibrational energies* are a third important indication of bond strength. We shall refer to these three quantities as *bond properties*. We now wish to point out some of the experimental methods by which these quantities may be measured.

All of the common spectroscopic methods used to measure bond properties involve impingement of light on the molecule under study:

$$M + h\nu \rightarrow M^*$$

Depending upon the energy, $h\nu$, of the photon that is absorbed, the molecule may undergo *rotational, vibrational,* or *electronic* transitions.

Let us examine the three types of spectroscopic transitions for a diatomic molecule. It can rotate end over end about any axis that is perpendicular to the internuclear axis, as shown in Figure 4-10(a). A diatomic molecule can also vibrate; see Figure 4-10(b). A physical picture of an electronic transition is not so straightforward as those for the rotational and vibrational

Figure 4-10 (a) Rotational motion of a diatomic molecule; (b) vibrational motion of a diatomic molecule.

motions. However, if we choose H_2 as our prototype molecule, we can imagine one of the bonding electrons being excited to a relatively non-bonding orbital. Such a transition could be pictured as

$$H : H + h\nu \rightarrow H \cdot \dot{H}$$

The total energy of a molecule depends on its rotational, vibrational, and electronic energy. Therefore, a reasonable approximation for the total molecular energy is

$$E_{total} = E_{rotational} + E_{vibrational} + E_{electronic}$$

Figure 4-11 shows a generalized energy-level diagram for a molecule. Two electronic levels, E_1 and E_2, are shown with their vibrational and rotational levels. As is apparent in the figure, separations between electronic energy levels usually are much larger than those between vibrational levels, which in turn are much larger than those between rotational levels.

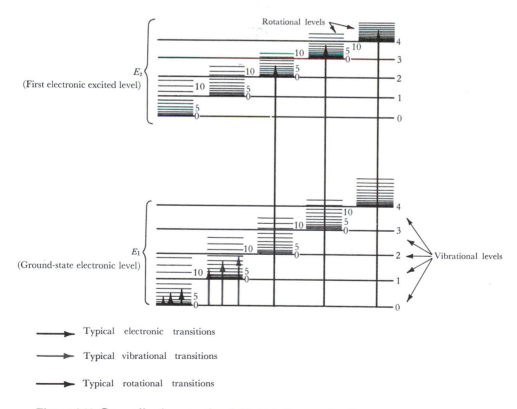

Figure 4-11 Generalized energy-level diagram for a molecule.

TABLE 4-2. ELECTROMAGNETIC RADIATION AND SPECTROSCOPY

Frequency, sec⁻¹	10^6	10^7	10^8	10^9	10^{10}	10^{11}
Wave number, cm⁻¹	3.3×10^{-5}	3.3×10^{-4}	0.0033	0.0333	0.333	3.33
Wavelength	300 m	30 m	3 m	30 cm	3 cm	0.3 cm
Energy	4000 erg mole⁻¹	0.004 joule mole⁻¹	0.04 joule mole⁻¹ 0.01 cal mole⁻¹	0.4 joule mole⁻¹ 0.1 cal mole⁻¹	4 joule mole⁻¹ 1 cal mole⁻¹	40 joule mole⁻¹ 10 cal mole⁻¹
Name	Radio (long-wave)	(short-wave)	(television and FM) (UHF)	Microwave	→ Infrared	(far IR)
Source and detector	(not used)	Vacuum tubes, wires, antenna, coil		Klystron, guide, cavity	Hot wire, etc.	

Frequency, sec⁻¹	10^{12}	10^{13}	10^{14}	10^{15}	10^{16}	10^{17}
Wave number, cm⁻¹	33	333	3333	33333	3.3×10^5	3.3×10^6
Wavelength	300 μ	30 μ	3 μ 30,000 Å	300 nm 3000 Å	30 nm 300 Å	3 nm 30 Å
Energy	0.1 kcal mole⁻¹ 0.16 kT	1 kcal mole⁻¹ 1.6 kT	10 kcal mole⁻¹ 16 kT	100 kcal mole⁻¹ 4 eV	1000 kcal mole⁻¹ 40 eV	400 eV
Name	Infrared (near IR)	(near IR)	→ Visible (red → blue)	Ultraviolet (near UV)		(far UV)
Source and detector	Lamp, prism, grating; phototube or photographic plate			Lamp, etc., grating, phototube		

Frequency, sec⁻¹	10^{18}	10^{19}	10^{20}
Wave number, cm⁻¹	3.3×10^7	3.3×10^8	3.3×10^9
Wavelength	3 Å	0.3 Å	0.03 Å
Energy	4 keV	40 keV	400 keV 0.4 MeV
Name	X rays		→ Gamma rays
Source and detector	X-ray tube, photographic plate		Nuclear reaction, counter

Table 4-2 shows the range of electromagnetic radiation, gives energies in all commonly used units, and indicates the kind of spectroscopy that is used in each range. Notice that rotational transitions require very little energy ($\sim$0.1–10 cm^{-1}) and correspond to absorption of radiation in the far-infrared to microwave regions. Vibrational transitions are more energetic ($\sim$100–4000 cm^{-1}) and correspond to absorption in the near-infrared and infrared regions. Electronic transitions are still more energetic ($\sim$ 10,000–100,000 cm^{-1}) and correspond to absorption in the visible and ultraviolet portion of the spectrum.

Let us examine the three types of spectroscopic transitions for the hydrogen molecule. As we have already said, it can rotate end over end about any axis that is perpendicular to the internuclear axis, as shown in Figure 4-10(a). The study of rotational transitions by far-infrared and microwave spectroscopy is used to obtain extremely precise data on bond lengths and bond angles. The bond length of H_2 has been determined to be 0.74116 Å.

Infrared and *Raman* spectroscopy are the experimental methods employed to study vibrational transition energies. In infrared spectroscopy a beam of infrared radiation, whose wavelength typically varies from 2.5 μ to 15 μ (wave number varies from 4000 cm^{-1} to 667 cm^{-1}), is passed through a sample of a compound; see Figure 4-12. A sodium chloride prism is used because quartz or glass is opaque to infrared light. The sodium chloride prism position and the slit widths determine the wavelength of the radiation that is allowed to reach the detector. The absorption of radiation at different wave

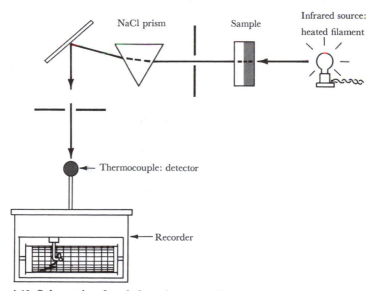

Figure 4-12 Schematic of an infrared spectrophotometer.

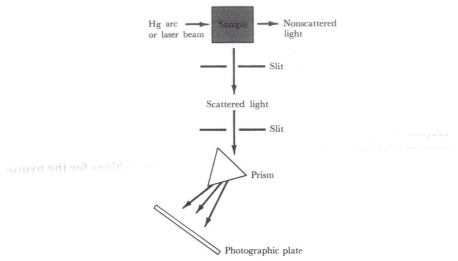

Figure 4-13 Schematic of a Raman spectrometer.

numbers corresponds to the excitation of molecules from a low (usually the lowest) vibrational energy level to the next higher vibrational energy level.

Absorbed radiation is identified by its wavelength (λ in Å, μ, or nm; 1μ = 10^3 nm = 10^4 Å = 10^{-4} cm), its frequency (ν is sec^{-1}), or its wave number ($\bar{\nu}$ in cm^{-1}). Radiation absorption is detected electronically and recorded in some suitable form as a graphical trace. A strong absorption throughout a narrow range of frequencies causes a sharp "peak" or "band" in the recorded spectrum. Absorption peaks are not always narrow and sharp because each vibrational energy level has superimposed upon it an array of rotational energy levels (Figure 4-11); therefore a particular vibrational transition really is the superposition of transitions from many vibrational-rotational levels.

Not all molecules respond to infrared radiation. In particular, molecules with certain elements of symmetry, such as homonuclear diatomic molecules, do not absorb infrared radiation. To assist in examining the vibrations of these molecules, Raman spectroscopy, named for the Indian scientist Chandrasekhara V. Raman, often can be used. The Raman spectrum results from the irradiation of molecules with light (usually in the visible region) of a known wavelength. A laser beam commonly is used to irradiate the sample in a Raman spectrometer, as shown in Figure 4-13. Radiation absorption is not detected directly. When irradiated with high-energy light, the molecules may add to, or extract from, the incident light small amounts of energy corresponding to the energy of the molecular vibration. The incident light is said to be scattered, rather than absorbed, and may be observed as light with a wavelength different from the incident light.

The Raman spectrum of H_2 exhibits a band at 4159.2 cm^{-1}. Since the vibrational energy depends upon the mass of the atoms, the D_2 molecule exhibits a band at lower energy (2990.3 cm^{-1}).

We now wish to examine in more detail the vibrational transition energy, and relate this energy not only to the masses of the atoms but also to the bond strength. If the molecular vibration can be described as a simple harmonic oscillator, the vibrational energy levels are given by

$$E_v = \left(v + \frac{1}{2}\right)h\nu \tag{4-16}$$

in which the only allowed values of the vibrational quantum number v are $v = 0, 1, 2, 3, \ldots$ and ν is the frequency of vibration (sec^{-1}). A potential

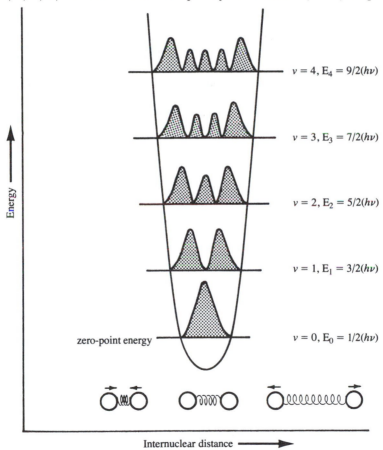

Figure 4-14 Potential energy curve for a diatomic molecule assuming that its vibrational motion can be described as a harmonic oscillator. The shaded region on each vibrational energy level represents the probability distribution of the vibrational wave function.

energy curve associated with the harmonic oscillator is shown in Figure 4-14; the vibrational energy levels are superimposed along with the calculated probability distribution in each vibrational energy level. Notice that this figure is somewhat analogous to Figure 4-2, except that in the former only the lowest vibrational energy level is depicted ($v = 0$).

There are several important features of the harmonic oscillator vibrational energy levels. First, the energy levels are evenly spaced, as demanded by Equation 4-16. Second, notice that even in the lowest energy level ($v = 0$) there is a *zero-point* energy: $E_0 = \frac{1}{2}h\nu$.* Third, the vibrational probability distributions indicate a large probability of finding the molecule at the equilibrium internuclear distance in the lowest vibrational level ($v = 0$). However, in the higher vibrational levels the probability is larger at the extremes of bond compression and expansion.

The harmonic oscillator potential curve compares well with a typical molecular potential curve near the bottom of the potential (Figure 4-15). However, the harmonic oscillator does not exhibit the correct dissociation at larger internuclear distances. Due to molecular dissociation, the vibrational energy levels are evenly spaced only near the bottom of the potential energy curve. The higher vibrational levels come successively closer together in energy.

Let us now examine the relationship between bond strength and vibrational energy. The connection between vibrational frequency and the bond force constant for the harmonic oscillator problem is

$$\nu = \left(\frac{1}{2\pi}\right)\sqrt{\frac{k}{\mu}} \tag{4-17}$$

In this equation k is the force constant of the bond (i.e., a measure of the force required to deform the bond by a given amount) and hence is a measure of the bond strength; μ is the reduced mass for the vibrating atoms. For a diatomic molecule AB in which the atomic masses differ, the reduced mass is given by $m_A m_B/(m_A + m_B)$.

From Equation 4-17, we can see that for a given force constant, a molecule with a larger reduced mass will have a lower frequency of vibration (and therefore a lower vibrational energy in wave numbers). We already have

---

*The zero-point energy for H_2 is about 6 kcal mole^{-1}. This can be calculated as follows: Experimentally, the bond dissociation takes place from the lowest vibrational energy level ($v = 0$) and is 103.24 kcal mole^{-1}. The hypothetical bond dissociation energy of H_2 from the *bottom* of the potential energy curve is 109.5 kcal mole^{-1}. The difference between these two values is $\frac{1}{2}h\nu$ and is the zero-point energy for H_2. In Section 4-3 we referred to the experimental bond energy (103.24 kcal mole^{-1}); in Section 4-2 we referred to the hypothetical value of 109.5 kcal mole^{-1}. The calculations made in Section 4-2 do not include the effect of the zero-point vibrational energy.

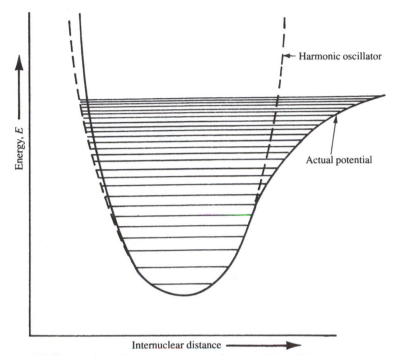

Figure 4-15 Comparison between the actual potential and the harmonic oscillator potential for a diatomic molecule.

pointed to such a case in comparing H_2 and D_2; these molecules have vibrational energies of 4159.2 cm^{-1} and 2990.3 cm^{-1}, respectively. The second important aspect of Equation 4-17 relates to comparing molecules of the same reduced mass but different force constants. Such an example is provided by H_2 and H_2^+; the vibrational energies are 4159.2 cm^{-1} and 2173 cm^{-1}, respectively. The corresponding force constants calculated from Equation 4-17 are 5.2 mdyn Å^{-1} and 1.4 mdyn Å^{-1}. We see that a larger frequency of vibration results in a larger bond force constant, corresponding to a stronger bond.

We turn next to electronic transitions. As stated previously, electronic transitions correspond to the absorption of large amounts of energy, compared to the energy absorbed in vibrational or rotational transitions. Electronic excitations usually are associated with absorption of visible and ultraviolet light. A schematic diagram of a visible and ultraviolet spectrophotometer is shown in Figure 4-16. A prism, or diffraction grating, and the width of the slit regulate the spread of wavelengths through the sample. The light incident on the sample is denoted I_0 and the transmitted light is denoted I. The absorption of light can be expressed quantitatively by the relationship

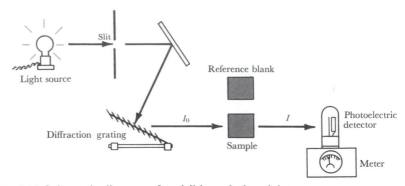

Figure 4-16 Schematic diagram of a visible and ultraviolet spectrometer.

known as *Beer's law*, $A = \varepsilon c l$, in which A is the absorbance ($\log I_0/I$), c is the concentration of the substance (in mole liter^{-1}), and l is the length (in cm) of the light path through the substance. The *molar extinction coefficient, ε,* is characteristic of the absorbing sample. Values for the molar extinction coefficient vary considerably from compound to compound and from peak to peak in the absorption spectrum of the same compound. The extent to which an electronic transition is "allowed" or "forbidden" by quantum-mechanical selection rules is reflected in the value of ε for the absorption peak accompanying the transition. Thus the experimentally determined ε value often can be extremely useful in assigning a given absorption peak to a particular type of electronic transition. If ε is in the range 10^3–10^5, the transition meets all the requirements of the quantum-mechanical selection rules, and the absorption is said to be "allowed."

The selection rules are only approximate when they describe a transition as "forbidden" because in actual practice an absorption band generally can be observed. Even so, the intensity of the band will be diminished significantly if it corresponds to an "orbitally forbidden" transition; in this situation ε usually is in the range 10^0–10^3.

Finally, extremely small values of ε (10^{-5}–10^0) are associated with an electronic transition in which the spin of one electron is required to change in going from the ground state to the excited state. Thus a "spin-forbidden" transition often is difficult to observe as an absorption band because of its extraordinarily weak intensity.

Let us consider some possible electronic transitions in the hydrogen molecule. We could imagine an electronic transition in which an electron is promoted from the $\sigma_g{}^b$ orbital to the $\sigma_u{}^*$ orbital:

$$(\sigma_g{}^b)^2 \rightarrow (\sigma_g{}^b)^1 \, (\sigma_u{}^*)^1$$

This transition is predicted to lead to an unstable or weakly bonded molecule, since the effect of the bonding electron is canceled by that of the anti-bonding electron.

There are many other possible electronic transitions. One example would be the promotion of one of the σ_g^b electrons to an empty π molecular orbital formed from the $2p$ orbitals on the hydrogen atoms:

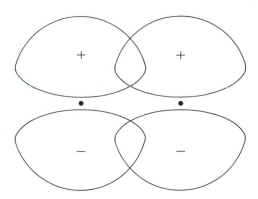

Such a transition can be designated as

$$(\sigma_g^b)^2 \rightarrow (\sigma_g^b)^1(\pi)^1$$

and usually is referred to as a $\sigma \rightarrow \pi$ electronic transition. This electronic transition occurs at about 100,000 cm^{-1} in the H_2 molecule.

There are two ways to present an electronic transition in a graphical fashion. In the first, a molecular-orbital energy-level diagram is drawn for H_2, and the $\sigma \rightarrow \pi$ electronic transition is depicted (Figure 4-17). Notice that this figure includes the π molecular orbital, which was not shown in Figure 4-8. The second method by which we can depict this electronic transition is to draw a potential energy curve for the molecule in the ground state and in the excited electronic state. This description is shown in Figure 4-18. Note that the ground-state molecule dissociates into two ground-state hydrogen atoms:

$$H_2 \rightarrow H(1s) + H(1s)$$

In contrast, the excited-state molecule (designated *) dissociates into one ground-state and one excited-state hydrogen atom:

$$H_2^* \rightarrow H(1s) + H(2p)$$

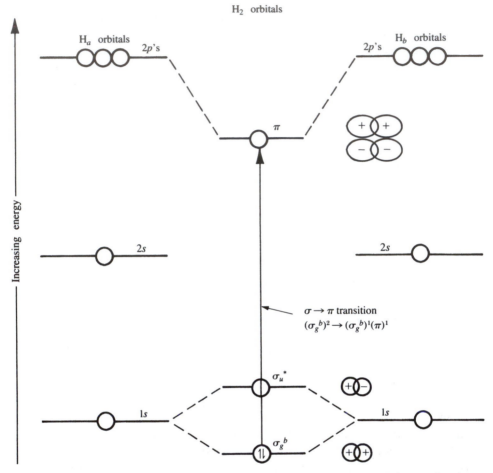

Figure 4-17 Illustration of a $\sigma \to \pi$ electronic transition in H_2. Other molecular orbitals are formed from the $2s$ and $2p$ atomic orbitals. These will be discussed in Section 4-6.

Up to this point we have assumed that the electronic transitions take place with no regard to the rotational and vibrational motions of the molecule. Actually, this is not true, and in Figure 4-18 we have also superimposed the vibrational energy levels on the potential energy curves of the electronic states. At normal temperatures, the ground-state molecule resides in the lowest vibrational energy level $(v''=0)$.* However, in undergoing electronic transitions, the excited-state molecule can reside in any number of vibrational energy levels, $v' = 0, 1, 2, 3, \ldots$, as shown in Figure 4-18. Hence,

*Vibrational quantum numbers of the ground-state molecule H_2 will be labeled v''; whereas those of H_2^* will be labeled v'.

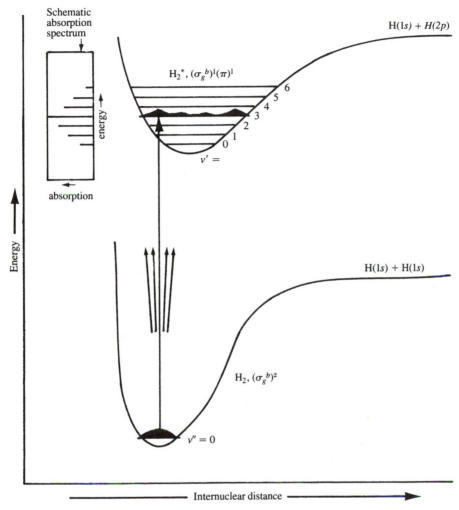

Figure 4-18 Relationship between the potential energy curves for H_2 and H_2^*. A schematic absorption spectrum is also indicated next to the H_2^* curve.

not one absorption but several absorptions will be exhibited, due to the vibrational levels in the excited state. Not all of these transitions will be of equal intensity. The transition of maximum intensity will occur to the level for which there is a maximum of *overlap* between the vibrational wave functions of the ground- and excited-state energy levels. Since the ground-state vibrational energy level has a maximum probability at the center of the bond, this will always be a transition that occurs vertically from that position. For the example shown in Figure 4-18, the maximum intensity band

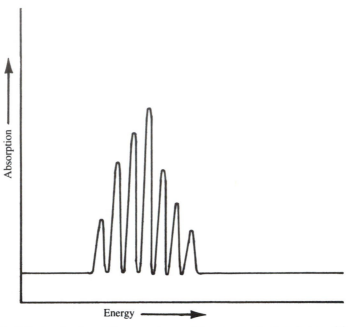

Figure 4-19 Schematic absorption spectrum of the $\sigma \rightarrow \pi$ electronic transition for H_2.

will occur for the transition $v'' = 0 \rightarrow v' = 3$. The other transitions will have lesser intensity. A schematic representation for this absorption spectrum also is indicated in Figure 4-18; the spectrum is sketched more completely in Figure 4-19. Each of the transitions is referred to as a *Franck-Condon transition,* and the relative intensities of the peaks is called the *Franck-Condon distribution.* The molar extinction coefficient, ε, can be calculated from the intensity of the peaks in Figure 4-19. Since this $\sigma \rightarrow \pi$ transition is "allowed," the corresponding ε value will be large.

4-5 Photoelectron Spectroscopy:
An Experimental Method of Studying Molecular Orbitals

The discussion in Section 4-3 showed that the molecular-orbital theory for simple homonuclear diatomic molecules agrees well with experimental bond energies and bond lengths. In Section 4-4 we showed how molecular spectroscopy can be used to determine bond properties.

In this section we will see that further evidence for the vailidity of the MO theory is obtained from another type of molecular spectroscopy called *photoelectron spectroscopy* (PES). This experimental technique can measure the energy required to eject electrons from each of the occupied molecular orbitals. The name of this technique indicates how it works: if monochromatic photons of sufficient energy, $h\nu$, are impinged on a gaseous sample, electrons with kinetic energy $\frac{1}{2}mv^2$ are ejected according to the Einstein photoelectric law (Section 1-6):

$$M + h\nu \rightarrow M^+ + e^-$$

$$h\nu = IE + \tfrac{1}{2}mv^2 \tag{4-18}$$

When PES is applied to molecules or atoms, only M^+ is produced, and M^{2+}, M^{3+}, . . . are not usually observed. In the experiment, the spectrometer detects the number of emitted electrons and their kinetic energies. Since $h\nu$ is a known constant, the IE can be calculated from Equation 4-18.

A schematic diagram of a photoelectron spectrometer is shown in Figure 4-20. A typical photon source for the study of valence electrons is the so-

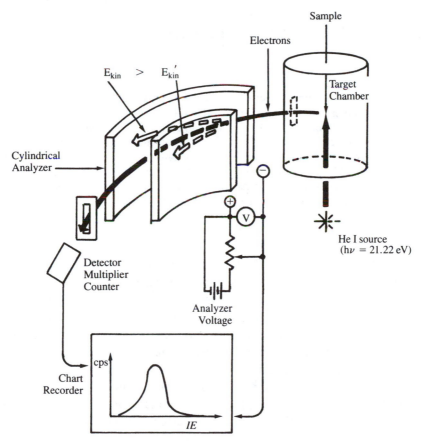

Figure 4-20 Schematic diagram of a photoelectron spectrometer. Adapted with permission from H. Bock and P.D. Mollère, *J. Chem. Educ.* 51: 506 (1974).

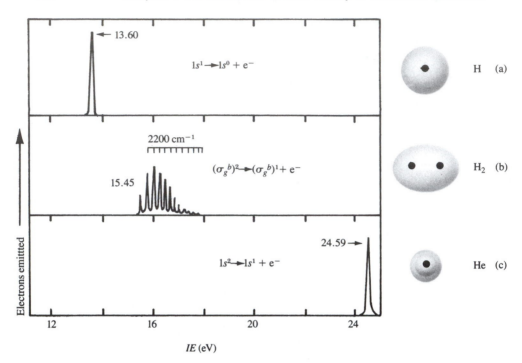

Figure 4-21 Schematic photoelectron spectra of (a) hydrogen atoms, (b) H_2 molecules, and (c) helium atoms. The sharp band in the H atom spectrum is due to ionization of the $1s^1$ electron at 13.60 eV. The band in the spectrum of H_2 molecules corresponds to electron ejection from the σ_g^b molecular orbital. The sharp peaks superimposed on the band are a result of vibrations of the ion H_2^+. This vibrational energy is about 2200 cm^{-1}. The sharp band at 24.59 eV in the He spectrum is due to ionization of one of the electrons from the $1s^2$ electron configuration. Adapted with permission from W.C. Price, *Endeavour* 26: 75 (1967).

called HeI source ($h\nu = 21.22$ eV). The photons strike the sample in the target chamber and a portion of the ejected electrons is passed through the cylindrical analyzer. The voltage on the analyzer plates is scanned in order to focus the electrons of different kinetic energy onto the detector. From the detector the signal is amplified and displayed on the chart recorder as "electrons emitted" per unit time (counts per second, cps) on the ordinate and ionization energy on the abscissa.

A schematic spectrum for a sample of H atoms is shown in Figure 4-21(a). Only one sharp peak (band) is present, which corresponds to the ionization energy of the $1s^1$ electron of the hydrogen atom at 13.60 eV:

$$H + h\nu \rightarrow H^+ + e^-$$

The ionization energy of the H atom is equal to the negative of the orbital energy as determined by the Bohr theory and the Schrödinger equation:

$$E = \frac{-k}{n^2} = -13.60 \text{ eV}$$

The simple statement that $IE_n = -E_n$ (Equation 1-11) is exactly true for one-electron atoms. For multielectron atoms and molecules, *Koopmans' theorem* states that the equation $IE_n = -E_n$ is still true, provided the orbitals in the ion M⁺ are unchanged from those of the neutral molecule or atom M. This statement is often called the "frozen orbital" approximation. A second approximation involved in Koopmans' theorem states that the electron correlation in the ion is unchanged from that in the neutral molecule. This assumption cannot be exactly true. Because the ion has one less electron than the neutral molecule, the amount of electron correlation in the ion will always be less than that in the molecule, even if the orbitals are frozen. Although the approximations involved in Koopmans' theorem show that we cannot *exactly* equate the ionization energy with the negative of the orbital energy, it is still a useful theorem in multielectron atoms and molecules. This is because the errors in the two approximations tend to cancel one another, thereby making the error in Koopmans' theorem relatively small.

The photoelectron spectrum of H_2 molecules is shown in Figure 4-21(b). Only one band is present, beginning at 15.45 eV and extending to nearly 18 eV with several regularly spaced peaks superimposed on it. This band corresponds to ionization of the $(\sigma_g^b)^2$ electrons of H_2. Because the two electrons are indistinguishable, it makes no difference which electron is ionized in forming H_2^+:

$$H_2(\sigma_g^b)^2 + h\nu \rightarrow H_2^+(\sigma_g^b)^1 + e^-$$

Notice that the $(\sigma_g^b)^2$ electrons of H_2 are more stable than the $1s^1$ electron of H by $15.45 - 13.60 = 1.85$ eV, which is in agreement with the prediction of Figure 4-8.

The sharp peaks superimposed on the band in Figure 4-21(b) are a consequence of the vibrational motion of the nuclei relative to each other. The occurrence of this "fine structure" can be understood in the same way as was discussed for the $H_2 \rightarrow H_2^*$ electronic transition (Figures 4-18 and 4-19). In fact, ionization is just a special type of electronic transition in which the electron is completely ejected from the molecule rather than promoted to an excited level. Figure 4-22 illustrates the potential energy curve of H_2 relative to H_2^+. As discussed previously, several vibrational

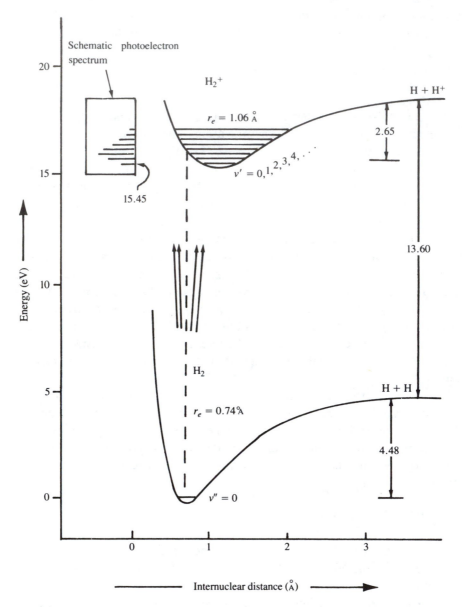

Figure 4-22 Energy curves for H_2 and H_2^+. The depth of each curve corresponds to the bond dissociation energy for that species. The energy difference between the minimum of each curve corresponds to the *IE* of H_2 (15.45 eV). The straight lines drawn "within" the H_2^+ curve illustrate the vibrational energy states ($v' = 0$, 1, 2, . . .) in which H_2^+ can be formed. A schematic photoelectron spectrum is sketched alongside the energy curve of H_2^+. The bond energies of H_2 (4.48 eV = 103 kcal mole^{-1}) and H_2^+ (2.65 eV = 61 kcal mole^{-1}) are also depicted. Adapted with permission from W.C. Price, *Endeavour* 26: 75 (1967).

energy levels exist "within" the potential energy curves of H_2 and H_2^+. Most molecules in their molecular ground state exist only in the lowest vibrational energy level, $v'' = 0$; hence we illustrate only this level for H_2. However, when H_2^+ is formed from H_2, it can be formed in several vibrational energy states, $v' = 0, 1, 2, 3, \ldots$, which are responsible for the fine structure of Figure 4-21(b).

The first peak in the photoelectron spectrum of H_2 occurs at 15.45 eV; this corresponds to ionization from the $v'' = 0$ vibrational energy level of H_2 to the $v' = 0$ vibrational energy level of H_2^+. This is called the *adiabatic* ionization energy. The most intense peak, however, is called the *vertical* ionization energy, because it corresponds to the ionization event in which H_2 and H_2^+ have the same bond length. The relative intensities of each peak within the band are determined by the overlap of the H_2 and H_2^+ vibrational wave functions. A detailed analysis of the relative intensities can show us the change in bond length that occurs upon ionization.

The energy difference between the peaks shown in Figure 4-21(b) is equal to $h\nu$ (see Equation 4-16) in which ν is the frequency of vibration in H_2^+. Vibrational energies are usually expressed in wave numbers, and for H_2^+ the measured value is about 2200 cm^{-1}. This shows that the H_2^+ bond is weaker than that of H_2, which exhibits a vibrational energy of about 4200 cm^{-1} (as determined by molecular spectroscopy, Section 4-4).

A third species that has been studied by photoelectron spectroscopy is He atoms; the spectrum is presented in Figure 4-21(c). The spectrum exhibits only one sharp band at 24.59 eV, which is due to the ionization process

$$He(1s^2) + h\nu \rightarrow He^+(1s^1) + e^-$$

We can consider the He atom to be formed by bringing the two protons of H_2 together. The $(\sigma_g^b)^2$ electrons of H_2 are converted into the $1s^2$ electrons of He in this hypothetical transformation. The experimental ionization energies indicate that the two electrons experience a greater attraction to the protons in He compared to H_2 by $24.59 - 15.45 = 9.14$ eV.

4-6 Molecules with s and p Valence Atomic Orbitals

Now we will discuss diatomic molecules that have $2s$ and $2p$ valence atomic orbitals. Our purpose is to determine which orbitals are involved in electron-pair bonds in a molecule such as N_2. We first need to specify a coordinate system for the general homonuclear diatomic molecule A_2, because the $2p$

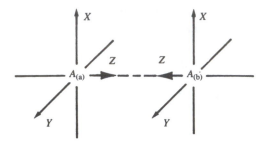

Figure 4-23 Coordinate system for an A_2 molecule.

orbitals have directional properties. The Z axis customarily is assigned to be the *unique molecular axis*, as shown in Figure 4-23. We will continue to follow the method of taking linear (i.e., additive and subtractive) combinations of appropriate atomic orbitals. We know from atomic structure that $2s$ orbitals have less energy than $2p$ orbitals have. In Figure 4-24 we see that there are three $2p$ orbitals of equal energy and one $2s$ orbital of less energy for each atom. We want to find the relative energies of the molecular orbitals for this system and to show how we describe the electronic structures for diatomic molecules involving $2s$ and $2p$ valence orbitals.

Sigma Orbitals

First consider the atomic $2s$ orbitals. The two molecular orbitals derived from $2s$ orbitals are similar to the H_2 molecular orbitals. The normalized wave functions are

$$\sigma_g^b(s) = \frac{1}{\sqrt{2}} (2s_a + 2s_b)$$

$$\sigma_u^*(s) = \frac{1}{\sqrt{2}} (2s_a - 2s_b)$$

The $\sigma_g^b(s)$ and $\sigma_u^*(s)$ molecular orbitals are shown at the top of Figure 4-25. The combination $2s_a + 2s_b$ concentrates electron density between the nuclei and is a bonding orbital. The combination $2s_a - 2s_b$ produces the relatively high-energy orbital that has a nodal plane in which there is no probability of finding the electron. Electron density is forced out of the bonding region, thus this is an antibonding orbital.

The three $2p$ orbitals are directed along the coordinate axes X, Y, and Z. As shown in Figure 4-23, the two sets of corresponding X and Y axes

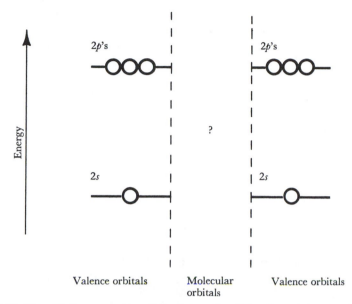

Figure 4-24 The relative energies of the two sets of the $2s$ and $2p$ atomic orbitals of elements in the second row of the periodic table.

are parallel and the Z axis is common to both nuclei. Thus there are two different types of p orbitals in a diatomic molecule. The two $2p_z$ orbitals are directed toward each other and overlap along the Z axis (Figure 4-25). The other p orbitals are not directed toward each other in this way. The two $2p_x$ and the two $2p_y$ orbitals do not overlap along the Z axis; rather they overlap above and below it, as shown in part (b) of the figure.*

The normalized wave functions for the z_a and z_b orbitals are

$$\sigma_g^b(z) = \frac{1}{\sqrt{2}}(z_a + z_b)$$

and

$$\sigma_u^*(z) = \frac{1}{\sqrt{2}}(z_a - z_b)$$

Notice in Figure 4-25 that the $\sigma_g^b(z)$ and $\sigma_u^*(z)$ orbitals are symmetric around the Z axis. For the bonding molecular orbital, electron density is greater in the internuclear region, but the antibonding molecular orbital

---

*In the following discussion, we abbreviate $2p_{z_a}$ as z_a, $2p_{z_b}$ as z_b, $2p_{x_a}$ as x_a, and so on.

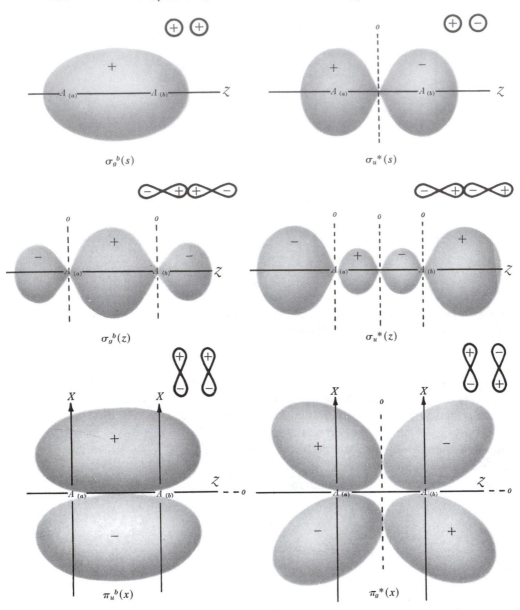

Figure 4-25 The six different kinds of molecular orbitals formed from the s, p_x, p_y, and p_z orbitals of two equivalent atoms in a diatomic molecule. The line drawn through the two nuclei is chosen as the Z axis. Plus and minus signs represent the signs of the wave function, not electrical charge. The atomic orbitals from which these molecular orbitals are obtained are shown, with their appropriate signs, at the upper right of each molecular orbital. The atomic orbitals used are s (top row), p_z (middle row), and p_x or the equivalent p_y (bottom row). Bonding orbitals are in the left column; antibonding orbitals are in the right column. Dashed lines designated "0" represent nodal planes of zero electron density.

reduces electron density in the internuclear region. The antibonding $z_a - z_b$ combination has a nodal plane halfway between the nuclei, just as $s_a - s_b$ has.

Pi Orbitals

Now we shall investigate the molecular orbitals formed from the $2p_x$ and $2p_y$ orbitals. The normalized wave functions are

$$\pi_u{}^b(x) = \frac{1}{\sqrt{2}}(x_a + x_b)$$

$$\pi_u{}^b(y) = \frac{1}{\sqrt{2}}(y_a + y_b)$$

$$\pi_g{}^*(x) = \frac{1}{\sqrt{2}}(x_a - x_b)$$

$$\pi_g{}^*(y) = \frac{1}{\sqrt{2}}(y_a - y_b)$$

Schematic representations of the $x_a + x_b$ and $x_a - x_b$ molecular orbitals are shown in Figure 4-25. The combination $x_a + x_b$ is a new type of molecular orbital. Since p orbitals have a node at the nucleus, the molecular orbital $x_a + x_b$ has a nodal plane that contains the internuclear line. Thus if we rotate the molecular orbital $x_a + x_b$ by 180° it changes sign, thereby becoming $-x_a - x_b$ (Figure 4-26). This type of molecular orbital is called a *pi* or π

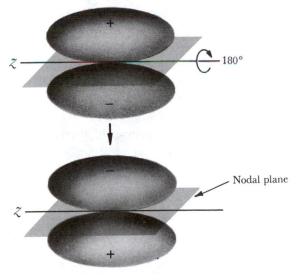

Figure 4-26 The overlap of two $2p_x$ orbitals gives rise to a π molecular orbital. A π molecular orbital changes sign when rotated 180° around the internuclear axis, but the electron distribution remains unchanged.

orbital. Pi molecular orbitals originate from combinations of parallel p orbitals that overlap above and below the internuclear line. The $x_a + x_b$ orbital concentrates electron density in the bonding region; thus it is a π bonding or π^b orbital. If we invert the π^b orbital through the center of symmetry we see that a sign change also occurs, just as for rotation by 180°. Therefore the π^b orbital is *ungerade* and we will designate it π_u^b. To complete the designation we say it is $\pi_u^b(x)$, that is, a combination of the two $2p_x$ atomic orbitals. It should be clear that an equivalent $y_a + y_b$ orbital exists, which we call $\pi_u^b(y)$. Thus there are two bonding π combinations that have the same shape and the same energy, and their orientations in space are mutually perpendicular.

Now let us consider the combinations $x_a - x_b$ and $y_a - y_b$, which have a nodal plane containing the internuclear axis. They are π antibonding orbitals (π^*) because they have reduced electron density (and an additional nodal plane perpendicular to the internuclear axis) between the nuclei. They are *gerade,* because they are symmetric with respect to inversion at the center of symmetry.

To summarize, we started with eight valence orbitals (one $2s$ and three $2p$ orbitals on each atom) and constructed eight molecular orbitals: $\sigma_g^b(s)$, $\sigma_u^*(s)$, $\sigma_g^b(z)$, $\sigma_u^*(z)$, $\pi_u^b(x)$, $\pi_u^b(y)$, $\pi_g^*(x)$, and $\pi_g^*(y)$. The relative energies of the molecular orbitals can be obtained from calculations and from experiment. The best experimental technique presently available to measure the energies of the occupied molecular orbitals is photoelectron spectroscopy. Experimental data for estimating relative energies of the unoccupied molecular orbitals can be obtained from absorption and emission spectroscopic measurements.

One of the fundamental features of the quantum theory is the view that orbitals of the same symmetry can mix or interact. This means that of the above eight molecular orbitals, $\sigma_g^b(s)$ can mix with $\sigma_g^b(z)$, and $\sigma_u^*(s)$ can mix with $\sigma_u^*(z)$. In other words, $2s$ and $2p_z$ orbitals can mix. This type of interaction often is called "s–p mixing." For the moment let us neglect such interactions. The resulting molecular-orbital energy-level diagram is shown in Figure 4-27. The expected order of orbital energy is

$$\sigma_g^b(s) < \sigma_u^*(s) < \sigma_g^b(z) < \pi_u^b(x) = \pi_u^b(y) < \pi_g^*(x) = \pi_g^*(y) < \sigma_u^*(z)$$

In the absence of s–p mixing, the $\sigma_g^b(z)$ orbital is predicted to be more stable than $\pi_u^b(x)$ and $\pi_u^b(y)$, since the $2p_z$ orbitals point directly toward each other and overlap more strongly than the π orbitals that result from parallel orbitals.

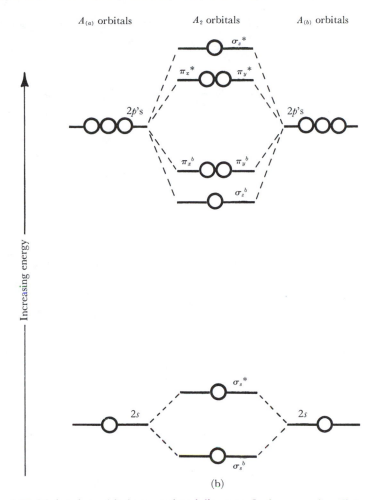

$A_{(a)}$ orbitals A_2 orbitals $A_{(b)}$ orbitals

Increasing energy

(b)

Figure 4-27 Molecular-orbital energy-level diagrams for homonuclear diatomic molecules without *s–p* mixing. This order of energy levels should be observed when the energy separation of the atomic *s* and *p* valence levels is relatively large. Referring to Table 4-4 we see that the large 2*s*–2*p* energy separation in O and F atoms makes the O_2 and F_2 molecules likely candidates for this scheme of energy levels.

s–p Sigma Mixing

The six molecular orbitals sketched in Figure 4-25 do not represent all the possible combinations of orbital overlap. Figure 4-28 shows that an overlap also can exist between s_a, z_b and s_b, z_a. There is no overlap, however, between the 2*s* orbitals and the p_x or p_y orbitals, since regions of equal positive and negative overlap give zero net overlap (Figure 4-28). The s_a, z_b and s_b, z_a overlap results in *s–p* mixing. The possibility of this mixing

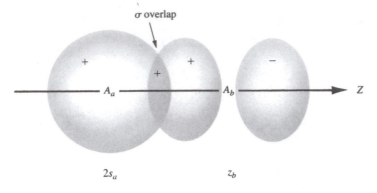

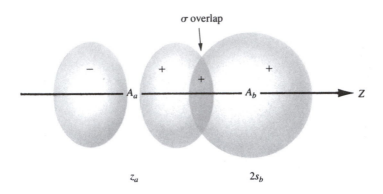

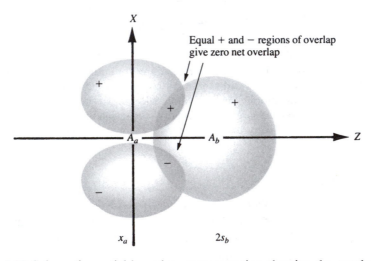

Figure 4-28 Schematic spatial boundary representation showing the overlap of $2s_a$ with z_b and $2s_b$ with z_a. It is this type of overlap that causes s–p mixing on atom A and on atom B. Notice that the p_x and p_y orbitals cannot participate in s–p mixing because the net overlap of either of these orbitals with an s orbital is zero.

TABLE 4-3. COMPARISON OF ORBITAL BONDING CHARACTER IN HOMONUCLEAR
DIATOMIC MOLECULES WITH AND WITHOUT $s–p$ MIXING

Orbital (no s–p mixing)	Bonding Character	Orbital (with s–p mixing)	Orbital Notation	Bonding Character
$\sigma_u^*(z)$	Antibonding	$\sigma_u^*(s) - \sigma_u^*(z)$	$2\sigma_u$	Antibonding
$\pi_g^*(x) = \pi_g^*(y)$	Antibonding	$\pi_g^*(x) = \pi_g^*(y)$	$1\pi_g$	Antibonding
$\pi_u^b(x) = \pi_u^b(y)$	Bonding	$\pi_u^b(x) = \pi_u^b(y)$	$1\pi_u$	Bonding
$\sigma_g^b(z)$	Bonding	$\sigma_g^b(s) - \sigma_g^b(z)$	$2\sigma_g$	Slightly bonding
$\sigma_u^*(s)$	Antibonding	$\sigma_u^*(s) + \sigma_u^*(z)$	$1\sigma_u$	Slightly antibonding
$\sigma_g^b(s)$	Bonding	$\sigma_g^b(s) + \sigma_g^b(z)$	$1\sigma_g$	Bonding

could have been determined by noting that $\sigma_g^b(s)$ and $\sigma_g^b(z)$ molecular orbitals derived in the absence of mixing have the same symmetry, σ_g, and therefore can mix to form $\sigma_g^b(s) + \sigma_g^b(z)$ and $\sigma_g^b(s) - \sigma_g^b(z)$.* The same considerations apply to $\sigma_u^*(s)$ and $\sigma_u^*(z)$ in forming $\sigma_u^*(s) + \sigma_u^*(z)$ and $\sigma_u^*(s) - \sigma_u^*(z)$ orbitals. The effect of this mixing on both the shapes and the energies of the orbitals is shown in Figure 4-29.

The two σ_g^b orbitals interact to produce two new molecular orbitals, one that is more stable and another less stable than the original molecular orbitals. The more stable orbital becomes even more bonding as a result of $s–p$ mixing. The less stable orbital becomes relatively nonbonding, whereas in the absence of $s–p$ mixing it was bonding. For the two σ_u^* orbitals, upon $s–p$ mixing, the lower orbital changes from antibonding to relatively non-bonding and the upper orbital becomes even more antibonding. The energy-level diagrams with and without $s–p$ mixing are identical except that the order of $\sigma_g^b(z)$ and $\pi_u^b(x) = \pi_u^b(y)$ are interchanged. Table 4-3 summarizes the bonding character of the molecular orbitals with and without $s–p$ mixing. Notice that $s–p$ mixing is impossible for the π orbitals, since they do not overlap with the s orbitals (Figure 4-28). Table 4-3 and Figure 4-29 use a simplified MO notation in which the orbitals of each symmetry are numbered sequentially, whatever their specific atomic-orbital composition or their character as bonding, nonbonding, or antibonding. Thus the lowest energy σ_g orbital is designated $1\sigma_g$, the next lowest σ_g orbital is $2\sigma_g$, and so on.

We now must answer the question as to which molecular-orbital diagram in Figure 4-29 is correct, the one that includes $s–p$ mixing or the one that neglects it. The final answer must be determined by experiment. For example, if there were no $s–p$ mixing, a molecule with eight valence electrons would

*The $\sigma_g^b(s)$ and $\sigma_g^b(z)$ orbitals need not mix in the 1:1 ratio shown here. In the more general case we form two molecular orbitals: $c_1\sigma_g^b(s) + c_2\sigma_g^b(z)$ and $c_2\sigma_g^b(s) - c_1\sigma_g^b(z)$, where $c_1 > c_2$.

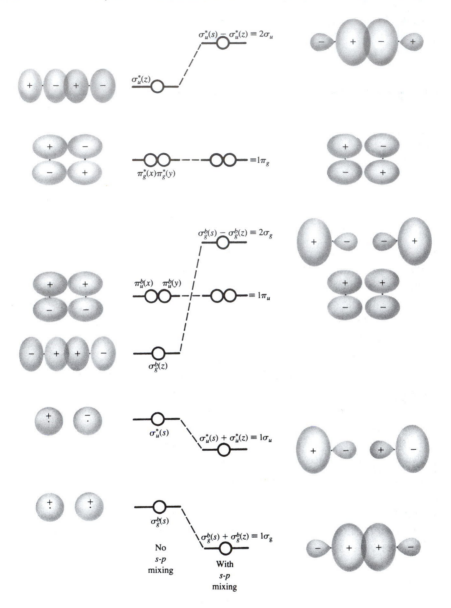

Figure 4-29 The effect of s–p mixing on the shapes and energies of the molecular orbitals for a homonuclear diatomic molecule. Notice that the $1\sigma_u$ and $2\sigma_g$ orbitals become relatively nonbonding, and that the energy order of $1\pi_u$ and $2\sigma_g$ interchanges. The energy-level scheme on the right has been established for most homonuclear diatomic molecules through detailed experiments involving magnetic and spectroscopic properties of molecules. The extent of s–p mixing becomes smaller as the energy separation of the valence s and p atomic levels increases.

have the ground-state electronic configuration $[\sigma_g^b(s)]^2[\sigma_u^*(s)]^2[\sigma_g^b(z)]^2$ $[\pi_u^b(x,y)]^2$, or using the simplified notation $(1\sigma_g)^2(1\sigma_u)^2(2\sigma_g)^2(1\pi_u)^2$. Such a configuration results in two unpaired electrons $[(1\pi_u)^2 = \pi_u(x)^1\pi_u(y)^1]$, thus the molecule would be paramagnetic. In contrast, the ground-state configuration in the presence of s–p mixing would be $(1\sigma_g)^2(1\sigma_u)^2(1\pi_u)^4$ with no unpaired electrons.

It is important not only to examine the experimental results, but also to understand theoretically when to expect s–p mixing. From quantum theory we know that the amount of s–p mixing will depend on two factors:

1. The magnitude of the s_a, z_b (or s_b, z_a) overlap (Figure 4-28): the greater the overlap, the more the s–p mixing.

2. The energy difference between $2s_a$ and z_b (or $2s_b$ and z_a): the smaller the energy difference, the greater the s–p mixing. Because of this factor

TABLE 4-4. VALENCE-ORBITAL IONIZATION ENERGIES (eV)

Atom	1s	2s	2p	3s	3p	4s	4p
H	13.6	—	—	—	—	—	—
He	24.6	—	—	—	—	—	—
Li	—	5.4	—	—	—	—	—
Be	—	9.3	—	—	—	—	—
B	—	14.0	8.3	—	—	—	—
C	—	19.4	10.6	—	—	—	—
N	—	25.6	13.2	—	—	—	—
O	—	32.3	15.8	—	—	—	—
F	—	40.2	18.6	—	—	—	—
Ne	—	48.5	21.6	—	—	—	—
Na	—	—	—	5.1	—	—	—
Mg	—	—	—	7.6	—	—	—
Al	—	—	—	11.3	5.9	—	—
Si	—	—	—	14.9	7.7	—	—
P	—	—	—	18.8	10.1	—	—
S	—	—	—	20.7	11.6	—	—
Cl	—	—	—	25.3	13.7	—	—
Ar	—	—	—	29.2	15.8	—	—
K	—	—	—	—	—	4.3	—
Ca	—	—	—	—	—	6.1	—
Zn	—	—	—	—	—	9.4	—
Ga	—	—	—	—	—	12.6	6.0
Ge	—	—	—	—	—	15.6	7.6
As	—	—	—	—	—	17.6	9.1
Se	—	—	—	—	—	20.8	10.8
Br	—	—	—	—	—	24.1	12.5
Kr	—	—	—	—	—	27.5	14.3

NOTE: The reference zero point is the ionized atom. Thus the corresponding valence-orbital energies are obtained simply by changing the sign of the ionization energy. For example, the 1s orbital energy of atomic H is −13.6 eV.

s–p mixing can be neglected in molecules that have a large energy separation between the valence s and p atomic levels, as shown in Figure 4-27. The valence-orbital ionization energies ($VOIEs$) for the atoms hydrogen through krypton are given in Table 4-4. The corresponding energies of the valence-electron orbitals, which we use to construct MO diagrams, are obtained simply by taking the negative of the $VOIEs$. Examination of the $2s$ and $2p$ $VOIE$ values for the second-row atoms shows that for B, C, and N the $2s$–$2p$ energy difference is 12.5 eV or less, whereas for O and F it is 16 eV or greater. Consequently, we expect s–p mixing to be relatively important in the homonuclear diatomic molecules B_2, C_2, and N_2 but negligible for O_2 and F_2.

4-7 Homonuclear Diatomic Molecules

Now we are able to discuss the electronic structures and bonding properties of all the possible homonuclear diatomic molecules of elements in the second row of the periodic table: Li_2, Be_2, B_2, C_2, N_2, O_2, F_2, and Ne_2. The bond lengths and bond energies of some important homonuclear diatomic molecules and ions are listed in Table 4-5. We also shall compare the results predicted by molecular-orbital theory to the photoelectron spectra of N_2, O_2, and F_2.

Lithium

A lithium atom has one $2s$ valence electron. In Li the 2s–2p energy difference is small, and the $\sigma_g^b(s)$ molecular orbital of Li_2 undoubtedly has considerable $2p$ character. The two valence electrons in Li_2 occupy the $\sigma_g^b(s)$ molecular orbital, thereby giving the ground-state electronic configuration $[(\sigma_g^b(s)]^2$ or $(1\sigma_g)^2$. In agreement with theory, experimental measurements show that a lithium molecule has no unpaired electrons. With two electrons in a bonding molecular orbital there is one net bond. The bond length of Li_2 is 2.67 Å, compared with 0.74 Å for H_2. The longer bond in Li_2 is consistent with the larger effective radius of the Li $2s$ orbital. The mutual repulsion of the two $1s$ electron pairs, an interaction not present in H_2, is also partly responsible for the longer bond in Li_2.

The bond energies of H_2 and Li_2 are 103.24 kcal mole^{-1} and 26.3 kcal mole^{-1}, respectively. On the average, the two $\sigma_g^b(s)$ electrons are much farther from the shielded nuclei in Li_2 than from the nuclei in H_2. At the longer distances the potential energy due to electron-nucleus attractions is smaller; thus the two $\sigma_g^b(2s_a + 2s_b)$ electrons do not bind Li_2 as strongly as do the two $\sigma_g^b(1s_a + 1s_b)$ electrons in H_2.

TABLE **4-5.** BOND PROPERTIES OF SOME HOMO-
NUCLEAR DIATOMIC MOLECULES AND IONS

Molecule	Bond length (Å)	Bond Dissociation Energy (kcal mole⁻¹)
Ag_2	—	38.7 ± 2.2
As_2	2.288	91.3
Au_2	2.472	53.9 ± 2.2
B_2	1.589	65.5 ± 5
Bi_2	—	46.6 ± 1.5
Br_2	2.2809	45.440 ± 0.003
C_2	1.2425	144
Cl_2	1.988	57.18 ± 0.006
Cl_2^+	1.8917	99.2
Cs_2	—	10.4
Cu_2	2.2195	47.3 ± 2.2
F_2	1.417	37
Ge_2	—	65
H_2	0.74116	103.24
H_2^+	1.06	61.06
He_2^+	1.080	55
I_2	2.6666	35.55
K_2	3.923	11.8
Li_2	2.672	26.3
N_2	1.0976	225.07
N_2^+	1.116	201.28
Na_2	3.078	17.3
O_2	1.20741	117.96
O_2^+	1.1227	—
O_2^-	1.26	93.9
O_2^{2-}	1.49	—
P_2	1.8937	114
Pb_2	—	23
Rb_2	—	11.3
S_2	1.889	100.69
Sb_2	2.21	71.3
Se_2	2.1663	77.6
Si_2	2.246	75
Sn_2	—	46
Te_2	2.5574	62.3

NOTE: The values of bond dissociation energy, *DE*, generally refer
to the $\Delta E°$ for the process $A_2(g) \rightarrow A(g) + A(g)$.

Beryllium

A beryllium atom has the valence electronic structure $2s^2$. The electronic
configuration of Be_2 would be $[\sigma_g^b(s)]^2[\sigma_u^*(s)]^2$ in the absence of s–p
mixing. This configuration gives no net bonds $[(2 - 2)/2] = 0$, suggesting
that Be_2 is an unstable molecule. As was the case with Li, however, the

$2s-2p$ energy difference is small and $s-p$ mixing is important. As a consequence of $s-p$ mixing, the electronic configuration is more appropriately
labeled $(1\sigma_g)^2(1\sigma_u)^2$. According to Table 4-3 and Figure 4-29 the $1\sigma_g$ orbital
is bonding and the $1\sigma_u$ orbital is slightly antibonding. The molecule Be_2
is therefore predicted to have a weak single bond. Accurate quantum-
mechanical calculations predict that there will be a small buildup of electron
density in the internuclear region but that this will be insufficient to offset
the repulsion of the two nuclei. Therefore the molecule is predicted to be
unstable in spite of the formation of a weak single bond. In fact, Be_2 has
been detected experimentally at low temperatures. The bond energy is only
about 1 kcal mole^{-1}, which is too weak for an ordinary chemical bond. Weak
bonds of this nature will be discussed in Section 7-4 under the general topic
of van der Waals forces.

Boron

Atomic boron has the valence electronic configuration $2s^2 2p^1$. For B_2 there
are six valence electrons to assign to molecular orbitals. The experimental
ground-state electronic configuration of B_2 is $(1\sigma_g)^2(1\sigma_u)^2(1\pi_u)^2$, which
corresponds to the ordering of levels shown in Figure 4-29, with the inclusion of $s-p$ mixing. The bonding consists of one σ bond due to $1\sigma_g$, one π
bond due to $1\pi_u$, and some slight σ antibonding due to $1\sigma_u$.

The example of B_2 illustrates that in the molecular-orbital theory there is
no precise method by which bond order (the number of bonds between two
atoms) can be defined. If we discount the σ antibonding character of $1\sigma_u$ we
obtain a double bond for B_2. However, if $1\sigma_u$ is counted as being completely
antibonding, we conclude that B_2 has only a single π bond since the bonding
character of $1\sigma_g$ would be canceled by the antibonding character of $1\sigma_u$,
thereby leaving only the $1\pi_u$ bond. We conclude that the bond order of B_2
is somewhere between one and two, depending on the extent of $s-p$ mixing.

The experimental B—B bond length of 1.59 Å is much shorter than the
Li—Li length of 2.67 Å for which a single bond was proposed. This fact
favors a double-bond representation for B_2, because the bond shortening
of $2.67 - 1.59 = 1.08$ Å is more than that expected as a result of change in
relative atomic radii only (Figure 2-1). Typical B—B single-bond lengths in
other compounds are 1.77 Å, again showing that the B—B bond length is
shortened in B_2. This is a further indication of some double-bond character
in B_2.

The bond energy of B_2 is 65 kcal mole^{-1}, which is considerably larger
than the value of 26 kcal mole^{-1} for Li_2. These data also are consistent with
the bond order of B_2 being greater than that of Li_2.

Carbon

Atomic carbon has the valence elctronic configuration $2s^2 2p^2$. For C_2 there are eight valence electrons, and the experimental ground-state electronic configuration is $(1\sigma_g)^2(1\sigma_u)^2(1\pi_u)^4$, which corresponds to the ordering of levels shown in Figure 4-29, with the inclusion of s–p mixing. The C_2 molecule absorbs light in the visible region of the spectrum, with maximum absorption at approximately 19,300 cm^{-1}. This absorption has been assigned to an electronic transition from the $1\pi_u$ orbital to the $2\sigma_g$ orbital, abbreviated $1\pi_u \rightarrow 2\sigma_g$. The ground- and excited-state configurations can be pictured as

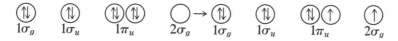

Thus the excited-state configuration $(1\sigma_g)^2(1\sigma_u)^2(1\pi_u)^3(2\sigma_g)^1$, which has two unpaired electrons, differs in energy from the ground-state configuration by only 19,300 cm^{-1}.

Like that of B_2, the bond order of C_2 cannot be predicted with certainty. The bonding consists of one σ bond due to $1\sigma_g$, two π bonds due to $1\pi_u$, and some slight σ antibonding due to $1\sigma_u$. If we discount the σ antibonding character of $1\sigma_u$, we obtain a triple bond for C_2. However, if we consider $1\sigma_u$ as being fully antibonding (as in the absence of s–p mixing, see Figure 4-29), we conclude that C_2 has only a double π bond. In this case the bonding characters of $1\sigma_g$ and $1\sigma_u$ would cancel each other, leaving only the four electrons in the $1\pi_u$ orbital. The bond order of C_2 therefore must be somewhere between two and three, depending on the extent of s–p mixing. The experimentally detemined bond energy of 144 kcal mole^{-1} and the bond length of 1.24 Å are compatible with a bond order of 2–3.

Nitrogen

Atomic nitrogen has the valence configuration $2s^2 2p^3$. The ground-state electronic configuration of N_2 has been determined experimentally to be $(1\sigma_g)^2(1\sigma_u)^2(1\pi_u)^4(2\sigma_g)^2$, which is consistent with the observed diamagnetism of this molecule. If one invokes the molecular-orbital scheme that includes s–p mixing (Figure 4-29), there are three net bonds for the nitrogen molecule. The triple bond in N_2 accounts for its unusual stability, its extraordinarily large bond energy of 225 kcal mole^{-1}, and its very short bond length of 1.10 Å. Recall that the Lewis structure of N_2 is $:N\equiv N:$, with three bonds and two lone pairs. The Lewis structure can now be analyzed in terms of MO theory. The two lone pairs correspond to the $(1\sigma_u)^2$ and $(2\sigma_g)^2$ configurations (Figure 4-29). The three bonds in the Lewis structure correspond to the two π bonds, $(1\pi_u)^4$, and one σ bond, $(1\sigma_g)^2$, proposed from MO theory.

Although one can draw a qualitative one-to-one correpondence between the molecular orbitals and the Lewis electron-dot structure, there is one important difference. An examination of the simple Lewis electron-dot structure implies that both lone-pair electrons have the same energy, while the molecular-orbital theory clearly shows that the $1\sigma_u$ and $2\sigma_u$ molecular orbitals have different energies.

The low-energy electronic transition $1\pi_u \rightarrow 2\sigma_g$ observed for C_2 is not possible in N_2 because $2\sigma_g$ is fully occupied in the ground state. As a result, N_2 does not absorb light in the visible region. The lowest electronic transition in N_2, $2\sigma_g \rightarrow 1\pi_g$, occurs at very high energy (approximately 70,000 cm^{-1}).

Oxygen

Atomic oxygen has the valence electronic configuration $2s^2 2p^4$. The experimentally determined electronic configuration of O_2 is $(1\sigma_g)^2 (1\sigma_u)^2 (2\sigma_g)^2 (1\pi_u)^4 (1\pi_g)^2$. Following Hund's rule, the electrons in $1\pi_g$ have the same spin in the ground state, thereby giving two unpaired electrons in O_2. The oxygen molecule is paramagnetic, which is in agreement with this electronic structural model. In this respect the molecular-orbital theory is superior to the simple Lewis picture $:\ddot{O}=\ddot{O}:$, which does not show that O_2 has two unpaired electrons.

Two net bonds (one σ and one π) are predicted for O_2. The bond energy of O_2 is 118 kcal mole^{-1} and the bond length is 1.21 Å. The change in bond length caused by changing the number of electrons in the $1\pi_g$ level of the O_2 system is instructive. The precise bond length of O_2 is 1.2074 Å. When an electron is added to the $1\pi_g$ level of O_2, thereby forming O_2^-, the bond length increases to 1.26 Å. Addition of a second electron to form O_2^{2-} increases the bond length to 1.49 Å. This is in agreement with the prediction of $1\frac{1}{2}$ bonds for O_2^- and 1 bond for O_2^{2-}.

Fluorine

Atomic fluorine has the valence electronic configuration $2s^2 2p^5$. The electronic configuration of F_2 is $(1\sigma_g)^2 (1\sigma_u)^2 (2\sigma_g)^2 (1\pi_u)^4 (1\pi_g)^4$, with no unpaired electrons and one net bond. This electronic structure is consistent with the diamagnetism of F_2, its 37 kcal mole^{-1} bond energy, and the F—F bond length of 1.42 Å.

Neon

A neon atom has the closed-shell electronic configuration $2s^2 2p^6$. The hypothetical Ne_2 molecule would have the configuration $(1\sigma_g)^2(1\sigma_u)^2(2\sigma_g)^2(1\pi_u)^4(1\pi_g)^4(2\sigma_u)^2$ and zero net bonds. There is no experimental evidence for the existence of a stable diatomic neon molecule.

Summary

The MO theory provides an excellent framework for the correlation of bond lengths and bond energies of diatomic molecules. We already have shown for the series H_2^+, H_2, He_2^+, and He_2 that as the bond order increases the bond length decreases and the bond energy increases. This relationship holds for comparisons of the bonds between atoms of approximately the same effective size. In addition to bond length and bond energy, the force constant provides a third measure of the bond order in a molecule. The quantitative correlations for B_2, C_2, N_2, O_2, and F_2 are

Bond order (number):*
$$B_2(1) < C_2(2) < N_2(3) > O_2(2) > F_2(1).$$

Bond energy (kcal mole^{-1}):
$$B_2(65) < C_2(144) < N_2(225) > O_2(118) > F_2(37).$$

Bond length (Å):
$$B_2(1.59) > C_2(1.24) > N_2(1.10) < O_2(1.21) < F_2(1.42).$$

Force constant (mdyn Å^{-1}):
$$B_2(3.5) < C_2(9.3) < N_2(22.4) > O_2(11.4) > F_2(4.5).$$

Plots of the bond orders, bond energies, bond lengths, and force constants of diatomic molecules of the second-row elements are shown in Figure 4-30.

4-8 Term Symbols for Linear Molecules

Electronic states of a linear molecule may be classified conveniently in terms of orbital and spin angular momentum, a classification analogous to the Russell-Saunders term-symbol scheme for atoms. The unique molecular axis in linear molecules is labeled the Z axis, and we are interested in the component of angular momentum about this axis. The *combining atomic*

---

*The tabulated bond orders are those calculated in the absence of s–p mixing.

TABLE 4-6. QUANTUM NUMBER
ASSIGNMENTS FOR MOLECULAR ORBITALS
IN LINEAR MOLECULES

Molecular Orbitals	m_l	λ	Atomic Orbitals
σ	0	0	s, p_z, d_{z^2}
π	± 1	1	p_x, p_y, d_{xz}, d_{yz}
δ	± 2	2	d_{xy}, $d_{x^2-y^2}$

orbitals in any given molecular orbital have the same m_l value; it is the m_l quantum number that indicates the component of angular momentum about the Z axis. Thus each different type of MO can be assigned a quantum number $\lambda = |m_l|$, as indicated in Table 4-6. The number of nodal planes containing the molecular axis is equal to the quantum number λ. Thus σ, π, and δ orbitals have zero, one, and two such nodal planes, respectively.

The term designations for a system of several electrons is arrived at by first determining the value of M_L, just as we did for atoms:

$$M_L = m_{l_1} + m_{l_2} + \cdots + m_{l_n}$$

We then apply the relationship $\Lambda = |M_L|$ and write the term symbol in the form

$$^{2S+1}\Lambda$$

in which S has the same significance as for atoms. The Λ state abbreviations correspond to the orbital designation λ. That is,

$$\Lambda = 0 \quad 1 \quad 2 \quad 3 \quad \cdots$$
$$\quad\quad \updownarrow \quad \updownarrow \quad \updownarrow \quad \updownarrow$$
$$\quad\quad \Sigma \quad \Pi \quad \Delta \quad \Phi \quad \cdots$$

In molecules possessing a center of symmetry, the g or u character of the term is indicated by an additional right subscript. This symmetry is easy to determine because for any two-wave functions the following "direct products" apply:

$$g \times g = g$$

$$g \times u = u$$

$$u \times u = g$$

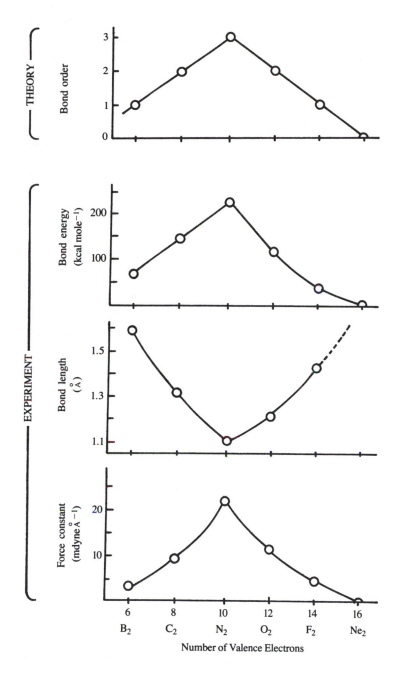

Figure 4-30 Trends in bond properties and predicted bond orders in first-row homonuclear diatomic molecules. Adapted with permission from G. C. Pimentel and R. D. Spratley, *Understanding Chemistry,* Holden-Day, Inc., San Francisco, 1971, p. 572.

Sigma states also must be distinguished as Σ^+ or Σ^- depending upon whether the wave function is symmetric or antisymmetric with respect to reflection in any plane containing the molecular axis. We shall not go into the details of assigning this $\pm$ symmetry. Suffice it to say that most wave functions give Σ^+ states. A Σ^- state can arise when two electrons with parallel spins reside in π or δ orbitals. For parallel spins, the two electrons reside separately in the degenerate pair of orbitals [for example, $(\pi_{+1})^1(\pi_{-1})^1$, or alternatively, $(\pi_x)^1(\pi_y)^1$]. In this case $\Lambda = 0$ is attained by cancellation of the orbital angular momenta of the individual π electrons. Consequently, the electron configurations π^2 or δ^2 always give rise to a $^3\Sigma^-$ state, along with other singlet states (see Example 4-2). These are the only Σ^- states that we shall encounter.

If spin-orbit coupling is important, the value of the total angular momentum, Ω, is sometimes also indicated as a right subscript:

$$^{2S+1}\Lambda_\Omega$$

The possible values of Ω are determined in the same way as the J values for atoms:

$$\Omega = \Lambda + S, \Lambda + S - 1, \ldots, |\Lambda - S|$$

The value of Ω seldom is given for Σ states or singlets, for which Ω simply is equal to S or Λ, respectively.

We shall give three examples to illustrate the procedure for determining the term symbols for linear molecules.

Example 4-1

Find the ground-state term of H_2. We proceed as follows:

1. Find M_L: The two electrons are placed in the $1\sigma_g$ MO, thereby giving the $(1\sigma_g)^2$ configuration. This is the most stable state of H_2. The MO is σ type, so each electron has $m_l = 0$. Then

$$M_L = m_{l_1} + m_{l_2} = 0 + 0 = 0$$

and the state is Σ.

2. Find M_S: Because both electrons have $m_l = 0$, they must have different m_s values (the Pauli principle). Thus,

$$M_S = m_{s_1} + m_{s_2} = \left(+\frac{1}{2}\right) + \left(-\frac{1}{2}\right) = 0$$

Because $M_S = 0$ we must have $S = 0$.

3. Determine the symmetry with respect to the inversion center. Because both electrons are in *gerade* orbitals we have $g \times g = g$. The correct term symbol therefore is $^1\Sigma_g^+$.

From the result for H_2, you may infer that filled molecular orbitals always give $M_L = 0$ and $M_S = 0$. Indeed this is so, because in filled orbitals every positive m_l value is matched with a canceling negative m_l value. The same is true for the m_s values; they come in $+\frac{1}{2}$, $-\frac{1}{2}$ pairs in filled orbitals. This information eliminates considerable work in arriving at the term symbols for states of molecules in which there are many electrons, because most of the electrons are paired in different molecular orbitals.

Example 4-2

Find the ground-state term of O_2. The electronic configuration of O_2 is $(1\sigma_g)^2(1\sigma_u)^2(2\sigma_g)^2(1\pi_u)^4(1\pi_g)^2$. All the orbitals are filled up to $1\pi_g$ and consequently give $M_L = 0$. The two electrons in $1\pi_g$ can be arranged as shown in Table 4-7.

There is a term with $M_L = +2, -2 (\Lambda = 2)$, and $M_S = 0$ $(S = 0)$; the term designation is $^1\Delta_g$. There is a term with $M_L = 0$ $(\Lambda = 0)$ and $M_S = +1, 0, -1$ $(S = 1)$; the term designation is $^3\Sigma_g^-$. This leaves one microstate unaccounted for — the one with $M_L = 0$ $(\Lambda = 0)$ and $M_S = 0$ $(S = 0)$; thus there is a $^1\Sigma_g^+$ term.

The ground state must be either $^1\Delta$, $^3\Sigma_g^-$, or $^1\Sigma_g^+$. According to Hund's first rule (Section 1-15), the ground state has the highest spin multiplicity; the ground state therefore is $^3\Sigma_g^-$. Spectroscopic evidence confirms the $^3\Sigma_g^-$ ground state for O_2.

TABLE 4-7. M_L, M_S VALUES FOR EXAMPLE 4-2

π_{+1}		↿⇂	↓	↑	↑	↓
π_{-1}	↿⇂		↑	↓	↑	↓
M_S	0	0	0	0	1	−1
M_L	−2	+2	0	0	0	0

Example 4-3

Let us now find the term symbol to be associated with two of the ionized states of N_2.

1. The electronic configuration of the first ionized state of N_2^+ is $(1\sigma_g)^2$ $(1\sigma_u)^2(1\pi_u)^4(2\sigma_g)^1$. All of the orbitals except $2\sigma_g$ are filled, consequently $M_L = 0$ and $M_S = 0$. Therefore, since m_l is zero for a σ electron, we have

$M_L = 0$ and $\Lambda = 0$. With only one unpaired electron, $M_S = \pm\frac{1}{2}$ and $S = \frac{1}{2}$. The full notation then is $^2\Sigma_g{}^+$.

2. The electronic configuration of the second ionized state of $N_2{}^+$ is $(1\sigma_g)^2(1\sigma_u)^2(1\pi_u)^3(2\sigma_g)^2$. The only orbital that is not completely filled is $1\pi_u$. The appropriate term symbol is more readily determined in this case from the number of vacancies or "holes" in the $1\pi_u$ orbital, which is only one. Therefore $M_L = \pm1$ ($\Lambda = 1$) and $M_S = \pm\frac{1}{2}$ ($S = \frac{1}{2}$), so the term symbol is $^2\Pi_u$. The reader should analyze this example in terms of the three π_u electrons, rather than the single vacancy, to be convinced that the same answer results.

From parts 1 and 2 of Example 4-3 it should be clear that whenever a closed-shell molecule is ionized, the term symbol of the ionized molecule corresponds to the symmetry of the orbital from which electron ejection took place. That is,

Orbital Ionized	Term Symbol
$2\sigma_g$	$^2\Sigma_g{}^+$
$1\pi_u$	$^2\Pi_u$
$1\sigma_u$	$^2\Sigma_u{}^+$

4-9 Photoelectron Spectra of N_2, O_2, and F_2

Now we will illustrate how photoelectron spectroscopy can verify the molecular-orbital schemes for N_2, O_2, and F_2. In Section 4-5 we discussed the application of photoelectron spectroscopy to H, H_2, and He. None of these species contains more than two electrons. Consequently, only one orbital is occupied and only one band appears in the photoelectron spectrum (Figure 4-21). For any multielectron molecule, several orbitals will be occupied. However, this does not unduly complicate the photoelectron spectrum because each occupied molecular orbital gives rise to a single band in the spectrum.

Nitrogen

Let us examine the photoelectron spectrum of N_2 in order to illustrate the close connection between this experiment and molecular-orbital theory. For N_2 the photoelectron spectrum (Figure 4-31) of the valence electrons exhibits four bands that are assigned to the energy-level sequence $(1\sigma_u)^2(1\sigma_u)^2(1\pi_u)^4(2\sigma_g)^2$. In the photoelectron spectrum, each band results from the ionization of a different orbital. These four ionization bands are described as follows:

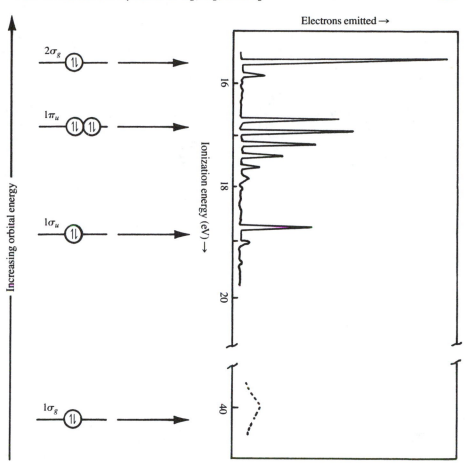

Figure 4-31 Relationship between the molecular-orbital energy-level diagram for N_2 and its photoelectron spectrum. Only the occupied molecular orbitals are shown. Note that the orbital energy scale increases in opposition to the ionization energy. This is because $IE_n = -E_n$ (Equation 1-21). Photoelectron spectrum adapted with permission from J.L. Gardner and J.A.R. Samson, *J. Chem. Phys.* 62: 1447 (1975).

$$N_2 + h\nu \rightarrow N_2^+ + e^-$$
$$(1\sigma_g)^2(1\sigma_u)^2(1\pi_u)^4(2\sigma_g)^2 \rightarrow [(1\sigma_g)^2(1\sigma_u)^2(1\pi_u)^4(2\sigma_g)^1, \, ^2\Sigma_g^+] + e^-$$
$$\rightarrow [(1\sigma_g)^2(1\sigma_u)^2(1\pi_u)^3(2\sigma_g)^2, \, ^2\Pi_u] + e^-$$
$$\rightarrow [(1\sigma_g)^2(1\sigma_u)^1(1\pi_u)^4(2\sigma_g)^2, \, ^2\Sigma_u^+] + e^-$$
$$\rightarrow [(1\sigma_g)^1(1\sigma_u)^2(1\pi_u)^4(2\sigma_g)^2, \, ^2\Sigma_g^+] + e^-$$

The energy level ordering for N_2 has the sequence $1\pi_u < 2\sigma_g$, which is in accord with Figure 4-29 when *s–p* mixing is included. Further evidence for

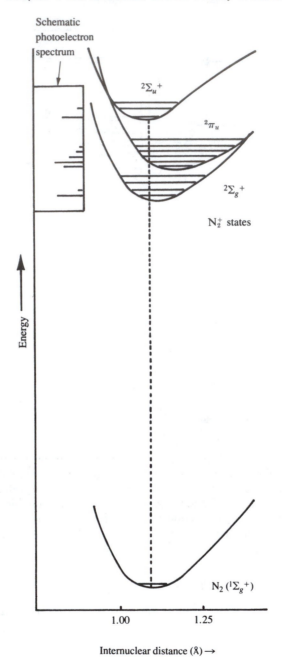

Figure 4-32 Potential energy curves for N_2 and three states of N_2^+. Notice that the equilibrium bond length of the $^2\Pi_u$ state of N_2^+ is considerably longer than that of N_2. This results in the extensive vibrational fine structure shown in the photoelectron spectrum for the band assigned to the $^2\Pi_u$ state. Adapted with permission from P.E. Cade, K.D. Sales, and A.C. Wahl, *J. Chem. Phys.* 44: 1973 (1966).

TABLE 4-8. BOND PROPERTIES OF N_2 AND THREE STATES OF N_2^+

Orbital Ionized	Vibrational Wavenumber (cm^{-1})	Bond Length (Å)	Bond Energy (kcal mole^{-1})
N_2 molecule	2330	1.0976	225.07
$N_2^+(2\sigma_g)$	2175	1.116	201.28
$N_2^+(1\pi_u)$	1873	1.176	—
$N_2^+(1\sigma_u)$	2373	1.075	—

s–p mixing is obtained from a study of the shapes of the bands. The bands assigned to $1\sigma_u$ and $2\sigma_g$ of N_2 are sharp, with little evidence of vibrational fine structure (Figure 4-31). This indicates that when a $1\sigma_u$ or $2\sigma_g$ electron is ionized, the bond length of N_2^+ is little changed from that of N_2; this can be so only if the ionized electron is nonbonding. Figure 4-29 shows that s–p mixing would cause the $1\sigma_u$ and $2\sigma_g$ orbitals to become mainly non-bonding. The $1\pi_u$ orbital is not affected by this s–p mixing, and remains strongly bonding. Experimental evidence for this bonding is provided by the vibrational fine structure on the $1\pi_u$ band. The photoelectron spectrum indicates that the triple bond in N_2 is due to one sigma bond ($1\sigma_g$) and the degenerate π bonds ($1\pi_u$).* This result is in complete agreement with the previously discussed prediction of molecular-orbital theory based on s–p mixing.

Let us examine in more detail the concept that extensive vibrational fine structure results if the electron is ejected from a bonding orbital (for example, $1\pi_u$), whereas there is little evidence of vibrational fine structure if the electron is ejected from a mainly nonbonding orbital (for example, $1\sigma_u$ and $2\sigma_g$). In Figure 4-32 we illustrate the formation of the $^2\Sigma_g^+$, $^2\Pi_u$, and $^2\Sigma_u^+$ ionic states of N_2^+ in terms of the potential energy curves. This diagram illustrates another way of looking at the situation shown in Figure 4-31. Since the N_2^+ bond length in the $^2\Sigma_g^+$ and $^2\Sigma_u^+$ states is little changed from that in N_2, the two bands assigned to these ionic states exhibit little vibrational fine structure. However, the N_2^+ bond length is considerably longer in the $^2\Pi_u$ state, and consequently the band assigned to this ionic state exhibits extensive vibrational structure. A longer bond in the $^2\Pi_u$ state implies that the $1\pi_u$ orbital is bonding, so that electron ejection from this orbital reduces the bond order in N_2 from 3 to $2\frac{1}{2}$ in the $^2\Pi_u$ state of N_2^+.

Spectroscopic measurements of the bond properties of N_2 and N_2^+ are given in Table 4-8. These properties have been obtained from the experi-

*No vibrational structure is evident on the photoelectron band assigned to the $1\sigma_g$ orbital in Figure 4-31. This is a result of poor instrumental resolution at ionization energy greater than 20 eV.

mentally determined potential energy curves shown in Figure 4-32. Notice
that the electronic states with shorter bond lengths exhibit larger vibrational
energies and larger dissociation energies, as expected.

It is interesting to compare the Lewis structure :N≡N: with the interpretation of bonding that we obtain from the photoelectron spectrum. An
analysis of the simple Lewis structure would indicate that the two lone
electron pairs have identical energy. By contrast, the molecular-orbital
theory (Figure 4-29) and the photoelectron spectrum (Figure 4-31) indicate
that the two lone electron pairs correspond predominantly to the $1\sigma_u$ and
$2\sigma_g$ orbitals, which are nondegenerate. This example shows that the concept
of orbital energy cannot be applied to the localized electron-pair bonds that
are represented in the Lewis structure.

Actually, molecular nitrogen presents a well-known example of what
often is called a "breakdown" in Koopmans' theorem. Accurate quantum-
mechanical calculations indicate that the order of orbitals in N_2 is

$$(1\sigma_g)^2(1\sigma_u)^2(2\sigma_g)^2(1\pi_u)^4 \qquad (4\text{-}19)$$

From experiments we know that electrons ionize from the $2\sigma_g$ orbital at
less energy than from the $1\pi_u$ orbital. Consequently, the lowest ionic state
of N_2^+ corresponds to the orbital configuration

$$(1\sigma_g)^2(1\sigma_u)^2(1\pi_u)^4(2\sigma_g)^1 \qquad (4\text{-}20)$$

Applying Koopmans' theorem, we conclude that the $2\sigma_g$ orbital is less
stable than the $1\pi_u$ orbital. When the ordering of orbitals in the ground-
state molecule is deduced to be different from that of the molecule-ion (compare Equations 4-19 and 4-20), we say that there is a breakdown in Koopmans'
theorem. This does not mean that either the calculation or the experiment
is incorrect. What it does mean is that the approximations involved in
Koopmans' theorem are not valid in this particular case. In fact, if calculations are carried out on the molecule-ion N_2^+ and the orbitals are allowed
to "relax" (i.e., vary) compared to those found for N_2, the experimental
and calculated results agree.

Oxygen and Fluorine

In Figure 4-33 we display the photoelectron spectra of O_2 and F_2, along with
that of N_2 for comparison purposes. The photoelectron spectrum of O_2 is
in accord with the energy sequence $(1\sigma_g)^2(1\sigma_u)^2(2\sigma_g)^2(1\pi_u)^4(1\pi_g)^2$. There
are three important features of this spectrum that we wish to examine. First,
although the oxygen atom has a higher Z_{eff} than the nitrogen atom, the first
IE of O_2 is less than that of N_2 because the $(1\pi_g)^2$ electrons are antibonding.

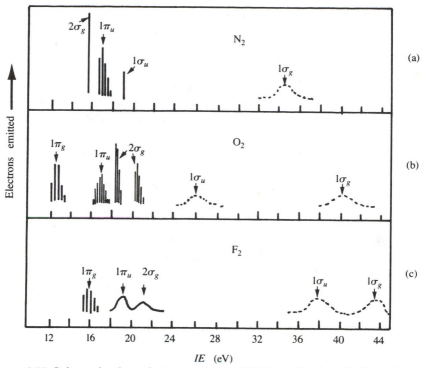

Figure 4-33 Schematic photoelectron spectra of (a) N_2 molecules, (b) O_2 molecules, and (c) F_2 molecules. Adapted with permission from W. C. Price in *Electron Spectroscopy: Theory, Techniques, and Applications,* vol. 1, edited by C.R. Brundle and A.D. Baker, New York: Academic, 1977, p. 151.

Second, the photoelectron spectrum reveals two bands due to ionization of the $2\sigma_g$ electrons. This is understood readily by examining the resulting electron configuration of O_2^+. If a spin-down electron is ionized from the $2\sigma_g$ orbital, the electron configuration of O_2^+ will contain three parallel unpaired electrons. However, if a spin-up electron is ionized, the result will be two spin-up electrons and one spin-down electron. These two electronic states will have different energies, and consequently two bands will appear in the photoelectron spectrum due to $2\sigma_g$ ionization. A pictorial representation of these electron configurations is

A similar double-band effect will occur upon ionization of a $1\pi_u$ electron. Close examination of the photoelectron band near 17 eV, with more detail than is shown in Figure 4-33(b), reveals that it is composed of two electronic states. A third important feature of the O_2 photoelectron spectrum, as compared to that of N_2, is that the sequence of the $1\pi_u$ and $2\sigma_g$ levels is interchanged. This indicates that $s–p$ mixing is not as important for O_2 as for N_2, which is not surprising in view of the larger $2s–2p$ energy separation for atomic oxygen as compared to atomic nitrogen (Table 4-4).

The photoelectron spectrum of F_2 [Figure 4-33(c)] exhibits the same orbital energy sequence as O_2 and corresponds to the molecular-orbital diagram with little $s–p$ mixing (Figure 4-27). The spectrum of F_2 is simpler than that of O_2, since F_2 has no unpaired electrons and therefore each ionization results in only one electronic state. Notice that each ionization energy of F_2 is larger than the corresponding ionization energy of O_2. This is due to the higher Z_{eff} of F, as compared to that of O.

The Photoelectron Spectra of N_2, O_2, and F_2 Core Electrons

Up to now we have been interested only in the molecular orbitals formed from the valence electrons and valence orbitals ($2s$ and $2p$). It should be obvious that in addition to the valence electrons, each of the molecules N_2, O_2, and F_2 has a total of four core electrons, $1s^2$ from each of the constituent atoms. These electrons can be ionized by using a higher-energy photon source. The molecular orbitals formed from the $1s$ atomic orbitals have the same form as those discussed for H_2^+ and H_2, namely

$$\psi(\sigma_g{}^b) = 1/\sqrt{2}(1s_a + 1s_b)$$
$$\psi(\sigma_u{}^*) = 1/\sqrt{2}(1s_a - 1s_b)$$

The energy *difference* between $\psi(\sigma_g{}^b)$ and $\psi(\sigma_u{}^*)$ in the second-row molecules N_2, O_2, and F_2 is negligible compared to that in H_2. In other words, the exchange integral β (Equation 4-12) is negligible for N_2, O_2, and F_2. We readily can understand that there will be almost no interaction between the core orbitals because of the small effective radius of the $1s^2$ charge distribution on each atom. Consequently the four electrons occupy the $\sigma_g{}^b(1s)$ and $\sigma_u{}^*(1s)$ molecular orbitals; however, only one band is observed in the photoelectron spectrum of the core electrons, because $\sigma_g{}^b(1s)$ and $\sigma_u{}^*(1s)$ have essentially the same energy. These core electrons are observed at 409.9 eV in N_2, 543.1 eV in O_2, and 696.70 eV in F_2. Note the increase in ionization energy throughout the series, due to an increase in Z_{eff}.

4-10 Homonuclear Diatomic Molecules of the Transition Elements

In their normal states, the transition elements exist as solids. However, at high temperatures it is possible to generate the diatomic M_2 molecules, just as C_2 can be generated by heating graphite. All of the first-row homonuclear diatomic transition element molecules have been observed experimentally. Several of the second- and third-row molecules also are known; for example, Y_2, Nb_2, Mo_2, Pd_2, Ag_2, La_2, and Au_2. The bond properties of Cu_2, Ag_2, and Au_2 are given in Table 4-5.

The molecular-orbital theory of the homonuclear diatomic transition metal molecules is more complicated than that of the second-row diatomic molecules discussed in Section 4-7. The complication arises from the introduction of the nd orbitals, in addition to the $(n+1)s$ and $(n+1)p$ orbitals. Just as for the second-row diatomic molecules, there is no one molecular-orbital scheme that can apply to all of the M_2 molecules. In particular, the relative energy of the $(n+1)s$, $(n+1)p$, and nd orbitals will vary, depending upon the element under consideration. Generally, on the left side of the transition series the energy level sequence is $(n+1)s \simeq nd < (n+1)p$, whereas on the right side the sequence is $nd < (n+1)s < (n+1)p$. Let us assume the latter energy-level sequence in order to simplify our discussion. Thus in Figure 4-34 we see that for a first-row transition element there are three $4p$ orbitals of equal energy, one $4s$ orbital of less energy, and five $3d$ orbitals lowest in energy for each atom. We want to find the relative energies of the molecular orbitals for this system and to show how we build the electronic structures for these M_2 molecules. We will further simplify the discussion by considering only $3d$–$3d$, $4s$–$4s$, and $4p$–$4p$ interactions. All other orbital mixing (s–p, s–d, and p–d) will be neglected. Obviously, the $4s$–$4s$ and $4p$–$4p$ interactions, to form σ and π molecular orbitals, will be similar to the $2s$–$2s$ and $2p$–$2p$ interactions discussed in Section 4-6 and shown in Figure 4-25.

We now focus our attention on the molecular orbitals formed from the $3d$–$3d$ interactions. Let us designate the line that connects the nuclei as the Z axis. There are two sigma molecular orbitals derived from linear combinations of the $3d_{z^2}$ orbitals:

$$\sigma_g^b(z^2) = 1/\sqrt{2}(z_a^2 + z_b^2)$$
$$\sigma_u^*(z^2) = 1/\sqrt{2}(z_a^2 - z_b^2)$$

The $\sigma_g^b(z^2)$ and $\sigma_u^*(z^2)$ molecular orbitals are shown at the top of Figure 4-35.

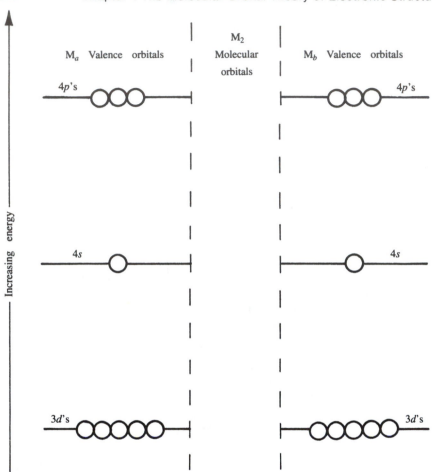

Figure 4-34 The relative energies of two sets of the $3d$, $4s$, and $4p$ atomic orbitals of elements in the first transition-metal series of the periodic table.

Now we shall investigate the pi molecular orbitals formed from the $3d_{xz}$ and $3d_{yz}$ orbitals. The normalized wave functions are

$$\pi_u^b(xz) = 1/\sqrt{2}(xz_a + xz_b)$$
$$\pi_u^b(yz) = 1/\sqrt{2}(yz_a + yz_b)$$
$$\pi_g^*(xz) = 1/\sqrt{2}(xz_a - xz_b)$$
$$\pi_g^*(yz) = 1/\sqrt{2}(yz_a - yz_b)$$

Schematic representation of these molecular orbitals are also shown in Figure 4-35. Obviously, these are all π molecular orbitals because they have a nodal plane that contains the internuclear line.

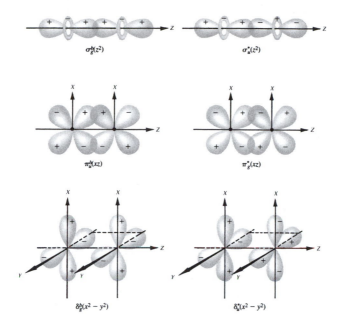

Figure 4-35 The six different kinds of molecular orbitals formed from the $3d$ orbitals of two equivalent atoms in a diatomic molecule. The line drawn through the two nuclei is chosen as the Z axis. Bonding orbitals are in the left column; antibonding orbitals are in the right column. Corresponding to the $\pi(xz)$ orbitals, there is an equivalent set of $\pi(yz)$ orbitals. Likewise, $\delta(xy)$ is equivalent to $\delta(x^2 - y^2)$.

Finally, we examine the molecular orbitals formed from the $3d_{xy}$ and $3d_{x^2 - y^2}$ orbitals. The normalized wave functions are

$$\delta_g{}^b(xy) = 1/\sqrt{2}(xy_a + xy_b)$$
$$\delta_g{}^b(x^2 - y^2) = 1/\sqrt{2}[(x^2 - y^2)_a + (x^2 - y^2)_b]$$
$$\delta_u{}^*(xy) = 1/\sqrt{2}(xy_a - xy_b)$$
$$\delta_u{}^*(x^2 - y^2) = 1/\sqrt{2}[(x^2 - y^2)_a - (x^2 - y^2)_b].$$

Schematic representations of the $xy_a + xy_b$ and $xy_a - xy_b$ molecular orbitals are shown in Figure 4-35; they represent a new type of molecular orbital since there are *two* nodal planes that contain the internuclear line. This type of molecular orbital is called a *delta* or δ *orbital*. Delta molecular orbitals originate from combinations of parallel d orbitals just as pi orbitals result from combinations of parallel p orbitals.

To summarize the $3d$–$3d$ interactions, we started with ten valence orbitals (five on each atom) and constructed ten molecular orbitals: $\sigma_g{}^b(z^2), \sigma_u{}^*(z^2),$

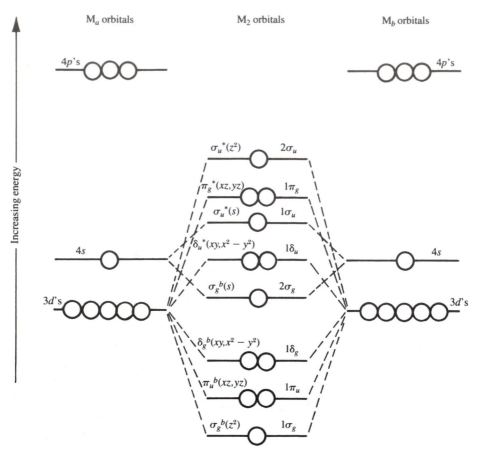

Figure 4-36 Molecular-orbital energy-level diagram for homonuclear diatomic molecules of the transition elements. Molecular orbitals formed from the $4p$ orbitals are not shown, since they remain unoccupied in these molecules.

$\pi_u{}^b(xz)$, $\pi_u{}^b(yz)$, $\pi_g{}^*(xz)$, $\pi_g{}^*(yz)$, $\delta_g{}^b(xy)$, $\delta_g{}^b(x^2 - y^2)$, $\delta_u{}^*(xy)$, and $\delta_u{}^*(x^2 - y^2)$.

A typical molecular-orbital energy-level diagram is shown in Figure 4-36. Notice that of the molecular orbitals derived from the $3d$ orbitals, the energy level sequence of the bonding combinations is

$$\sigma_g{}^b(z^2) < \pi_u{}^b(xz) = \pi_u{}^b(yz) < \delta_g{}^b(xy) = \delta_g{}^b(x^2 - y^2)$$

This energy level sequence is expected, since the amount of orbital overlap follows the order $\sigma > \pi > \delta$. As a general rule, the larger the amount of atomic-orbital overlap, the more stable the molecular orbital will be. Figure 4-36 also illustrates that the $\sigma_g{}^b(s)$ and $\sigma_u{}^*(s)$ orbitals are likely to be in the same energy region as the molecular orbitals formed from the $3d$ orbitals. The σ and π molecular orbitals formed from the $4p$–$4p$ interactions are not shown in Figure 4-36, since they are so high in energy as to be unimportant in bonding the M_2 molecules. Now we will apply this molecular-orbital scheme to three molecules: V_2, Nb_2, and Cu_2.

The V_2 Molecule

The vanadium atom has five valence electrons with the electronic configuration $[Ar]3d^34s^2$. Thus the V_2 molecule has ten valence electrons. Calculations indicate that eight of the ten valence electrons remain unpaired in the V_2 molecule, so the resulting electronic configuration is

$$[\sigma_g^b(z^2)]^2 \; [\pi_u^b(xz)]^1 \; [\pi_u^b(yz)]^1 \; [\delta_g^b(xy)]^1 \; [\delta_g^b(x^2-y^2)]^1$$
$$[\sigma_g^b(s)]^1 \; [\delta_u^*(xy)]^1 \; [\delta_u^*(x^2-y^2)]^1 \; [\pi_g^*(xz)]^1$$

If we recognize that the $4s$ and $3d_{z^2}$ orbitals can mix due to their common σ_g^b character, we can write the electronic configuration in the standard nomenclature

$$(1\sigma_g)^2 \; (1\pi_u)^2 \; (1\delta_g)^2 \; (2\sigma_g)^1 \; (1\delta_u)^2 \; (1\pi_g)^1$$

It is difficult to state precisely what the predicted bond order is for V_2. Analysis of this simplified bonding scheme reveals that there are seven bonding and three antibonding electrons. However, calculations indicate that only the $1\sigma_g$ molecular orbital is strongly bonding; we can say only that the bond order is somewhere between one and two.

The Nb_2 Molecule

The Nb_2 molecule is the congener of V_2. Therefore, we once again have ten valence electrons. In contrast to the V_2 molecule, however, calculations indicate that the Nb_2 molecule has no unpaired electrons, so the resulting electronic configuration is $(1\sigma_g)^2(1\pi_u)^4(1\delta_g)^4$. This results in a formal bond order of *five* or a *pentuple* bond! The two δ bonds are expected to be relatively weak due to the small overlap between the δ orbitals. Consequently, we would not expect the bond dissociation energy for Nb_2 to be double or triple that of V_2. In agreement with this prediction, calculations indicate that the bond dissociation energy for Nb_2 is only about one and one-half times larger than that for V_2.

The Cu₂ Molecule

The copper atom has the electronic configuration $[\text{Ar}]3d^{10}4s^1$. Because the $3d$ orbitals are filled, their contribution to the bonding will be negligible in Cu_2. The lack of bonding contribution on the part of the $3d$ orbitals results because for each filled $\sigma_g{}^b$, $\pi_u{}^b$, and $\delta_g{}^b$ bonding molecular orbital, there also will be a filled $\sigma_u{}^*$, $\pi_g{}^*$, and $\delta_u{}^*$ antibonding molecular orbital. Consequently, the $3d^{10}$ electrons are unable to contribute to the bonding in Cu_2, just as the $2p^6$ electrons cannot contribute for Ne_2.

The $3d^{10}$ electrons therefore can be treated as part of the core and we need consider only the $4s^1$ valence electron in our bonding scheme. The resulting electronic configuration for Cu_2 is $[\sigma_g{}^b(s)]^2$, just as it would be for Li_2 or K_2.

4-11 Heteronuclear Diatomic Molecules

We now turn to a detailed molecular-orbital treatment of *heteronuclear* diatomic molecules. For any general heteronuclear molecule, AB, the molecular orbitals can be classified only according to rotational symmetry (σ or π). Inversion symmetry is no longer applicable, because the two halves of the molecule are not identical and hence there is no center of symmetry. We will discuss three molecules — HF, CO, and BF — in order to illustrate the general features of bonding in heteronuclear molecules. In addition, the bond properties of several second-row diatomic molecules will be compared with those of CO and BF, as will that of the first-row transition-metal monoxide, ScO.

Hydrogen Fluoride

When combining valence orbitals of different atoms it is helpful to know the relative energies of the orbitals. These can be estimated as the negative of the valence-orbital ionization energies given in Table 4-4. A hydrogen atom has a $1s$ valence orbital and a fluorine atom has $2s$ and $2p$ valence orbitals. The orbital energies are

Hydrogen: $1s$, -13.6 eV

Fluorine: $2s$, -40.2 eV; $2p$, -18.6 eV

Since the $2p$ orbitals in fluorine are much closer in energy to the $1s$ orbital in hydrogen, we will consider, as a first approximation, only the interaction of the $1s$ orbital of hydrogen with the fluorine $2p$ orbitals, as shown in Figure 4-37. The Z axis corresponds to the internuclear line, and we see

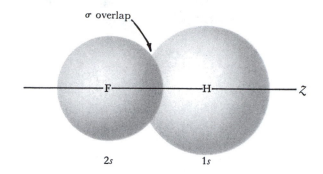

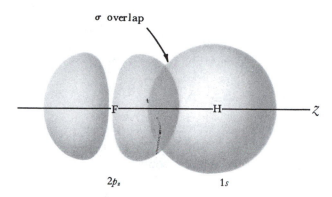

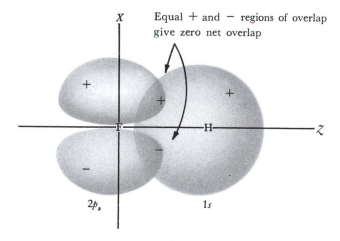

Figure 4-37 Overlap of the hydrogen $1s$ orbital with the valence orbitals of fluorine. The net overlap of a $2p_x$ or $2p_y$ orbital of fluorine with the hydrogen $1s$ orbital is zero, and thus these two p orbitals cannot be used to form molecular orbitals.

that the $1s(H)$ and $2p_z(F)$ orbitals overlap. Therefore the bonding orbital is represented by the combination of these orbitals:

$$\sigma^b = c_1[1s(H)] + c_2[2p_z(F)]$$

In the equation, the coefficients c_1 and c_2 give the relative contributions of the hydrogen $1s$ and the fluorine $2p_z$ orbitals in the σ^b molecular orbital. In this case the combining valence orbitals have different contributions in the molecular orbital, because they have different energies. As explained in Chapter 2, electrons naturally will be "pulled" toward the atom that has the larger electronegativity, which therefore furnishes the lower-energy valence orbital. In the molecular-orbital formulation for HF, c_2 is much greater than c_1 in the bonding orbital, thus contributing significantly more to the lower-energy fluorine $2p_z$ orbital. There are virtually two electrons in the fluorine $2p_z$ orbital in the ground state.

The antibonding orbital in HF can be written

$$\sigma^* = c_3[1s(H)] - c_4[2p_z(F)]$$

The antibonding molecular orbital is not occupied by electrons in HF in the ground state, but it could be occupied in certain excited electronic states. The coefficient c_4 gives the relative contribution of the fluorine $2p_z$ orbital in the antibonding molecular orbital. Electron density in σ^* will be reduced between the nuclei. Furthermore, since the bonding orbital has "used" most of the fluorine $2p_z$ orbital, electron density will be forced mainly into the domain of the hydrogen nucleus. This means that c_4 will be considerably less than c_3; that is, an electron would be associated with the higher-energy $1s(H)$ orbital if it were found in the σ^* molecular orbital.

The fluorine $2p_x$ and $2p_y$ orbitals are suitable for π molecular orbitals. However, atomic hydrogen has only a $1s$ valence orbital, which is involved solely in σ bonding. The $1s(H)$ orbital has zero net overlap with the $2p_x$ and $2p_y$ orbitals (Figure 4-37), so the fluorine $2p_x$ and $2p_y$ orbitals are nonbonding in HF.

The molecular orbitals for HF, and their relative energies, are shown in Figure 4-38. The valence orbitals of fluorine are on the right in the diagram, with the $2p$ energy level above the $2s$ level. On the left, the hydrogen $1s$ energy level is placed higher than the fluorine $2p$ level, in agreement with their known relative energies. The σ^b and σ^* molecular orbitals are in the center. The σ^b molecular orbital is lower in energy than the fluorine $2p_z$ orbital, and the diagram illustrates that σ^b has a large component of the fluorine $2p_z$ orbital and a small component of the hydrogen $1s$ orbital. The σ^* molecular orbital is higher in energy than the hydrogen $1s$ orbital, and

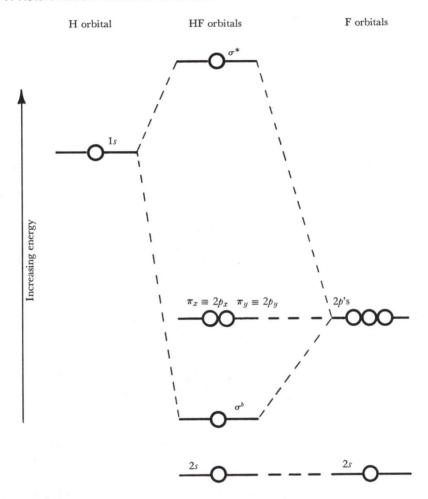

H orbital   HF orbitals   F orbitals

Figure 4-38 Relative energies of atomic and molecular orbitals in HF. The energy of an electron in an atomic hydrogen $1s$ orbital is -13.6 eV (the first ionization energy of H is $+13.6$ eV); the energy of the $2p$ orbitals of F is -18.6 eV (first valence-orbital ionization energy of F is $+18.6$ eV). The diagram is not to scale: The $2s$ orbital is actually much lower in energy than shown.

the diagram shows that σ^* is composed mainly of the hydrogen $1s$ orbital. The fluorine $2p_x$ and $2p_y$ orbitals are shown in the molecular-orbital column as π-type molecular orbitals. They are nonbonding because hydrogen has no p valence orbitals.

There are eight valence electrons to occupy the molecular orbitals for HF (Figure 4-38). Seven are from fluorine $(2s^22p^5)$ and one is from hydrogen $(1s^1)$. Therefore the ground-state electronic structure of HF is

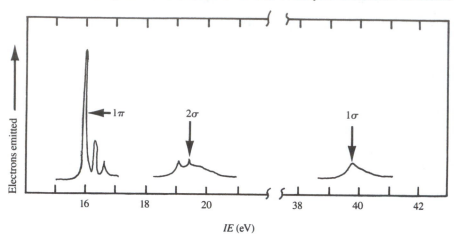

Figure 4-39 Photoelectron spectrum of HF. Adapted with permission from H.J. Lempka, T.R. Passmore, and W.C. Price, *Proc. Roy. Soc.* A304: 53 (1968).

$(2s)^2(\sigma^b)^2(2p_x)^2(2p_y)^2$. In addition to the one σ bond there are three lone pairs, a configuration which corresponds to the Lewis structure (H—$\ddot{\text{F}}$:).

The lowest molecular orbital will not be 100 percent fluorine $2s$, because the fluorine $2s$ and hydrogen $1s$ orbitals have a net overlap (Figure 4-37). To give a completely general description of the electronic structure of HF we can use the standard molecular orbital notation $(1\sigma)^2(2\sigma)^2(1\pi)^4$; the only orbital that contributes significantly to the bonding is 2σ.

Let us compare the molecular-orbital diagram shown in Figure 4-38 with the experimental photoelectron spectrum of HF shown in Figure 4-39. Indeed there is a good correlation, in that three bands are observed, corresponding to ionization from the 1σ, 2σ, and 1π molecular orbitals. These ionization processes can be depicted as

$$(1\sigma)^2(2\sigma)^2(1\pi)^4 \rightarrow (1\sigma)^2(2\sigma)^2(1\pi)^3 + e^-$$

$$(1\sigma)^2(2\sigma)^2(1\pi)^4 \rightarrow (1\sigma)^2(2\sigma)^1(1\pi)^4 + e^-$$

$$(1\sigma)^2(2\sigma)^2(1\pi)^4 \rightarrow (1\sigma)^1(2\sigma)^2(1\pi)^4 + e^-$$

Although the overall correlation between experiment and theory is good, there is a discrepancy between the observed and predicted value of the first ionization energy of HF. The molecular orbital diagram shown in Figure 4-38 indicates that the 1π orbital of HF will have an energy identical to that of the fluorine $2p$ orbital. However, the experimental results show that the first ionization energy of HF (16.05 eV) is lower than the *VOIE* of fluorine $2p$ (18.6 eV) by more than 2 eV. The explanation for this discrepancy lies in the fact that the 2σ molecular orbital has a higher electron density in the vicinity of the fluorine nucleus than of the hydrogen nucleus. It follows that there is a separation of charge in the ground state of HF, hydrogen having a partial positive charge $(\delta+)$ and fluorine a partial negative charge $(\delta-)$: $H^{\delta+}F^{\delta-}$. This negative charge will tend to destabilize the π electrons, which are localized on the fluorine atom, thereby making it easier to ionize them.

An extreme situation would exist if both 2σ electrons had their charge density concentrated completely around the fluorine atom. In that case an HF molecule would be composed of H^+ and F^- ions. Recall that a molecule that can be formulated accurately as an ion pair is described as an ionic molecule. Ion pairing is encountered in a diatomic molecule only if the valence orbital of one atom has much less energy than the valence orbital of the other. An example is NaF. The difference between the ionization energies of the Na $3s$ orbital and the F $2p$ orbital is 13.5 eV. In HF the energy difference between the H $1s$ orbital and the F $2p$ orbital is only 5.0 eV (Table 4-4). Therefore the HF molecule is not so ionic as NaF, but we say that HF has partial ionic character.

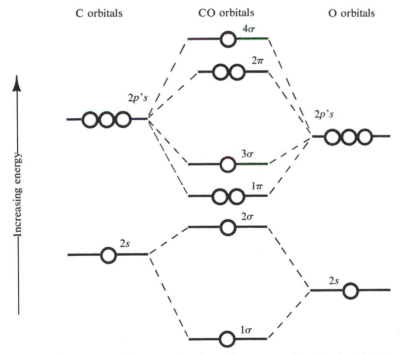

Figure 4-40 Relative orbital energies in carbon monoxide. Notice that the atomic orbitals belonging to the O atom are more stable than those belonging to the C atom. To simplify the diagram, not all correlation lines have been drawn. Each of the 1σ, 2σ, 3σ, and 4σ molecular orbitals can contain atomic-orbital contributions from s_a, z_a, s_b, and z_b.

The fluorine $1s$ core photoelectron spectrum of HF provides further evidence for the charge separation $H^{\delta+}F^{\delta-}$. Only one band is expected from the fluorine $1s$ core electrons, and it is observed at 694.25 eV. This is 2.55 eV lower than the fluorine $1s$ ionization in F_2 (696.70 eV). Once again we see that the negative charge on the fluorine atom tends to destabilize the electrons that are localized there, making it easier to ionize them.

Carbon Monoxide

In carbon monoxide, both atoms of the heteronuclear diatomic molecule have s and p orbitals. Figure 4-40 illustrates the molecular-orbital energy-level diagram. For carbon monoxide, with ten valence electrons, the predicted electronic configuration is $(1\sigma)^2(2\sigma)^2(1\pi)^4(3\sigma)^2$.

The s and p orbitals of O are placed lower than the s and p orbitals of C, in agreement with the $VOIE$s (Table 4-4). The σ and π bonding and antibonding orbitals for a heteronuclear molecule such as carbon monoxide are formed in the same manner as for a homonuclear molecule. However, just as we discussed for HF, the orbital coefficients of the valence orbitals for the more electronegative atom will be larger in the more stable molecular orbitals. The less electronegative atom will have larger coefficients for its orbitals in the less stable molecular orbitals. For CO, this means that the electrons in the more stable orbitals have a larger electron density near the more electronegative atom, O. In the higher-energy antibonding orbitals the electrons have a larger electron density near the less electronegative atom, C. Spatial representations of the molecular orbitals for CO are given in Figure 4-41.

Boron Monofluoride

We choose BF as our next example of a heteronuclear diatomic molecule. In the BF molecule, there is a much larger disparity between the $VOIE$s of the two atoms than there is for CO, due to the large electronegativity difference between B and F. Figure 4-42 presents a molecular-orbital energy-level diagram for BF. Notice that the F_{2p} orbitals are more stable than the B_{2s} orbitals, in agreement with the $VOIE$s (Table 4-4). Consequently, there will be extensive mixing between these orbitals. Although the electronic configuration for BF is the same as that for CO $[(1\sigma)^2(2\sigma)^2(1\pi)^4(3\sigma)^2]$, the atomic-orbital character of the 3σ orbital is considerably different from that of CO. As a result of s–p mixing, the 3σ orbital of CO is composed mainly of $2p_{z0}$ (Figure 4-41), whereas for BF the 3σ orbital is predominantly $2s_B$ in character (Figure 4-42).

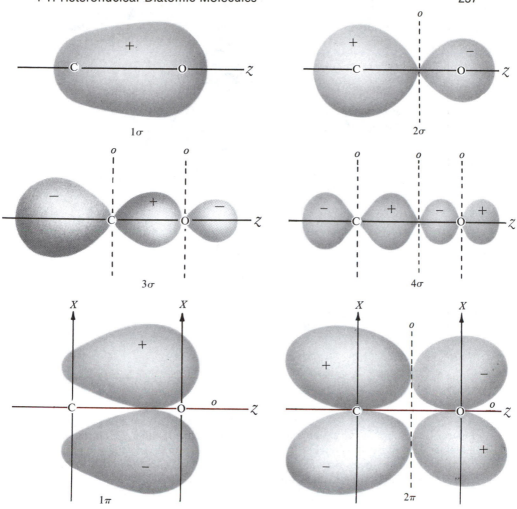

Figure 4-41 Spatial representation of molecular orbitals for the CO molecule. Dashed lines designated "0" represent nodal planes of zero electron density.

Bond Properties of Other Heteronuclear Diatomic Molecules and Ions

The bond properties of other heteronuclear diatomic molecules and ions are listed in Table 4-9. Several examples are discussed in the following paragraphs.

BO, CN, *and* CO⁺ *(Nine Valence Electrons)*

Of these molecules, the molecular-orbital diagram of CN should correspond to Figure 4-40, since the electronegativity difference between the atoms is

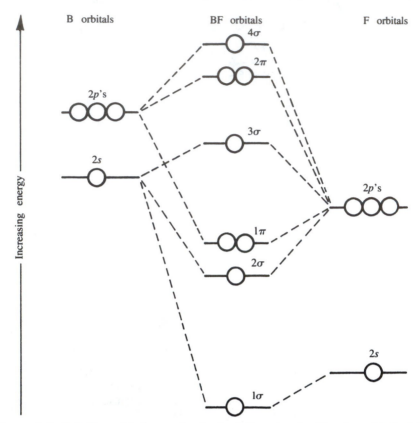

Figure 4-42 Relative orbital energies in the BF molecule. The $2p$ orbital of F is more stable than the $2s$ orbital of B. To simplify the diagram, only the important correlation lines have been drawn. Each of the 1σ, 2σ, 3σ, and 4σ molecular orbitals can contain atomic-orbital contributions from s_a, z_a, s_b, and z_b.

small, while the MO diagram of BO should correspond to Figure 4-42. The ground-state configuration of all three molecules is predicted to be $(1\sigma)^2(2\sigma)^2(1\pi)^4(3\sigma)^1$. Although the photoelectron spectra of BO and CN have not yet been determined, there is another experimental technique called *electron spin resonance* that allows one to determine the orbital character of the unpaired electron. These experiments indicate that in CN the unpaired electron resides in an orbital that is mainly carbon $2p_z$ in character, whereas in BO it is predominantly boron $2s$ in character. These results are in agreement with the above discussion concerning the character of the 3σ orbital.

Each of the BO, CN, and CO^+ molecular species has one more electron than the C_2 molecule. The additional electron is found in the 3σ orbital.

TABLE 4-9. BOND PROPERTIES OF SOME HETERONUCLEAR DIATOMIC
MOLECULES AND IONS

Molecule	Bond Length (Å)	Bond Dissociation Energy, (kcal mole^{-1})	Molecule	Bond Length (Å)	Bond Dissociation Energy, (kcal mole^{-1})
AsN	1.620	115	KH	2.244	43
AsO	1.623	113	KI	3.0478	77.2
BF	1.262	131	LiBr	2.1704	101
BH	1.2325	70	LiCl	2.018	113.25
BN	1.281	92	LiF	1.5639	135.8
BO	1.2043	191.2 ± 2.3	LiH	1.5953	56
BaO	1.940	130.4 ± 6	LiI	2.3919	81
BeF	1.3614	135.9 ± 2.3	MgO	1.749	81
BeH	1.297	53	NH	1.045	85
BeO	1.3308	106.1 ± 2.3	NH$^+$	1.081	—
BrCl	2.138	52.1	NO	1.1508	162
BrF	1.7555	55	NO$^+$	1.0619	—
CF	1.2718	131	NP	1.4910	—
CH	1.1202	80	NS	1.495	115
CN	1.1719	188	NS$^+$	1.25	—
CN$^+$	1.1727	—	NaBr	2.502	88
CN$^-$	1.14	—	NaCl	2.3606	98.5
CO	1.1283	255.8	NaF	1.9260	113.9
CO$^+$	1.1152	192.4	NaH	1.8873	47
CP	1.5583	122.1 ± 5	NaI	2.7115	69
CS	1.5349	173.6 ± 3.5	NaK	—	14.3
CSe	1.66	138 ± 5	NaRb	—	13.8
CaO	1.822	91.32 ± 1.4	OH	0.9706	101.5
ClF	1.6281	60.3	OH$^+$	1.0289	101.0
CsBr	3.072	91.5	PH	1.4328	—
CsCl	2.9062	101.7	PN	1.4869	174.6
CsF	2.345	122	PO	1.473	124
CsH	2.494	42	RbBr	2.9448	90.9
CsI	3.315	75.4	RbCl	2.7868	102.8
GeO	1.650	157	RbF	2.2704	119.5
HBr	1.4145	86.5	RbH	2.367	39
HBr$^+$	1.459	—	RbI	3.1769	77.7
HCl	1.2744	102.2	SO	1.4810	123.66
HCl$^+$	1.3153	108.3	SbO	1.848	74
HF	0.91680	135.1	SiF	1.6008	129.5
HI	1.6090	70.5	SiH	1.5201	74
HS	1.3503	81.4	SiN	1.575	104
IBr	2.485	41.90	SiO	1.5097	182.8
ICl	2.32070	49.63	SiS	1.929	148
IF	1.908	45.7	SnH	1.785	74
KBr	2.8207	91.4	SnO	1.838	126.5
KCl	2.6666	100.8	SnS	2.209	110.3
KF	2.1715	118.9	SrO	1.9199	99.2

The experimental bond lengths all are shorter than that of C_2 (1.2425 Å):
1.204 Å for BO, 1.172 Å for CN, and 1.115 Å for CO^+. This bond shortening
cannot be accounted for simply by the change in atomic radius (Figure 2-1);
consequently, the additional 3σ electron must have some bonding character,
as depicted in Figure 4-41. Also, the bond energies are greater than the bond
energy of C_2 (144 kcal mole^{-1}): 191 kcal mole^{-1} for BO, 188 kcal mole^{-1} for
CN, and 192 kcal mole^{-1} for CO^+.

NO^+, CO, and CN^- *(Ten Valence Electrons)*

The NO^+, CO, and CN^- molecular species are isoelectronic with N_2. From
the configuration $(1\sigma)^2(2\sigma)^2(1\pi)^4(3\sigma)^2$ we predict one σ and two π bonds.
The bond lengths of NO^+, CO, and CN^- increase with increasing negative
charge, being 1.062 Å for NO^+, 1.128 Å for CO, and 1.14 Å for CN^-. Comparing CO with BO and NO^+ with CO^+ (species of like charge), we see that
the bond length of CO is less than that of BO and that of NO^+ is less than
that of CO^+, as we expect. The bond energy of CO is 255.8 kcal mole^{-1},
which is greater than the bond energy of 225 kcal mole^{-1} for N_2.

NO *(Eleven Valence Electrons)*

The electronic configuration of NO is $(1\sigma)^2(2\sigma)^2(1\pi)^4(3\sigma)^2(2\pi)^1$. Since
the eleventh electron is in a π^* orbital the bond order is two and one-half,
which is one half less than for NO^+. The bond length of NO is 1.151 Å, longer
than both the CO and NO^+ bonds. The bond energy of NO is 162 kcal mole^{-1},
which is considerably less than the bond energy of CO.

Some Heteronuclear Transition-Metal Molecules

A typical molecular-orbital energy-level diagram for a first-row transition-
metal monoxide molecule, such as ScO, is presented in Figure 4-43. Except
for the introduction of the transition-metal $3d$ orbitals, this diagram is very
similar to the energy-level scheme presented in Figure 4-42 for BF. In both
ScO and BF there is a large electronegativity difference between the two
atoms in the diatomic molecule. The molecular-orbital energy-level sequence
is $1\sigma < 2\sigma < 1\pi < 1\delta < 2\pi < 3\sigma(3d) < 4\sigma(4s) < 3\pi < 5\sigma$. Also shown in
Figure 4-43 is the effect of $4s$–$3d_{z^2}$ orbital mixing on the energy-level sequence.
Such mixing causes the 3σ and 4σ molecular orbitals to move apart in
energy. As a consequence, the 3σ molecular orbital can become more stable
than the 1δ molecular orbital, so the revised energy-level scheme is

$$1\sigma < 2\sigma < 1\pi < 3\sigma < 1\delta < 2\pi < 4\sigma < 3\pi < 5\sigma$$

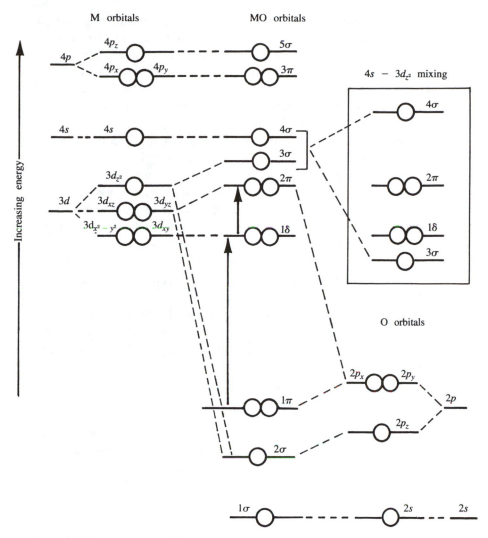

Figure 4-43 Relative orbital energies in a general transition-metal monoxide molecule, MO. Notice that the atomic-orbital energies of the oxygen atom are much more stable than those of the metal atom. The metal atomic-orbital energies are placed in the sequence $3d < 4s < 4p$, and the $4p$ orbitals are assumed to be so high in energy that their contribution to the bonding is negligible. The 1σ molecular orbital is predominantly oxygen $2s$ in character, and the 3π and 5σ molecular orbitals are predominantly metal $4p$ in character. The effect of $4s$–$3d_{z^2}$ orbital mixing is shown in the inset. Note that the 3σ molecular orbital can become more stable than 1δ as a consequence of $4s$–$3d_{z^2}$ mixing. The arrows indicate possible electronic transitions. For example, in TiO we could have a $1\delta \rightarrow 2\pi$ electronic transition, and because both of these molecular orbitals are predominantly $3d$ in character, this is called a d-d transition. If a transition occurs from a predominantly oxygen molecular orbital (1π or 2σ) to a predominantly metal molecular orbital (1δ or 2π), the transition is designated a ligand-to-metal transition.

Figure 4-43 shows that in the absence of $4s-3d_{z^2}$ orbital mixing, the 1δ, 2π, and 3σ molecular orbitals are predominantly metal $3d$ in character. Removal of these orbitals from the energy-level scheme shown in the figure results in a diagram similar to that presented in Figure 4-42 for a hetero-nuclear diatomic molecule. As we shall see shortly, however, experimental results on transition-metal monoxide molecules agree with the energy-level scheme that includes $4s-3d_{z^2}$ orbital mixing.

Let us apply the energy-level scheme shown in Figure 4-43 to scandium monoxide, ScO. The scandium atom has three valence electrons and there-fore is isovalent with boron. Thus the general bonding description for ScO is closely related to that for BO. There is an even larger energy difference, however, between the scandium and oxygen valence orbitals in ScO than between the boron and oxygen orbitals in BO, for the first ionization energy of Sc is only 6.54 eV, whereas boron's is 8.298 eV. As a consequence, we expect ScO to be more ionic than BO. This conclusion also could have been arrived at by noting that scandium is less electronegative than boron (Table 2-4).

The ScO molecule has nine valence electrons. Three of these are con-tributed by $Sc(3d^1 4s^2)$ and six by $O(2s^2 2p^4)$. According to the energy-level scheme presented in Figure 4-43, in the absence of $4s-3d_{z^2}$ orbital mixing the ground-state electronic configuration of ScO is

$$(1\sigma)^2 (2\sigma)^2 (1\pi)^4 (1\delta)^1 (3\sigma)^0$$

However, the experimental ground-state electronic configuration indicates that the unpaired electron is in a σ molecular orbital, so the energy-level sequence, including $4s-3d_{z^2}$ mixing, is better written as

$$(1\sigma)^2 (2\sigma)^2 (1\pi)^4 (3\sigma)^1 (1\delta)^0$$

Simplified atomic-orbital overlaps for several of the molecular orbitals are shown in Figure 4-44. Notice that a σ molecular orbital has no nodes con-taining the internuclear axis, whereas a π molecular orbital has one such node, and a δ molecular orbital has two. Thus the 1δ molecular orbitals are completely nonbonding in ScO, and consists of the degenerate pair of orbitals $3d_{x^2 - y^2}$ and $3d_{xy}$.

---

*Notice that the metal valence orbitals are in the energy-level sequence $3d < 4s < 4p$. This energy-level sequence does not necessarily indicate the *order* of orbital occupation in the free metal atom, because the order will also be governed by the effects of electron–electron repul-sion. As explained in Section 1-14, the $3d$ orbital of scandium is more stable than the $4s$ orbital. However, the ground-state electronic configuration of the scandium atom is $[Ar]3d^1 4s^2$, not $[Ar]3d^2 4s^1$ or $[Ar]3d^3 4s^0$. This configuration is a result of decreased electron–electron repul-sion in the $4s$ orbital.

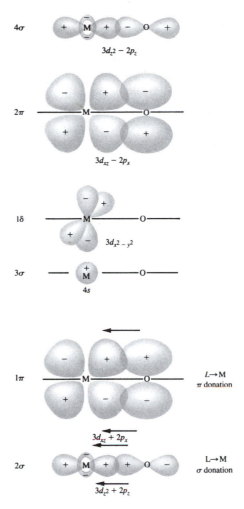

Figure 4-44 A schematic representation of the atomic-orbital overlap of metal-atom 3d orbitals with oxygen-atom 2p orbitals. Overlap diagrams of the 1σ, 5σ, and 3π molecular orbitals are not drawn, because these orbitals remain predominantly atomic in character. To simplify the diagram, atomic-orbital mixing of the $3d_{z^2}$ and 4s orbitals on the metal atom and $2s$–$2p_z$ on the oxygen atom are not represented. The 3σ molecular orbital is shown as mainly metal 4s, as predicted by electron-spin resonance experiments on ScO and the results of quantum-mechanical calculations.

In the completely ionic formulation, $Sc^{2+}O^{2-}$, the 1σ, 2σ, and 1π molecular orbitals correspond to the oxygen atomic orbitals, and the 3σ molecular orbital is mainly Sc 4s or $3d_{z^2}$. We could write the electronic configuration of ScO, giving the predominant atomic character, as

$$(\sigma_{2s})^2(\sigma_{2p_z})^2(\pi_{2p_{x,y}})^4(\sigma_{4s\text{-}3d_{z^2}})^1$$

and we would obtain the ionic formulation $Sc^{2+}O^{2-}$. However, if one of the σ orbitals (for example, σ_{2p_z}) becomes involved in the bonding by overlapping with the $3d_{z^2}$ orbital, as shown in Figure 4-44 for the 2σ molecular orbital, we would obtain a single bond with the correponding Lewis electron-dot structure $\cdot\overset{+}{Sc}$—$\ddot{\overset{-}{O}}$: . Such an orbital interaction is called *ligand-to-metal* σ *donation*. The $\pi_{2p_{x,y}}$ orbital also can exhibit bonding (Figure 4-44, 1π molecular orbital); in that case we obtain a triple bond, $\cdot Sc\equiv\overset{+}{O}$: . This mechanism of allowing the ligand $\pi_{2p_{x,y}}$ orbital to donate electron density into a d_π orbital on the metal ($3d_{xz,yz}$) is called *ligand-to-metal* ($L \rightarrow M$) π *bonding*. As is shown in Figure 4-43, this type of interaction stabilizes the 1π molecular orbital, but destabilizes the predominantly metal ($3d$) 2π molecular orbital.

   The actual type of bonding in ScO, whether ionic, single bond, or triple bond, can be determined only by recourse to experiments and calculations. Quantum-mechanical calculations indicate that ScO can be considered a triply bonded molecule, but with all three bonds strongly polarized toward the oxygen atom. Consequently, the molecule has partial ionic character $Sc^{\delta+}O^{\delta-}$. Experimentally, the best evidence for covalent character in ScO comes from electron-spin resonance, which shows that the unpaired electron in ScO is in an orbital (3σ) consisting mainly of Sc $4s$ character. If the ionic formulation $Sc^{2+}O^{2-}$ were correct, the odd electron would be Sc $3d$, because Sc^{2+} has the electronic configuration $[Ar]3d^1$ (Section 1-14).

   The electronic configurations of other transition-metal monoxides also can be understood on the basis of Figure 4-43. The experimental ground-state electronic configuration of TiO, which has one more valence electron than ScO, is

$$(1\sigma)^2(2\sigma)^2(1\pi)^4(3\sigma)^1(1\delta)^1$$

Evidently the 3σ and 1δ molecular orbitals of TiO are sufficiently close in energy that the electronic configuration with two unpaired electrons ($3\sigma^1 1\delta^1$) is more stable than the spin-paired configuration ($3\sigma^2$). However, the experimental ground-state electronic configuration for the isoelectronic ScF molecule is

$$(1\sigma)^2(2\sigma)^2(1\pi)^4(3\sigma)^2(1\delta)^0$$

   The experimental ground-state electronic configurations, bond lengths, and bond-dissociation energies for several diatomic transition-metal oxides, halides, and hydrides are tabulated in Table 4-10. Very few of the molecules listed in the table have been studied by photoelectron spectroscopy. Their optical absorption and emission spectra, however, have been studied exten-

TABLE 4-10. BOND PROPERTIES OF SOME DIATOMIC TRANSITION METAL HYDRIDE, HALIDE, AND OXIDE MOLECULES

Molecule	Ground State	Electron Configuration*	Bond Length (Å)	Bond Dissociation Energy (kcal mole^{-1})
CrH	$^6\Sigma$	$\sigma^1\delta^2\pi^2$	1.656	—
MnH	$^7\Sigma$	$\sigma^1\delta^2\pi^2\sigma^1$	1.722	—
CoH	$^3\Phi_4$	$\sigma^2\delta^3\pi^3$	1.542	—
NiH	$^2\Delta_{5/2}$	$\sigma^2\delta^3\pi^4$	1.475	—
PtH	$^2\Delta_{5/2}$	$\sigma^2\delta^3\pi^4$	1.528	—
CuH	$^1\Sigma^+$	$\sigma^2\delta^4\pi^4$	1.463	63.0 ± 1.5
AgH	$^1\Sigma^+$	$\sigma^2\delta^4\pi^4$	1.6179	53.2 ± 1.5
AuH	$^1\Sigma^+$	$\sigma^2\delta^4\pi^4$	1.5237	74.3 ± 3.0
ScO	$^2\Sigma^+$	σ^1	1.668	161.0
YO	$^2\Sigma^+$	σ^1	1.790	168.5
LaO	$^2\Sigma^+$	σ^1	1.825	192.5
TiO	$^3\Delta$	$\sigma^1\delta^1$	1.620	167.38 ± 2.30
ZrO	$^1\Sigma^+$	σ^2	1.711	181 ± 5
HfO	$^1\Sigma^+$	σ^2	1.724	182.6 ± 6
VO	$^4\Sigma^-$	$\sigma^1\delta^2$	1.589	147.5 ± 4.5
NbO	$^2\Delta$	δ^3	1.691	180.0 ± 2.5
TaO	$^2\Delta$	δ^3	1.687	197 ± 12
CuO	$^2\Pi$	$\sigma^2\delta^4\pi^3$	—	62.7 ± 3
ScF	$^1\Sigma^+$	σ^2	1.788	—
YF	$^1\Sigma^+$	σ^2	1.925	—
LaF	—	—	2.026	—
CuF	$^1\Sigma^+$	$\sigma^2\delta^4\pi^4$	1.749	88 ± 10
CuCl	$^1\Sigma^+$	$\sigma^2\delta^4\pi^4$	2.050	88 ± 6
CuBr	$^1\Sigma^+$	$\sigma^2\delta^4\pi^4$	—	79 ± 6
CuI	$^1\Sigma^+$	$\sigma^2\delta^4\pi^4$	2.337	<75
AgF	$^1\Sigma^+$	$\sigma^2\delta^4\pi^4$	—	—
AgCl	$^1\Sigma^+$	$\sigma^2\delta^4\pi^4$	2.2808	78
AgBr	$^1\Sigma^+$	$\sigma^2\delta^4\pi^4$	2.3922	73
AgI	$^1\Sigma^+$	$\sigma^2\delta^4\pi^4$	2.544	60
AuCl	$^1\Sigma^+$	$\sigma^2\delta^4\pi^4$	—	—

*This electron configuration refers only to the orbitals that are mainly metal in character, that is, the 3σ, 1δ, and 2π orbitals shown in the inset of Figure 4-43.

sively. As Figure 4-43 shows, there are two general types of absorption transitions. In the first, an electron is excited from a predominantly metal molecular orbital to a higher-energy molecular orbital, which also is predominantly metal in character. If the two relevant molecular orbitals have mainly $3d$ character, the absorptions are called *d-d transitions*. The $1\delta \rightarrow 2\pi$ transition in TiO would be an example of a *d-d* transition. In the second type of transition, an electron is excited from an orbital that is predominantly ligand in character to a higher, mainly metal, orbital. Such a transition is referred to as a *ligand-to-metal* transition or a *charge-transfer* transition. In ScF, for example, we could have a transition $1\pi \rightarrow 1\delta$. This transition effectively transfers one electron from the fluorine atom to the scandium atom, because 1π is predominantly fluorine $2p_{x,y}$ and 1δ is mainly scandium $3d_{xy,x^2-y^2}$. In terms of the Lewis electron-dot structure this transition corresponds to

$$:\text{Sc}\!-\!\ddot{\text{F}}: \rightarrow :\overset{-}{\text{Sc}}\!-\!\overset{+}{\text{F}}:$$

QUESTIONS AND PROBLEMS

1. The ground state of H_2 has the molecular-orbital configuration $(\sigma_g{}^b)^2$. In addition to the ground state there are excited states that have the following configurations:

    Predict which of these states would be highest in energy and which would be lowest. Explain your reasoning. Would you expect the lowest excited state of H_2 to be paramagnetic or diamagnetic?

2. Calculate the energies of the $\sigma_g{}^b$ and $\sigma_u{}^*$ orbitals for $H_2{}^+$, including the overlap integral S. Show that $\sigma_u{}^*$ is destabilized more than $\sigma_g{}^b$ is stabilized if the overlap is not zero but has some positive value.

3. Notice from Table 4-5 that Cl_2 has a longer, weaker bond than $Cl_2{}^+$. Use molecular-orbital theory to predict the bond order for each of these molecules.

4. The mercurous ion is found as the diatomic species $Hg_2{}^{2+}$. Ignore the filled d orbitals and describe the bonding in this ion in terms of molecular-orbital theory. The ion exhibits strong absorption in the ultraviolet region. To what electronic transition can this absorption be attributed?

5. The photoelectron spectrum of C_2 has not yet been measured. Sketch a predicted spectrum, based on the molecular-orbital energy-level diagram and the electronic configuration of the C_2 molecule. Ignore the $1s$ core orbitals in your treatment. Assign each band to an orbital ionization process. Which bands do you predict will exhibit extensive vibrational fine structure?

6. Find the ground-state term for (a) B_2; (b) F_2; (c) C_2; (d) S_2; (e) BeF; (f) BeO.

7. The core electron ($1s$) photoelectron spectrum of O_2 exhibits two bands at 543.1 eV and 544.2 eV. The corresponding spectra of F_2 and N_2 exhibit only one band each, at 696.70 eV and 409.9 eV, respectively. Explain why the O_2 spectrum exhibits two bands.

8. The 8 valence electrons in the C_2 molecule comprise a double bond and two lone electron pairs, $:C{=}C:$. The 12 valence electrons of Mo_2 comprise a sextuple bond with no lone electron pairs, $Mo{\equiv}\!\!{\equiv}Mo$. Explain why lone pairs are required in the bonding description of C_2 (it is not $C{\equiv}C$) but that no lone pairs are required for Mo_2.

9. Use the molecular-orbital theory for homonuclear diatomic transition-metal molecules to study the bond properties of Cr_2 and Mo_2. Assume that each of the electrons is spin paired so that the molecule attains a closed-shell $^1\Sigma_g^+$ ground state. What is the predicted bond order for these molecules?

10. Draw a molecular-orbital diagram for LiH and write the electronic configuration. Indicate the relative energies of the Li $2s$ and H $1s$ atomic orbitals by reference to the *VOIE* values in Table 4-4. The first ionization energy of LiH has not been measured, but calculation predicts it to be about 8 eV. To support this value, which atom in LiH must have a partial negative charge?

11. The OH radical has been observed in interstellar space. Formulate its electronic structure in terms of molecular-orbital theory, using only the $2p$ oxygen and the $1s$ hydrogen orbitals. What type of molecular orbital contains the unpaired electron? Is this orbital associated with both oxygen and hydrogen atoms, or is it localized on a single atom? Would you expect the lowest electronic transition of OH to occur at lower, or higher, energy than that of OH^-? Briefly explain your answer.

12. Formulate the bonding in the hydrogen-halide molecules HF, HCl, HBr, and HI in terms of molecular-orbital theory. The first ionization energies of these molecules are 16.05, 12.74, 11.67, and 10.38 eV, respectively. Plot these molecular ionization energies versus the ionization energy of the halogen atoms given in Table 2-1. Also plot the molecular ionization energies versus the *VOIE* of the valence p orbital for F, Cl, and Br (Table 4-4). Explain the results of these plots. Can you predict a value for the *VOIE* of the iodine $5p$ orbital based on these plots?

13. The first ionization energy of the OH radical is 13.2 eV, and that of HF is 16.05 eV. The difference between these values is almost identical to

the difference between the $2p$ *VOIE* of oxygen and fluorine atoms (Table 4-4). Explain this observation on the basis of molecular-orbital theory.

14. Consider the molecule NF and the ions NF^+ and NF^-. Write the Lewis structure and the molecular-orbital description of the ground state for each species. Determine which of the three species would be paramagnetic, and tell how many unpaired electrons there would be in each paramagnetic molecule. Predict bond orders for all three species.

15. Discuss the electronic structure of the SO molecule in terms of the molecular-orbital theory for heteronuclear diatomic molecules. How many unpaired electrons do you predict for the ground state?

16. What is the molecular-orbital configuration of the ground state of the CN molecule? How many unpaired electrons does CN have? The molecule has an absorption band in the near-infrared region (at 9000 cm^{-1}), which is due to an electronic transition. Suggest an assignment for this absorption band. The CN^- ion does not absorb in the near-infrared region. Is this observation consistent with your transition assignment? If not, it would be a good idea to reconsider!

17. The bond dissociation energy of CF^+ is 180 kcal mole^{-1}, whereas that of CF is only 131 kcal mole^{-1}. (a) Draw Lewis structures to explain this difference in bond energy. (b) Use molecular-orbital theory to explain the difference in bond energy for CF and CF^+.

18. The isoelectronic molecules NO and CF have bond dissociation energies of 162 and 131 kcal mole^{-1}, respectively. (a) Explain this difference on the basis of Lewis structures. (b) Use molecular-orbital theory to explain why CF has a weaker bond than NO.

19. The first ionization energies of BF, CO, and N_2 are 11.06 eV, 14.01 eV, and 15.57 eV, respectively. Explain the increase in ionization energy for this isoelectronic series on the basis of atomic-orbital composition of the highest occupied molecular orbital.

20. The first ionization energies of the isovalent molecules CO, SiO, and GeO are 14.01, 11.58, and 11.10 eV, respectively. Explain the decrease in ionization energy through the series.

21. The ionization energy of the 11 electron molecule NO (9.26 eV) is considerably less than that of the 10 electron molecule CO (14.01 eV). Explain this fact on the basis of the electronic configurations for these molecules.

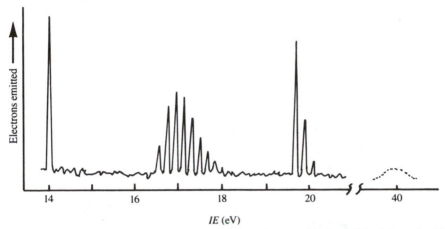

Figure 4-45 The photoelectron spectrum of the valence orbitals of CO. Adapted with permission from J.L. Gardner and J.A.R. Samson, *J. Chem. Phys.* 62: 1447 (1975).

22. The electron affinity of a molecule can be defined in the same way as for atoms (Chapter 2). In terms of molecular-orbital theory, explain why the electron affinities of O_2 and NO are small (approximately 10 and 21 kcal mole^{-1}, respectively) while those of CN and C_2 are large (88.1 and 81.6 kcal mole^{-1}, respectively).

23. The photoelectron spectrum of the valence orbitals of carbon monoxide is shown in Figure 4-45. Indicate the orbital assignment for each band in the spectrum. Compare this spectrum with that observed for N_2 (Figure 4-31). Explain why the spectra are so similar in terms of the number of bands and the extent of vibrational fine structure on each band.

24. The photoelectron spectrum of the core electrons of ClF exhibits a band at 694.54 eV due to F($1s$) electrons, and at 279.23 eV due to Cl($2s$) electrons. The corresponding ionization energies in F_2 and Cl_2 are 696.70 eV and 278.66 eV. Using this information, decide which representation is more correct: $Cl^{\delta+}F^{\delta-}$ or $Cl^{\delta-}F^{\delta+}$.

25. The experimental ground-state electronic configuration of ScO is . . . $3\sigma^1$ and of ScF . . . $3\sigma^2$ (see Figure 4-43). The corresponding equilibrium bond lengths are 1.668 Å and 1.791 Å, respectively. What does this indicate about the bonding or antibonding character of the 3σ molecular orbital? Do you expect the bond length in ScO$^+$ to be longer or shorter than that in ScO? Explain.

26. Predict the ground-state electronic configuration of ScH and VH.

Suggestions for Further Reading

Ballhausen, C.J., and Gray, H.B., *Molecular Orbital Theory,* Menlo Park, Calif.: Benjamin, 1964.

Bock, H., and Mollère, P.D., "Photoelectron Spectra," *J. Chem. Educ.* 51: 506 (1974).

Cade, P.E., Sales, K.D., and Wahl, A.C., "Electronic Structure of Diatomic Molecules. III. A. Hartree-Fock Wavefunctions and Energy Quantities for N_2 $(X, {}^1\Sigma_g^+)$ and $N_2(X\ {}^2\Sigma_g^+,\ A\ {}^2\Pi_u,\ B\ {}^2\Sigma_u^+)$ Molecular Ions," *J. Chem. Phys.* 44: 1973 (1966).

Carlson, K.D., and Claydon, C.R., "Electronic Structure of Molecules of High Temperature Interest," in *Advances in High Temperature Chemistry,* vol. 1, edited by L. Eyring, New York: Academic, 1967, p. 43.

Cartmell, E., and Fowles, G.W.A., *Valency and Molecular Structure,* 4th ed., London: Butterworths, 1977.

Chang, R., *Basic Principles of Spectroscopy,* New York: McGraw-Hill, 1971.

Cheetham, C.J., and Barrow, R.F., "The Spectroscopy of Diatomic Transition Element Molecules," in *Advances in High Temperature Chemistry,* vol. 1, edited by L. Eyring, New York: Academic, 1967, p.7.

Cohen, I., and Carlson, K.D., "Density Distributions and Chemical Bonding in Diatomic Molecules of the Transition Metals," *J. Phys. Chem.* 73: 1365 (1969).

Crooks, J.E, *The Spectrum of Chemistry,* New York: Academic, 1978.

Dunn, T.M., "Coordination Chemistry," in *Physical Chemistry: An Advanced Treatise,* vol. 5, edited by H. Eyring, New York: Academic, 1970, p. 205.

Eland, J.H.D., *Photoelectron Spectroscopy,* London: Butterworths, 1974.

Gray, H.B., *Chemical Bonds,* Menlo Park, Calif.: Benjamin, 1973.

Gray, H.B., *Electrons and Chemical Bonding,* Menlo Park, Calif.: Benjamin. 1965.

Herzberg, G., *Molecular Spectra and Molecular Structure* I. *Diatomic Molecules,* 2nd ed., Princeton: Van Nostrand, 1950.

Hollenberg, J.L., "Energy States of Molecules," *J. Chem. Educ.* 47: 2 (1970).

Karplus, M., and Porter, R.N., *Atoms and Molecules: An Introduction for Students of Physical Chemistry,* Menlo Park, Calif.: Benjamin, 1970.

Klotzbücher, W., and Ozin, G.A., "Diniobium, Nb_2, and Dimolybdenum, Mo_2. Synthesis, Ultraviolet-Visible Spectra and Molecular Orbital Investigations of Nb_2 and Mo_2. Spectral and Bonding Comparison with V_2 and Cr_2," *Inorg. Chem.* 16: 984 (1977).

Lowe, J.P., *Quantum Chemistry,* New York: Academic, 1979.

McWeeny, R., *Coulson's Valence,* New York: Oxford, 1979.

Murrell, J.N., Kettle, S.F.A., and Tedder, J.M., *The Chemical Bond,* New York: Wiley, 1978.

Norman, J.G., Jr., Kolari, H.J., Gray, H.B., and Trogler, W.C., "Electronic Structure of $Mo_2(O_2CH_4)_4$, Mo_2^{4+}, and Mo_2," *Inorg. Chem.* 16: 987 (1977).

Pimentel, G.C., and Spratley, R.D., *Understanding Chemistry,* San Francisco: Holden-Day, 1971.

Robiette, A.G., "The Variation Theorem Applied to H_2^+," *J. Chem. Educ.* 52: 95 (1975).

Turner, D.W., Baker, C., Baker, A.D., and Brundle, C.R., *Molecular Photoelectron Spectroscopy,* New York: Wiley-Interscience, 1970.

Weltner, W., Jr., "The Matrix-Isolation Technique Applied to High Temperature Molecules," in *Advances in High Temperature Chemistry,* vol. 2, edited by L. Eyring, New York: Academic, 1969, p. 85.

5 Electronic Structures, Photoelectron Spectroscopy, and the Frontier-Orbital Theory of Reactions of Polyatomic Molecules

The molecular-orbital method that we applied to diatomic molecules provides a logical starting point for understanding polyatomic systems. The most general method for constructing molecular wave functions for polyatomic molecules is to use unhybridized atomic orbitals in linear combinations. Electrons in these molecular orbitals are not localized between two atoms of a polyatomic molecule; rather, they are *delocalized* among several atoms. The resulting description is conceptually very different from the Lewis structure, in which two electrons between two atoms are equivalent to one chemical bond.

An alternative method for dealing with complex molecules is to use *localized* two-atom orbitals, which we discussed in Chapter 3 as part of the valence bond and hybrid-orbital descriptions of chemical bonding. For many molecules the localized-orbital method provides a simple framework for the discussion of ground-state properties, particularly molecular geometry. However, for an understanding of the spectroscopic aspects of molecules (e.g., excited states and ionization energies), the delocalized molecular-orbital approach is simpler. Since much of our discussion in this chapter will center on spectroscopic aspects of polyatomic molecules, we will emphasize the delocalized molecular-orbital method.

5-1 The Simplest Polyatomic Molecule, H_3^+

The H_3^+ ion has only two valence electrons and thus is the simplest possible polyatomic molecule. It has been found by mass spectrometry in electrical discharges through hydrogen gas. The structure of H_3^+ has been determined experimentally to be an equilateral triangular configuration of the nuclei. The H—H distance along each side of the triangle is approximately 1 Å.

There is a simple way in which we could have predicted this structure by considering the formation of H_3^+ as resulting from the reaction of a proton with the hydrogen molecule:

$$H^+ + H_2 \rightarrow H_3^+$$

Recall that the electron density in the H_2 molecule is somewhat concentrated between the two nuclei. The proton is an *electrophile* (Greek *philos*: love; hence electrophile—"lover" of electrons) and it will attack the H_2 molecule at the position of highest electron density, namely at the center of the H—H bond:

$$\begin{array}{c} H^+ \\ \downarrow \\ H-H \end{array}$$

Such an attack would result in a *triangular* H_3^+ ion which no single Lewis structure can describe because there is a possibility of three bonds but only two electrons. Instead, we can write resonance structures I, II, and III, which taken together could be written as structure IV:

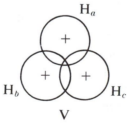

Structure IV represents a three-center two-electron bond.

If we use the molecular-orbital method we can generate one delocalized molecular orbital, V, by taking the bonding combination of all three hydrogen 1s orbitals, $\frac{1}{\sqrt{3}}(1s_a + 1s_b + 1s_c)$:

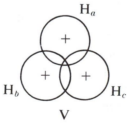

This molecular orbital is of the σ^b type and can contain the two electrons of H_3^+. By using delocalized molecular orbitals we have dispensed with the need for resonance structures, because the molecular orbital shown in V represents the same electronic structure as the superposition of structures I, II, and III presented in IV.

Now that we understand the type of orbital in which the two electrons of H_3^+ reside, we should inquire about the strength of the bonds in H_3^+. The exact dissociation energy of H_3^+ has not been determined experimentally, but accurate quantum-mechanical calculations predict it to be 203 kcal mole^{-1} for the process $H_3^+ \rightarrow 2H + H^+$. This is surprisingly large.

It is interesting to compute the energy of the reaction $H^+ + H_2 \rightarrow H_3^+$. The energy released in this reaction will be due to the *delocalization* of the two electrons from the two-center H_2 molecule to the three-center H_3^+ ion.

We know that $2H + H^+ \rightarrow H_3^+ + 203$ kcal mole^{-1} and $H_2 + 103$ kcal mole^{-1} $\rightarrow 2H$. By adding these two reactions we find that the energy released in the reaction

$$H^+ + H_2 \rightarrow H_3^+$$

is $(203 - 103) = 100$ kcal mole^{-1}. This energy is practically as large as the 103 kcal mole^{-1} released in the reaction $H + H \rightarrow H_2$. This example shows how important delocalization of electron pairs is for producing stability in molecules.

5-2 Delocalized Molecular Orbitals for BeH₂ and H₂O

One of the simplest molecules that we can use to illustrate the delocalized molecular-orbital method is BeH_2, for which we will assume the linear structure predicted from the VSEPR method (Section 2-15). As for a diatomic molecule, we label the molecular axis the Z axis (the H_a—Be—H_b line), shown in Figure 5-1. We form the molecular orbitals for BeH_2 by using the $2s$ and $2p$ valence orbitals of beryllium and the $1s$ valence orbitals of H_a and H_b. We obtain the correct linear combinations for the bonding molecular orbitals by writing the combinations of $1s_a$ and $1s_b$ that match the algebraic signs on the lobes of the central beryllium atom's $2s$ and $2p_z$ orbitals, respectively. This procedure gives a bonding orbital that concentrates electron density between the nuclei. The proper combinations of the hydrogen atom $1s$ orbitals with the beryllium atom valence orbitals are called *symmetry adapted linear combinations*, SALCs. Because the $2s$ orbital does not change sign over its spherical surface, the normalized SALC $\frac{1}{\sqrt{2}}(1s_a + 1s_b)$ is appropriate for one bonding molecular orbital. The $2p_z$ orbital has a plus lobe along the $+Z$ axis and a minus lobe along $-Z$. Thus the appropriate SALC of H orbitals for the second bonding molecular orbital is $\frac{1}{\sqrt{2}}(1s_a - 1s_b)$, as shown in Figure 5-2.

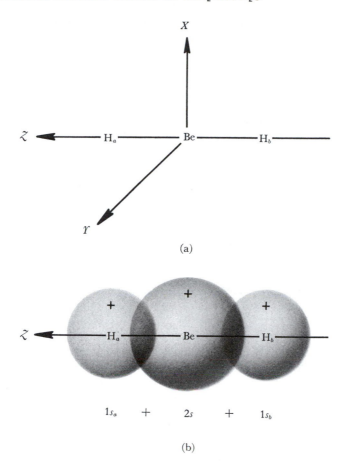

(a)

(b)

Figure 5-1 Linear polyatomic molecule: (a) coordinate system for BeH_2; (b) overlap of the hydrogen $1s$ orbitals with the beryllium $2s$ orbital.

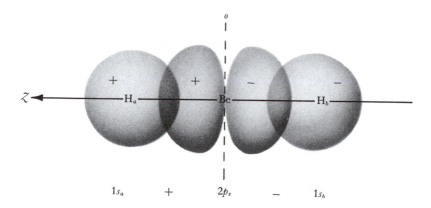

Figure 5-2 Overlap of the hydrogen $1s$ orbitals with the beryllium $2p_z$ orbital.

We can describe the two σ^b molecular orbitals by the following molecular wave functions:

$$\sigma_g{}^b(s) = c_1 2s + c_2\left[\frac{1}{\sqrt{2}}(1s_a + 1s_b)\right]$$

$$\sigma_u{}^b(z) = c_3 2p_z + c_4\left[\frac{1}{\sqrt{2}}(1s_a - 1s_b)\right]$$

The corresponding antibonding molecular orbitals, $\sigma_g{}^*(s)$ and $\sigma_u{}^*(z)$ will have nodes between the Be and the two H nuclei. That is, we will combine the beryllium $2s$ orbital with $-\frac{1}{\sqrt{2}}(1s_a + 1s_b)$ and the beryllium $2p_z$ orbital with the $-\frac{1}{\sqrt{2}}(1s_a - 1s_b)$. Therefore the two σ^* molecular orbitals are

$$\sigma_g{}^*(s) = c_5 2s - c_6\left[\frac{1}{\sqrt{2}}(1s_a + 1s_b)\right]$$

and

$$\sigma_u{}^*(z) = c_7 2p_z - c_8\left[\frac{1}{\sqrt{2}}(1s_a - 1s_b)\right]$$

Because the beryllium $2s$ and $2p_z$ orbitals are much higher in energy than the hydrogen $1s$ orbitals (H is more electronegative than Be; see also the valence-orbital ionization energies in Table 4-4), we can confidently assume that the electrons in the bonding orbitals of BeH_2 tend to concentrate around the hydrogen nuclei.

In the antibonding orbitals an electron is forced into the vicinity of the beryllium nucleus. In other words, calculation of the coefficients of the valence orbitals in the molecular-orbital combinations would reveal that c_2 and c_4 are greater than c_6 and c_8, respectively.

The $2p_x$ and $2p_y$ beryllium orbitals are not used in bonding because they are π orbitals in a linear molecule and because hydrogen has no valence orbitals capable of forming π molecular orbitals. Spatial representations of the BeH_2 molecular orbitals are shown in Figure 5-3.

The molecular-orbital energy-level scheme for BeH_2, shown in Figure 5-4, is constructed as follows. The valence orbitals of the beryllium atom are indicated at the left of the diagram, with the lower-energy $2s$ orbital below the $2p$ orbitals. The $1s$ orbitals of the two hydrogen atoms are placed on the right of the diagram. The $1s$ orbitals of hydrogen are placed lower than either the $2s$ or $2p$ orbitals of beryllium because of the difference in valence-orbital ionization energy of the orbitals (Table 4-4). The molecular orbitals — bonding, nonbonding, and antibonding — are placed in the middle of the diagram.

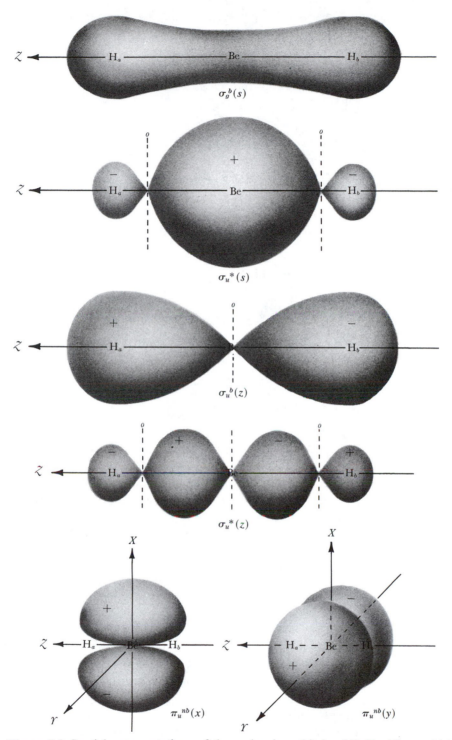

Figure 5-3 Spatial representations of the molecular orbitals of BeH_2. The $\pi_u^{nb}(x)$ and $\pi_u^{nb}(y)$ orbitals shows at the bottom are not used in bonding.

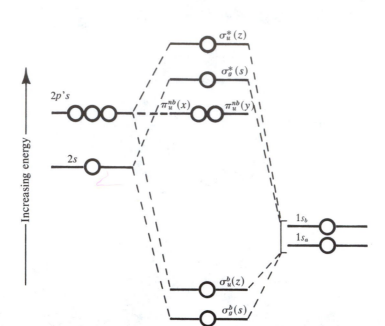

Figure 5-4 Relative orbital energies in BeH_2.

As usual, the bonding levels are lower in energy than the individual atomic orbitals, and the antibonding levels are correspondingly higher in energy. The $2p_x$ and $2p_y$ Be orbitals are nonbonding because they have zero net overlap with the hydrogen $1s$ orbitals. Consequently their energy does not change in our approximation scheme.

The energy-level sequence that we find in Figure 5-4 is as follows:

$$\sigma_g^b(s) < \sigma_u^b(z) < \pi_u^{nb}(x,y) < \sigma_g^*(s) < \sigma_u^*(z)$$

Nodes: 0   1   1   2   3

From the spatial representation shown in Figure 5-3 we can determine the number of nodes present in each molecular orbital, and we have indicated these beneath each orbital symbol. It is clear that in general, the energy increases with the number of nodes in the molecular orbital.

The ground state of BeH_2 is found by placing the valence electrons in the most stable molecular orbitals of Figure 5-4. There are four valence electrons, two from beryllium ($2s^2$) and two from the two hydrogen atoms. Therefore the ground-state electronic configuration is $[\sigma_g^b(s)]^2[\sigma_u^b(z)]^2$ or, using the simplified notation, $(1\sigma_g)^2(1\sigma_u)^2$.

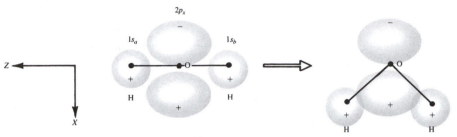

Figure 5-5 Schematic demonstration of the increase in overlap and bonding character that occurs between hydrogen $1s$ orbitals and the oxygen $2p_x$ orbital when water is bent in the XZ plane.

In this description of the electronic structure of BeH_2 the two electron-pair bonds are delocalized over the three atoms. According to the molecular-orbital method, the four electrons are distributed in two orbitals that do not have the same energy. This is contrary to what one would predict using the simple hybrid-orbital description wherein all four electrons appear to be equivalent in energy since the two bonds are identical in H—Be—H. This example further emphasizes a point made previously for N_2 (Section 4-9). It is worthwhile repeating that the concept of orbital energy cannot be applied to localized hybrid-orbital bonds. According to Koopmans' theorem, the ionization energy of an orbital is approximately equal to the negative of the calculated orbital energy, but this statement applies only to *delocalized* molecular orbitals. Although the photoelectron spectrum of BeH_2 has not been obtained, the molecular-orbital energy-level diagram presented in Figure 5-4 shows that two bands are expected, because of ionization from the $\sigma_g^b(s)$ and $\sigma_u^b(z)$ molecular orbitals.

Like BeH_2, the water molecule contains two H atoms, but H_2O has eight valence electrons compared to only four for BeH_2. Reference to Figure 5-4 suggests that if H_2O were a linear molecule, H—O—H, its electronic configuration would be

$$[\sigma_g^b(s)]^2[\sigma_u^b(z)]^2[\pi_u^{nb}(x,y)]^4$$

However, we can easily show that the water molecule would not be stable in this linear configuration. Suppose the molecule bends as shown in Figure 5-5. As the hydrogen atoms move toward the positive X axis, the $2p_x$ orbital and the hydrogen $1s$ orbitals overlap. This overlap causes the nonbonding $\pi(x)$ molecular orbital to become $\sigma(x)$ bonding, lowering its energy. Meanwhile the $\pi(y)$ orbital remains nonbonding, as in the linear configuration, and its energy remains unchanged because its zero net overlap with the hydrogen $1s$ orbitals is maintained during the bending process.

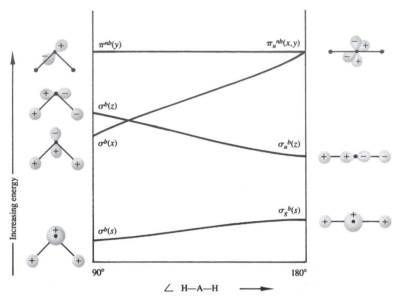

Figure 5-6 Atomic-orbital overlap and qualitative molecular-orbital energies of linear and bent AH_2 molecules. Changes in shape that cause an increase in overlap will lower the molecular-orbital energy level. Note that in the bent AH_2 molecule, s–p_x mixing could occur at the A atom. To simplify the diagram, we have not represented this s–p mixing. We have also not represented the two antibonding molecular orbitals in this figure.

Before asserting that the water molecule is bent by the stabilization of the $\sigma(x)$ orbital, we should examine the changes in energy of all the occupied molecular orbitals as the H_2O molecule bends. Figure 5-6 shows the change in energy of each of these molecular orbitals with variation in bond angle. Diagrams such as Figure 5-6 are called *Walsh correlation diagrams*, after the British chemist Arthur D. Walsh, who published an extensive account involving their use in 1953.

Consider a general triatomic molecule AH_2. Figure 5-6 shows that the $\sigma_g^b(s)$ orbital is slightly stabilized by the bending process. Since the central atom s orbital is spherical, there is no decrease in A-H overlap due to the bending of the AH_2 molecule. However, the bending process will cause an *increase* in the $1s$–$1s$ overlap between the two hydrogen atoms. This will serve to stabilize the $\sigma_g^b(s)$ molecular orbital slightly as the molecule bends.

In contrast to the $\sigma_g^b(s)$ orbital, the $\sigma_u^b(z)$ orbital is destabilized. The schematic orbital diagrams presented in Figure 5-6 show that this destabilization is a result of *decreased* overlap between the A ($2p_z$) orbital and the H ($1s$) orbitals. We can understand this decrease in overlap with bending by noting that the angular dependence of the $2p_z$ orbital involves a $\cos \theta$ term, where θ is the angle between the A-H internuclear line and the Z axis. Since $\cos \theta$ reaches a maximum for $\theta = 0°$, the $2p_z$–$1s$ overlap will obviously be largest for a linear AH_2 molecule.

We can achieve a general understanding of these orbital energy trends by noting that the movement of nuclei toward a nodal surface tends to *destabilize* the corresponding molecular orbital, whereas movement of nuclei away from a nodal surface tends to *stabilize* the corresponding molecular orbital. This rule provides a simple means of predicting the variation of orbital energy with bond angle.

We can use the Walsh diagram shown in Figure 5-6 to predict the molecular geometry of any general AH_2 molecule. We simply state that the most stable geometry will be that corresponding to the greatest stabilization for all of the occupied molecular orbitals. We therefore predict that a molecule with only four valence electrons, such as BeH_2, will be linear. However, in a molecule with eight valence electrons, such as H_2O, the stabilization of $\sigma(x)$ will cause the molecule to bend.

Experimental observation has shown that H_2O is angular, with a bond angle of 104.5°. Recall that in Section 2-15 we used the VSEPR method to predict the water molecule to be angular. Since predicting molecular shape using the molecular-orbital theory, as we have just done, is much more tedious than using the VSEPR method, we will generally employ the molecular-orbital theory only to discuss the spectroscopic properties of molecules, and not to determine molecular shape.

A complete molecular-orbital energy-level diagram for H_2O is presented in Figure 5-7. The ground-state electronic configuration of water is

$$[\sigma^b(s)]^2[\sigma^b(z)]^2[\sigma^b(x)]^2[\pi^{nb}(y)]^2$$

The orbitals can no longer be labeled *gerade* or *ungerade* because an angular molecule has no center of inversion. We see that the eight valence electrons occupy four molecular orbitals and that each orbital has a different energy. Consequently, using molecular-orbital theory, we predict that the photoelectron spectrum of H_2O should exhibit four bands, as experiment has indeed shown that it does (Figure 5-8). Not only does molecular-orbital theory predict the correct number of bands, but it also explains their shapes. The first band at 12.61 eV exhibits little vibrational structure, as would be expected for ionization from the nonbonding $\pi(y)$ orbital. However, the bands assigned to electron ejection from the $\sigma^b(x)$ and $\sigma^b(z)$ orbitals exhibit extensive vibrational fine structure, as would be expected for electron ejection from bonding orbitals.

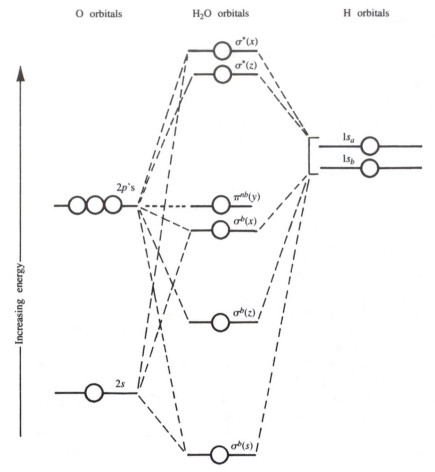

Figure 5-7 Relative orbital energies in H_2O.

It is important to recognize that whereas the simple Lewis electron-dot structure indicates only two types of electrons (bond pairs and lone pairs), the molecular-orbital theory indicates four different energy levels for the eight electrons. Once again we see that the concept of orbital energy applies only to the delocalized molecular orbitals, not to the localized hybrid orbitals.

5-3 Delocalized Molecular Orbitals for BH_3 and NH_3

We now wish to discuss the molecular-orbital theory for the trihydrides BH_3 and NH_3. The borane molecule has six valence electrons, and ammonia has eight.

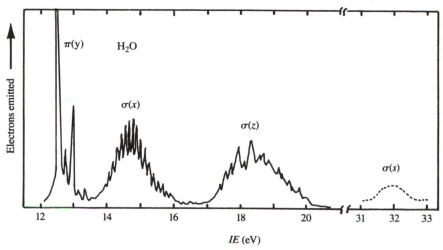

Figure 5-8 Photoelectron spectrum of H_2O. The first band corresponds to ionization from the nonbonding $\pi(y)$ orbital; in consequence, little vibrational structure is evident. Ionization from the bonding $\sigma(x)$ and $\sigma(z)$ orbitals, however, produces extensive vibrational structure. Adapted with permission from A.W. Potts and W.C. Price, *Proc. Roy. Soc. Lond.* A326: 181 (1972).

The Borane Molecule

According to the VSEPR method, BH_3 should have a trigonal-planar structure. We shall see shortly that the molecular-orbital method also predicts trigonal-planar geometry for BH_3. We choose the molecular Z axis to lie along the threefold axis, as shown in Figure 5-9.

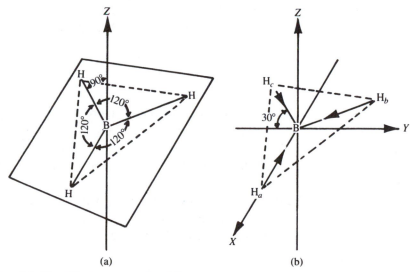

(a)　　　　　　　　　　(b)

Figure 5-9 Coordinate system for BH_3.

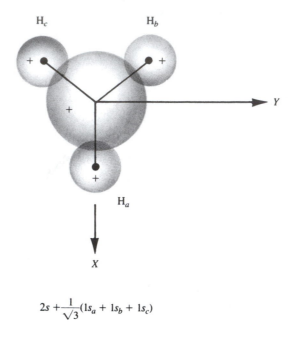

$$2s + \frac{1}{\sqrt{3}}(1s_a + 1s_b + 1s_c)$$

Figure 5-10 Overlap of the boron $2s$ orbital with the $1s$ orbitals of the hydrogen atoms.

We form the molecular orbitals for BH_3 by using the $2s$ and $2p$ boron valence orbitals and the $1s$ valence orbitals of H_a, H_b, and H_c. Our first step is to construct the correct symmetry-adapted linear combinations (SALCs) of the hydrogen $1s$ orbitals just as we did for BeH_2. Since the boron $2s$ orbital does not change sign over its spherical surface, the combination

$$\frac{1}{\sqrt{3}}(1s_a + 1s_b + 1s_c) \tag{5-1}$$

is appropriate for one bonding molecular orbital (Figure 5-10). The $2p_z$ orbital has a plus lobe along $+Z$ and a minus lobe along $-Z$. Because the BH_3 molecule lies in the X-Y plane, the hydrogen $1s$ orbitals have zero net overlap with the boron $2p_z$ orbital, and the latter is simply a π nonbonding orbital.

We are left to describe the correct SALCs to overlap with the boron $2p_x$ and $2p_y$ orbitals. If we place H_a along the X axis, as shown in Figure 5-11, the $1s_a$ orbital will have zero net overlap with the boron $2p_y$ orbital. The

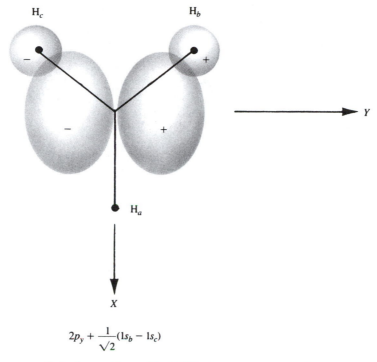

$$2p_y + \frac{1}{\sqrt{2}}(1s_b - 1s_c)$$

Figure 5-11 Overlap of the boron $2p_y$ orbital with the $1s$ orbitals of the hydrogen atoms.

correct SALC for boron $2p_y$ therefore is

$$\frac{1}{\sqrt{2}}(1s_b - 1s_c) \tag{5-2}$$

The boron $2p_x$ orbital is shown in Figure 5-12. The SALC $(1s_a - 1s_b - 1s_c)$ correctly overlaps the lobes of $2p_x$. However, there is a minor complication: The overlaps of $1s_a$, $1s_b$, and $1s_c$ with $2p_x$ are not identical. Specifically, $1s_a$ points directly at the positive lobe of $2p_x$, while $1s_b$ and $1s_c$ are 60° displaced from a comparable overlap with the negative lobe. At the end of this section (Exercise 5-1) we illustrate a method for determining the coefficients c_1, c_2, and c_3 in the SALC $c_1 1s_a - c_2 1s_b - c_3 1s_c$. The correct result is

$$\sqrt{2/3}(1s_a - \tfrac{1}{2}1s_b - \tfrac{1}{2}1s_c)$$

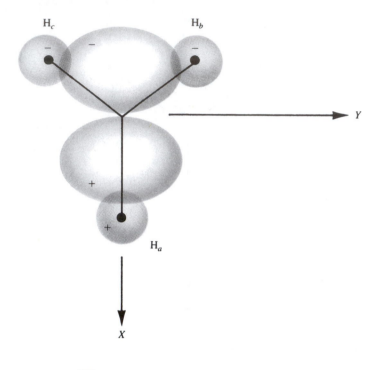

$$2p_x + \sqrt{2/3}\,[1s_a - \tfrac{1}{2}(1s_b) - \tfrac{1}{2}(1s_c)]$$

Figure 5-12 Overlap of the boron $2p_x$ orbital with the $1s$ orbitals of the hydrogen atoms.

Now that we have the correct SALCs we can describe the three σ^b molecular orbitals by the following molecular wave functions:

$$\sigma_s^b = c_1 2s + c_2\left[\frac{1}{\sqrt{3}}(1s_a + 1s_b + 1s_c)\right]$$

$$\sigma_y^b = c_3 2p_y + c_4\left[\frac{1}{\sqrt{2}}(1s_b - 1s_c)\right] \tag{5-3}$$

$$\sigma_x^b = c_5 2p_x + c_6[\sqrt{2/3}\,(1s_a - \tfrac{1}{2}1s_b - \tfrac{1}{2}1s_c)] \tag{5-4}$$

The corresponding antibonding molecular orbitals will have nodes between the B and the three H nuclei. They can be formed by taking the subtractive combination of the SALCs with the boron valence orbitals. Therefore the three σ^* molecular orbitals are

$$\sigma_s{}^* = c_7 2s - c_8\left[\frac{1}{\sqrt{3}}(1s_a + 1s_b + 1s_c)\right] \tag{5-5}$$

$$\sigma_y{}^* = c_9 2p_y - c_{10}\left[\frac{1}{\sqrt{2}}(1s_b - 1s_c)\right] \tag{5-6}$$

$$\sigma_x{}^* = c_{11} 2p_x - c_{12}\left[\sqrt{2/3}\,(1s_a - \tfrac{1}{2}1s_b - \tfrac{1}{2}1s_c)\right] \tag{5-7}$$

Altogether we have constructed three σ^b and three σ^* molecular orbitals. Because there are seven valence atomic orbitals (four from boron and one from each hydrogen atom) there must be seven molecular orbitals. In fact, the seventh is the simplest of all, corresponding to the boron $2p_z$ orbital only. This orbital will be π nonbonding, because it changes sign upon the operation of reflection in the molecular plane and has zero overlap with the hydrogen $1s$ orbitals: $\pi_z{}^{nb} = 2p_z$.

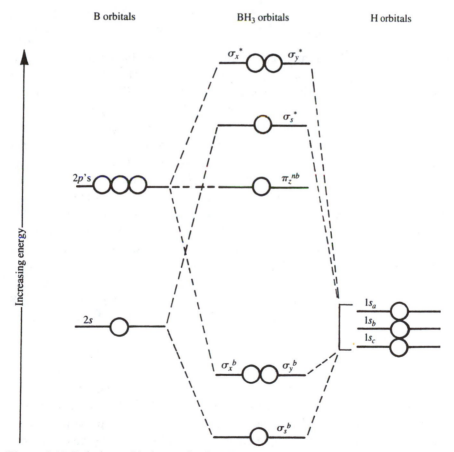

Figure 5-13 Relative orbital energies in BH₃.

Now that we know the forms of the seven molecular orbitals we can draw an energy-level diagram for BH_3, as in Figure 5-13. The σ_x^b and σ_y^b molecular orbitals are degenerate in trigonal-planar molecules such as BH_3. The σ_x^* and σ_y^* molecular orbitals are also degenerate. Because this is by no means obvious from Equations 5-3 through 5-7, we shall show that this is so in Exercise 5-2 at the end of this section.

There are six valence electrons in BH_3. Placing these electrons in the most stable molecular orbital of Figure 5-13, we see that the ground-state electronic configuration is $(\sigma_s^b)^2(\sigma_{x,y}^b)^4$. In the molecular-orbital description of the electronic structure of BH_3, the three electron-pair bonds are delocalized over the three atoms. The photoelectron spectrum of BH_3 has not been reported. By using molecular-orbital theory we predict that only two bands should be observed, one band due to electron ejection from the degenerate orbitals σ_x^b and σ_y^b and one band due to σ_s^b.

The Ammonia Molecule

Next we consider the ammonia molecule, NH_3. As we have noted, ammonia has eight valence electrons, compared to only six for BH_3. According to the molecular-orbital energy-level diagram employed for BH_3 (Figure 5-13), the NH_3 molecule would have the electronic configuration $(\sigma_s^b)^2(\sigma_{x,y}^b)^4(\pi_z^{nb})^2$ if it were trigonal-planar. However, it is easy to see (in Figure 5-14) that the π_z^{nb} molecular orbital becomes bonding if the molecule distorts to trigonal pyramidal. Also shown in the figure are the shapes of the σ_s^b, σ_x^b, and σ_y^b molecular orbitals in the trigonal-pyramidal shape. It can readily be appreciated that bending the molecule from trigonal planar to trigonal pyramidal will slightly stabilize the σ_s^b molecular orbital, whereas it will destabilize the σ_x^b and σ_y^b molecular orbitals, since they will overlap less with the relevant $2p_x$ and $2p_y$ orbitals on the central atom. The Walsh correlation diagram is also shown in Figure 5-14. The net result of our analysis is that we predict the ammonia molecule, with eight valence electrons, to be trigonal pyramidal due to the conversion of the π_z^{nb} orbital to σ_z^b. This prediction agrees with experiment and with the prediction based on the VSEPR method discussed in Section 2-15.

The photoelectron spectrum of NH_3 is presented in Figure 5-15; we see that three bands are observed, exactly as predicted by molecular-orbital theory. The first band in the photoelectron spectrum exhibits extensive vibrational structure. The vibrational spacing is 900 cm^{-1} and corresponds to the energy required to bring the molecule from a trigonal-pyramidal shape in NH_3 to one that is trigonal-planar in NH_3^+. This result agrees well with the character of the σ_z^b orbital, since it is this orbital that causes NH_3 to be pyramidal, rather than planar, like BH_3. Evidently the effect of the remaining

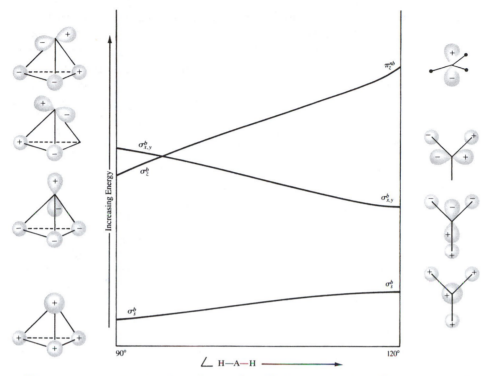

Figure 5-14 Atomic-orbital overlap and qualitative molecular-orbital energies of trigonal-planar and trigonal-pyramidal AH_3 molecules. Electron occupation of the σ_s^b, σ_x^b, and σ_y^b orbitals favors the trigonal-planar shape (for example, BH_3). Molecules with eight valence electrons, such as NH_3, require occupation of the π_z^{nb} orbital. The latter orbital favors a trigonal-pyramidal shape, because it then becomes σ_z^b.

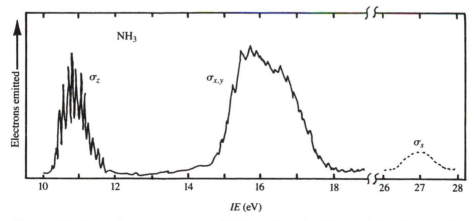

Figure 5-15 Photoelectron spectrum of NH_3. Adapted with permission from A.W. Potts and W.C. Price, *Proc. Roy. Soc. Lond.* A326: 181 (1972).

$\sigma_z{}^b$ electron in $NH_3{}^+$ is not sufficient to cause the molecule to distort to the pyramidal shape. The problem of determining molecular shape for the odd-electron isoelectronic series $BeH_3{}^{2-}$, $BH_3{}^-$, CH_3, and $NH_3{}^+$ was discussed in Section 2-15, where it was noted that the trend toward planarity increased throughout this series, with $BeH_3{}^{2-}$ strongly pyramidal and $NH_3{}^+$ strictly planar.

Exercise 5-1. Determine the coefficients c_1, c_2, and c_3 in the SALC $c_1 1s_a - c_2 1s_b - c_3 1s_c$, which will have the proper symmetry to overlap with the boron $2p_x$ orbital, as shown in Figure 5-12 for BH_3.

Solution. We must recognize three constraints that apply to any proper wave function, such as the SALC under consideration here. These conditions have already been discussed in Section 3-6. The first is the normalization condition, which states (neglecting overlap) that $c_1{}^2 + c_2{}^2 + c_3{}^2 = 1$. The second is that the symmetry of the molecule requires that $c_2 = c_3$. The third condition comes from the unit-orbital contribution, requiring that the squared contribution of any given orbital summed over all SALCs must be one. Considering the unit-orbital contribution for $1s_a$, we calculate a contribution of $(1/\sqrt{3})^2 = \frac{1}{3}$ from the first SALC (Equation 5-1), zero from the second SALC (Equation 5-2), and $c_1{}^2$ from the third SALC. Consequently $\frac{1}{3} + c_1^2 = 1$, and we obtain $c_1 = \sqrt{\frac{2}{3}}$. Now we can determine c_2 and c_3 from their constrained equality and the normalization condition:

$$c_1{}^2 + c_2{}^2 + c_3{}^2 = 1$$

$$c_1{}^2 + 2c_2{}^2 = 1$$

$$\tfrac{2}{3} + 2c_2{}^2 = 1$$

$$c_2{}^2 = \tfrac{1}{6}$$

$$c_2 = \sqrt{\tfrac{1}{6}} = \tfrac{1}{2}\sqrt{\tfrac{2}{3}}$$

Therefore the SALC that overlaps with the boron $2p_x$ orbital is

$$\sqrt{\tfrac{2}{3}}(1s_a - \tfrac{1}{2}1s_b - \tfrac{1}{2}1s_c).$$

Exercise 5-2. Show that the orbitals σ_x and σ_y are degenerate in trigonal-planar molecules.

Solution. The total overlap of the SALC $\sqrt{\tfrac{2}{3}}(1s_a - \tfrac{1}{2}1s_b - \tfrac{1}{2}1s_c)$ with $2p_x$ is labeled $S(\sigma_x)$; the total overlap of the SALC $\dfrac{1}{\sqrt{2}}(1s_b - 1s_c)$ with $2p_y$ is labeled $S(\sigma_y)$. A direct overlap, such as the overlap between $1s_a$ and $2p_x$ (Figure 5-12), is labeled $S(p_\sigma,s)$. To evaluate $S(\sigma_x)$ and $S(\sigma_y)$ in terms of $S(p_\sigma,s)$, we use the following calculations:

$$S(\sigma_x) = \sqrt{\tfrac{2}{3}} \int (2p_x)(1s_a - \tfrac{1}{2}1s_b - \tfrac{1}{2}1s_c) \, d\tau$$

$$= \sqrt{\tfrac{2}{3}}[S(p_\sigma,s) + \tfrac{1}{2}\cos 60° \, S(p_\sigma,s) + \tfrac{1}{2}\cos 60° \, S(p_\sigma,s)]$$

$$= \sqrt{\tfrac{2}{3}}(\tfrac{3}{2})[S(p_\sigma,s)]$$

$$= \sqrt{\tfrac{3}{2}}S(p_\sigma,s)$$

$$S(\sigma_y) = \frac{1}{\sqrt{2}} \int (2p_y)(1s_b - 1s_c) \, d\tau$$

$$= \frac{1}{\sqrt{2}}[\cos 30° \, S(p_\sigma,s) + \cos 30° \, S(p_\sigma,s)]$$

$$= \frac{1}{\sqrt{2}}\left(\frac{\sqrt{3}}{2} + \frac{\sqrt{3}}{2}\right)[S(p_\sigma,s)]$$

$$= \sqrt{\tfrac{3}{2}}S(p_\sigma,s)$$

Because the overlaps are the same for σ_x and σ_y and because the combining boron and hydrogen valence orbitals have the same initial energies, it follows that σ_x and σ_y are degenerate in trigonal-planar molecules. It is worth pointing out that σ_x and σ_y remain degenerate as long as the three-fold axis is maintained. Hence, even if the molecule is distorted from trigonal planar to trigonal pyramidal, the degeneracy remains.

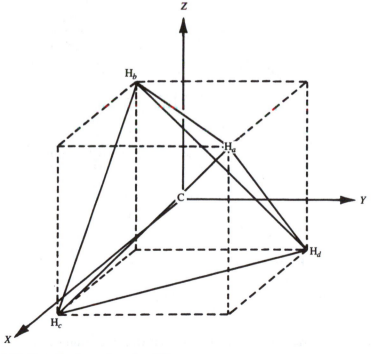

Figure 5-16 Coordinate system for CH_4.

5-4 Delocalized Molecular Orbitals for CH_4

The methane molecule, CH_4, is isoelectronic with the ammonia molecule because both have eight valence electrons. The tetrahedral structure of methane is shown in Figure 5-16. We place the carbon in the center of the cube, and the hydrogens at opposite corners, to define a regular tetrahedron. We choose the center of the cube as the origin of the rectangular coordinate system, with the X, Y, and Z axes perpendicular to the faces. We must use all of the carbon valence orbitals, $2s$, $2p_x$, $2p_y$, and $2p_z$, to form an adequate set of σ molecular orbitals.

We must now construct a SALC of the hydrogen $1s$ orbitals to match each of the carbon valence orbitals. The SALC $\frac{1}{2}(1s_a + 1s_b + 1s_c + 1s_d)$ is appropriate to match the carbon $2s$ orbital, as shown in Figure 5-17. The bonding and antibonding molecular orbitals are

$$\sigma_s{}^b = c_1 2s + c_2[\tfrac{1}{2}(1s_a + 1s_b + 1s_c + 1s_d)]$$

$$\sigma_s{}^* = c_3 2s - c_4[\tfrac{1}{2}(1s_a + 1s_b + 1s_c + 1s_d)]$$

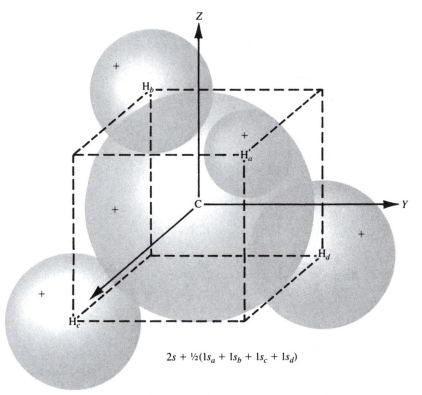

$$2s + \tfrac{1}{2}(1s_a + 1s_b + 1s_c + 1s_d)$$

Figure 5-17 Overlap of the carbon $2s$ orbital with the $1s$ orbitals of the hydrogen atoms in CH_4.

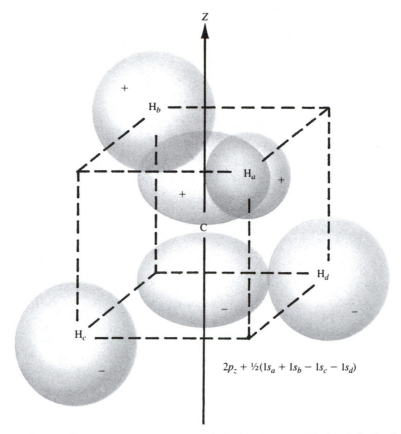

$$2p_z + \tfrac{1}{2}(1s_a + 1s_b - 1s_c - 1s_d)$$

Figure 5-18 Overlap of the carbon $2p_z$ orbital with the $1s$ orbitals of the hydrogen atoms in CH_4.

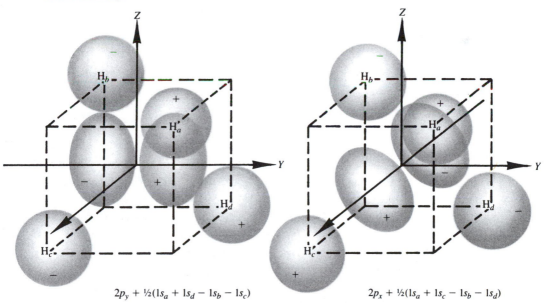

$$2p_y + \tfrac{1}{2}(1s_a + 1s_d - 1s_b - 1s_c)$$

$$2p_x + \tfrac{1}{2}(1s_a + 1s_c - 1s_b - 1s_d)$$

Figure 5-19 Overlap of the carbon $2p_x$ and $2p_y$ orbitals with the $1s$ orbitals of the hydrogen atoms in CH_4.

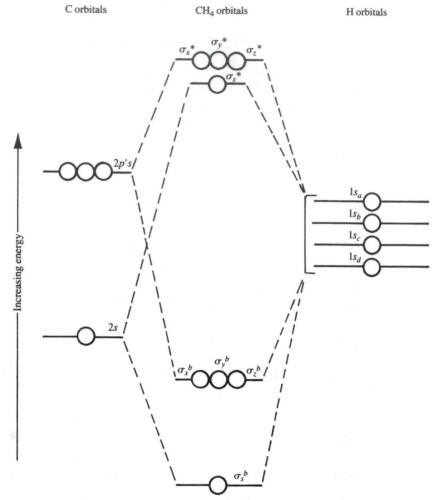

Figure 5-20 Relative orbital energies in CH_4.

The overlap of the four $1s$ orbitals with the carbon $2p_z$ orbital is shown in Figure 5-18. Hydrogen orbitals $1s_a$ and $1s_b$ overlap the plus lobe, and orbitals $1s_c$ and $1s_d$ overlap the minus lobe. Thus the proper combination is $\frac{1}{2}(1s_a + 1s_b - 1s_c - 1s_d)$. The $2p_x$ and $2p_y$ carbon orbitals overlap the four hydrogen orbitals in the same way as $2p_z$ does. This is shown in Figure 5-19. The SALCs are $\frac{1}{2}(1s_a + 1s_d - 1s_b - 1s_c)$ with $2p_y$, and $\frac{1}{2}(1s_a + 1s_c - 1s_b - 1s_d)$ with $2p_x$. The molecular orbitals are

$$\sigma_z^b = c_5 2p_z + c_6[\tfrac{1}{2}(1s_a + 1s_b - 1s_c - 1s_d)]$$

$$\sigma_z^* = c_7 2p_z - c_8[\tfrac{1}{2}(1s_a + 1s_b - 1s_c - 1s_d)]$$

$$\sigma_y^b = c_9 2p_y + c_{10}[\tfrac{1}{2}(1s_a + 1s_d - 1s_b - 1s_c)]$$

$$\sigma_y{}^* = c_{11}2p_y - c_{12}[\tfrac{1}{2}(1s_a + 1s_d - 1s_b - 1s_c)]$$
$$\sigma_x{}^b = c_{13}2p_x + c_{14}[\tfrac{1}{2}(1s_a + 1s_c - 1s_b - 1s_d)]$$
$$\sigma_x{}^* = c_{15}2p_x - c_{16}[\tfrac{1}{2}(1s_a + 1s_c - 1s_b - 1s_d)]$$

The molecular-orbital energy-level scheme for CH_4 is shown in Figure 5-20. The σ_x, σ_y, σ_z orbitals have the same overlap in a tetrahedral molecule, and are degenerate in energy. This is clear from the overlaps shown in Figures 5-18 and 5-19. The ground-state electronic configuration is

$$(\sigma_s{}^b)^2(\sigma_{x,y,z}{}^b)^6$$

Thus there are four delocalized σ bonds in CH_4. The molecular-orbital energy-level diagram presented in Figure 5-20 predicts that only two bands should be observed in the photoelectron spectrum, as experiment confirms (Figure 5-21).

5-5 Photoelectron Spectra for the Isoelectronic Sequence Ne, HF, H₂O, NH₃, and CH₄

We now wish to summarize the electronic structure of atoms, diatomic molecules, and polyatomic molecules by presenting the photoelectron spectra of

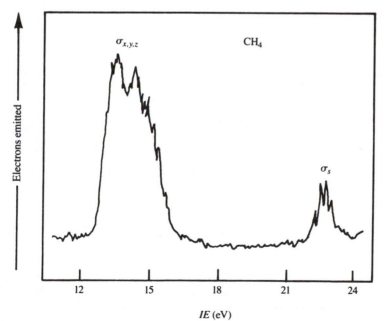

IE (eV)

Figure 5-21 Photoelectron spectrum of CH_4. Adapted with permission from A.W. Potts and W.C. Price, *Proc. Roy. Soc. Lond.* A326: 165 (1972).

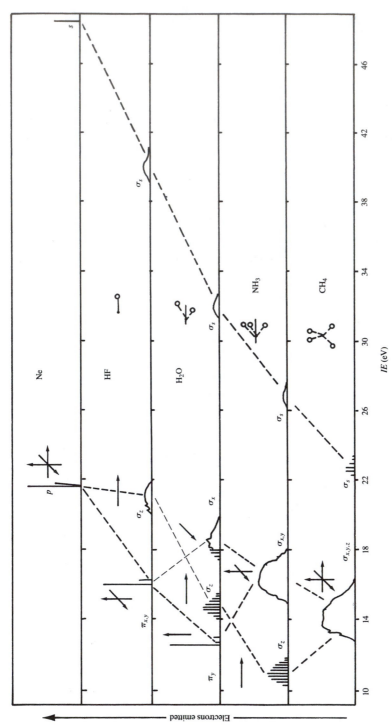

Figure 5-22 Schematic photoelectron spectra of the Ne atom and the isoelectronic molecules HF, H_2O, NH_3, and CH_4. Notice that we have interchanged the labels σ_z and σ_x for H_2O in this figure, compared to Figure 5-8. This is because in Figure 5-8 the twofold axis was the X axis, whereas in this figure the twofold axis is the Z axis. Adapted with permission from A.W. Potts and W.C. Price, *Proc. Roy. Soc. Lond.* A326: 181 (1972).

molecules isoelectronic with neon (Figure 5-22). These molecules are HF, H_2O, NH_3, and CH_4. The correlation of the spectra shown in Figure 5-22 was first discussed by the British scientist William C. Price. We can consider all of these molecules as being formed hypothetically by proton "withdrawal" from the neon nucleus.

Notice first that in its photoelectron spectrum neon exhibits only two sharp bands, which correspond to the valence electronic configuration $2s^2 2p^6$. If we focus our attention first on the band assigned to the neon 2s orbital, we see that each of the molecules also has a nondegenerate orbital that consists predominantly of central-atom 2s character. This orbital exhibits a smooth and continuous increase in energy (decrease in ionization energy) as protons are withdrawn from Ne to CH_4.

Next we turn to the behavior of the orbitals related to the $2p^6$ subshell of neon. This subshell is split into $(2\sigma)^2(1\pi)^4$ in HF. Note that relative to the $2p^6$ subshell of neon, the 1π orbital of HF is more destabilized than the 2σ orbital. This is so because the 2σ orbital is bonding, whereas the 1π orbital is nonbonding. In proceeding from HF to H_2O, the 1π orbital is further split into the π_y and σ_x orbitals. The σ_x orbital is actually stabilized relative to the π_x component of HF. This is because σ_x in H_2O has bonding character (Figure 5-5), whereas in HF the π_x component is nonbonding. Further proton withdrawal in proceeding from H_2O to NH_3 results in a coalescence of the π_y and σ_x orbitals of H_2O, forming the $\sigma_{x,y}$ orbital of NH_3. Finally, CH_4 is formed by withdrawing a proton from the nitrogen nucleus of NH_3 along the direction of the lone-pair orbital. This withdrawal should stabilize the σ_z lone-pair orbital of NH_3, as is observed. In the CH_4 molecule, the σ_z and $\sigma_{x,y}$ orbitals of NH_3 become degenerate in forming the $\sigma_{x,y,z}$ orbitals.

5-6 Delocalized Molecular Orbitals for CO_2 and XeF_2

At first sight it may seem strange to discuss such diverse molecules as carbon dioxide and xenon difluoride in the same section of this chapter. Yet both of these species are symmetric, linear, triatomic B—A—B molecules. The CO_2 molecule contains 16 valence electrons, and XeF_2 contains 22 valence electrons. From the point of view of simple MO theory, these two molecules differ only in the number of occupied molecular orbitals. Now we determine the forms of these molecular orbitals.

Molecular Orbitals for CO_2

Let us first treat the CO_2 molecule, realizing that the general molecular-orbital energy-level scheme also will be applicable to XeF_2. The CO_2 mole-

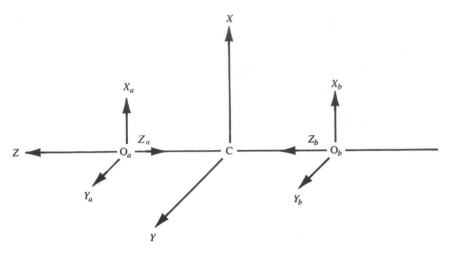

Figure 5-23 Coordinate system for CO_2.

cule is shown in Figure 5-23, drawn against our standard coordinate system. This molecule is an example of a linear triatomic molecule in which all three atoms have ns and np valence orbitals. The $2s$ and $2p_z$ carbon orbitals are used for bonding, along with the $2p_z$ orbitals on each oxygen atom.* The orbitals are the same as for BeH_2, except that now the oxygen atoms use mainly the $2p_z$ orbitals instead of the $1s$ valence orbitals used by the hydrogen atoms. The σ wave functions are

$$\sigma_g^{nb}(O_{2S}) = \frac{1}{\sqrt{2}}(2s_a + 2s_b)$$

$$\sigma_u^{nb}(O_{2S}) = \frac{1}{\sqrt{2}}(2s_a - 2s_b)$$

$$\sigma_g^b(s) = c_1 2s + c_2\left[\frac{1}{\sqrt{2}}(2p_{z_a} + 2p_{z_b})\right]$$

$$\sigma_g^*(s) = c_3 2s - c_4\left[\frac{1}{\sqrt{2}}(2p_{z_a} + 2p_{z_b})\right]$$

$$\sigma_u^b(z) = c_5 2p_z + c_6\left[\frac{1}{\sqrt{2}}(2p_{z_a} - 2p_{z_b})\right]$$

$$\sigma_u^*(z) = c_7 2p_z - c_8\left[\frac{1}{\sqrt{2}}(2p_{z_a} - 2p_{z_b})\right]$$

*The oxygen valence orbitals are $2s$ and $2p$. Thus a much better approximate σ MO scheme would include both $2s$ and $2p_z$ oxygen orbitals. For simplicity, however, we use only the $2p_z$ oxygen orbitals in forming the σ MOs. Later, we will discuss the effect of $s-p$ mixing of the oxygen orbitals.

The π molecular orbitals are made up of the $2p_x$ and $2p_y$ valence orbitals of the three atoms. Let us derive the $\pi(x)$ orbitals for CO_2. There are two different SALCs of the oxygen $2p_x$ orbitals:

$$\text{SALC } (\pi_u): \quad \frac{1}{\sqrt{2}}(2p_{x_a} + 2p_{x_b})$$

and

$$\text{SALC } (\pi_g): \quad \frac{1}{\sqrt{2}}(2p_{x_a} - 2p_{x_b})$$

The combination $\frac{1}{\sqrt{2}}(2p_{x_a} + 2p_{x_b})$ overlaps the carbon $2p_x$ orbital, as shown in Figure 5-24, because the correct symmetry match is found for π_u.

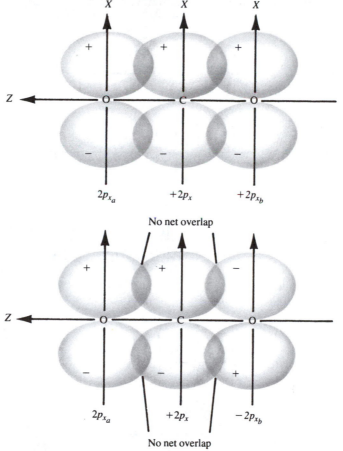

Figure 5-24 Overlap of the $2p_x$ orbitals of the carbon atom and the two oxygen atoms.

Since X and Y are equivalent, we obtain

$$\pi_u{}^b(x) = c_9 2p_x + c_{10}\left[\frac{1}{\sqrt{2}}(2p_{x_a} + 2p_{x_b})\right]$$

$$\pi_u{}^b(y) = c_{11} 2p_y + c_{12}\left[\frac{1}{\sqrt{2}}(2p_{y_a} + 2p_{y_b})\right]$$

$$\pi_u{}^*(x) = c_{13} 2p_x - c_{14}\left[\frac{1}{\sqrt{2}}(2p_{x_a} + 2p_{x_b})\right]$$

$$\pi_u{}^*(y) = c_{15} 2p_y - c_{16}\left[\frac{1}{\sqrt{2}}(2p_{y_a} + 2p_{y_b})\right]$$

In contrast, the π_g SALC $\left[\frac{1}{\sqrt{2}}(2p_{x_a} - 2p_{x_b})\right]$ has zero net overlap with the carbon $2p_x$ orbital (see Figure 5-24), and therefore is nonbonding in the molecular-orbital scheme. We could have known that the overlap was zero, because the carbon $2p_x$ orbital has π_u symmetry, whereas $\left[\frac{1}{\sqrt{2}}(2p_{x_a} - 2p_{x_b})\right]$

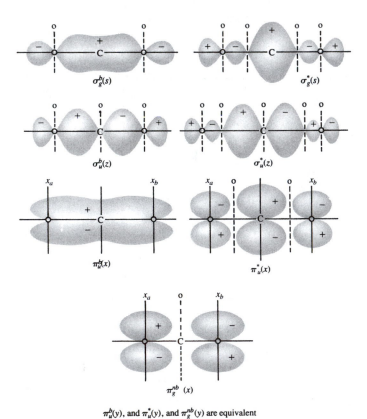

$\sigma_g^b(s)$ $\sigma_g^*(s)$

$\sigma_u^b(z)$ $\sigma_u^*(z)$

$\pi_u^b(x)$ $\pi_u^*(x)$

$\pi_g^{nb}(x)$

$\pi_u^b(y)$, and $\pi_u^*(y)$, and $\pi_g^{nb}(y)$ are equivalent

to $\pi_u^b(x)$, $\pi_u^*(x)$, and $\pi_g^{nb}(x)$.

Figure 5-25 Spatial representations of the molecular orbitals of CO_2.

has π_g symmetry. Orbitals must have the same symmetry to overlap or interact. We have, then, the normalized wave functions

$$\pi_g^{nb}(x) = \frac{1}{\sqrt{2}} (2p_{x_a} - 2p_{x_b})$$

and

$$\pi_g^{nb}(y) = \frac{1}{\sqrt{2}} (2p_{y_a} - 2p_{y_b})$$

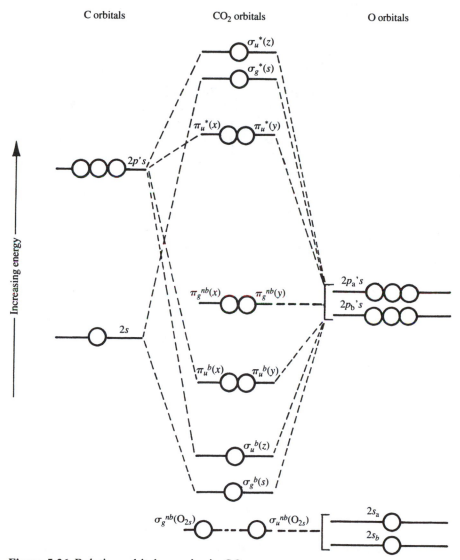

Figure 5-26 Relative orbital energies in CO_2.

Spatial representations of the MOs for CO_2 are shown in Figure 5-25. The MO energy-level scheme for CO_2 is given in Figure 5-26. Note that the oxygen orbitals are more stable than the carbon orbitals, in accord with the valence-orbital ionization energies presented in Table 4-4. There are 16 valence electrons (C is $2s^2 2p^2$, O is $2s^2 2p^4$) to place in the levels shown in the scheme. The ground-state electronic configuration of CO_2 therefore is

$$[\sigma_g^{nb}(O_{2s})]^2 [\sigma_u^{nb}(O_{2s})]^2 [\sigma_g^b(s)]^2 [\sigma_u^b(z)]^2 [\pi_u^b(x,y)]^4 [\pi_g^{nb}(x,y)]^4$$

There are four electrons in σ^b orbitals and four electrons in π^b orbitals. Thus we have two σ bonds and two π bonds for CO_2, in good agreement with the Lewis structure $:\overset{..}{O}{=}C{=}\overset{..}{O}:$.

The Photoelectron Spectrum of CO_2

The molecular-orbital energy-level diagram of CO_2 (Figure 5-26) predicts that the photoelectron spectrum of the valence electrons should exhibit six bands. Furthermore, it predicts that the second, third, and fourth bands should exhibit vibrational fine structure, because these bands correspond to the bonding molecular orbitals σ_g^b, σ_u^b, and π_u^b. The experimental spectrum shown in Figure 5-27 indicates that only the second band exhibits vibrational fine structure. This band is assigned to ionization from the $\pi_u^b(x,y)$ molecular orbital. The fact that the third and fourth bands (assigned to ionization from the $\sigma_g^b(s)$ and $\sigma_u^b(z)$ orbitals) do not show vibrational fine structure implies that these orbitals are nonbonding.

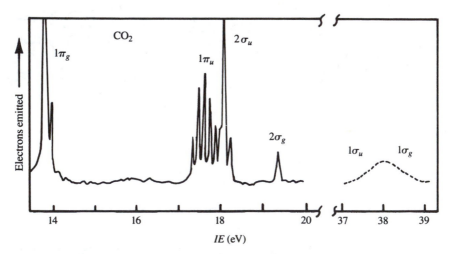

Figure 5-27 Photoelectron spectrum of CO_2. Adapted with permission from D.W. Turner, C. Baker, A.D. Baker, and C.R. Brundle, *Molecular Photoelectron Spectroscopy,* © 1970 by John Wiley & Sons (London), p. 103.

To understand the nonbonding character of the $\sigma_g^b(s)$ and $\sigma_u^b(z)$ orbitals, we must include the oxygen $2s$ orbitals in the bonding description. We assumed previously that the oxygen $2s$ orbitals did not interact with the carbon valence orbitals, and consequently the two combinations of oxygen $2s$ orbitals were designated $\sigma_g^{nb}(O_{2s})$ and $\sigma_u^{nb}(O_{2s})$. If we take account of the fact that all orbitals of σ_g symmetry can mix with each other and that all orbitals of σ_u symmetry can likewise mix together, we effectively introduce $2s$–$2p_z$ mixing on the oxygen atoms. The effect of this s–p mixing on the orbital energy and orbital character of the σ orbitals is shown in Figure 5-28. We see that the most stable orbitals, which were previously assumed to be nonbonding oxygen $2s$ orbitals, now become bonding. In contrast, the orbitals previously designated $\sigma_g^b(s)$ and $\sigma_u^b(z)$ now become relatively nonbonding.* Using simplified notation we now designate the ground-state electronic configuration of CO_2 as

$$(1\sigma_g)^2(1\sigma_u)^2(2\sigma_g)^2(2\sigma_u)^2(1\pi_u)^4(1\pi_g)^4$$

It is understood that the bonding orbitals are $(1\sigma_g)^2$, $(1\sigma_u)^2$, and $(1\pi_u)^4$. The $2\sigma_g$, $2\sigma_u$, and $1\pi_g$ orbitals are nonbonding, in agreement with the lack of vibrational structure observed for the bands assigned to these orbitals in the photoelectron spectrum.

Molecular Orbitals for XeF_2

The xenon difluoride molecule is linear and has 22 valence electrons. If we employ the xenon $5s$ and $5p$ valence orbitals, and the $2s$ and $2p$ valence orbitals on each fluorine atom, we can construct an energy-level scheme for XeF_2 that is analogous to that of CO_2 given in Figure 5-26. We may neglect

---

*This s–p mixing is entirely analogous to the s–p mixing discussed in Section 4-6 for diatomic molecules. In order to illustrate this analogy, let us examine carefully the mixing of $\sigma_g^{nb}(O_{2s})$ and $\sigma_g^b(s)$ to form $1\sigma_g$ and $2\sigma_g$ (see Figure 5-28). This mixing can be described quantitatively as

$$1\sigma_g = c_1\sigma_g^{nb}(O_{2s}) + c_2\sigma_g^b(s)$$

and

$$2\sigma_g = c_1\sigma_g^b(s) - c_2\sigma_g^{nb}(O_{2s})$$

This construction of orbitals is required in order to ensure that $1\sigma_g$ and $2\sigma_g$ will be orthogonal to each other.

In the absence of s–p mixing we have $c_1 = 1$ and $c_2 = 0$, so that $1\sigma_g = \sigma_g^{nb}(O_{2s})$ and $2\sigma_g = \sigma_g^b(s)$. Figure 5-28 displays a situation in which $c_1 \simeq c_2$ so that sp hybrids are formed on the oxygen atoms; this results in the $2\sigma_g$ orbital being slightly bonding. However, based on our analysis so far, there is nothing to prevent the $2\sigma_g$ orbital from being slightly antibonding:

The latter situation can occur if $c_2 > c_1$. Our main concern here is not whether the $2\sigma_g$ orbital is slightly bonding or antibonding, but rather that it is *mainly* nonbonding.

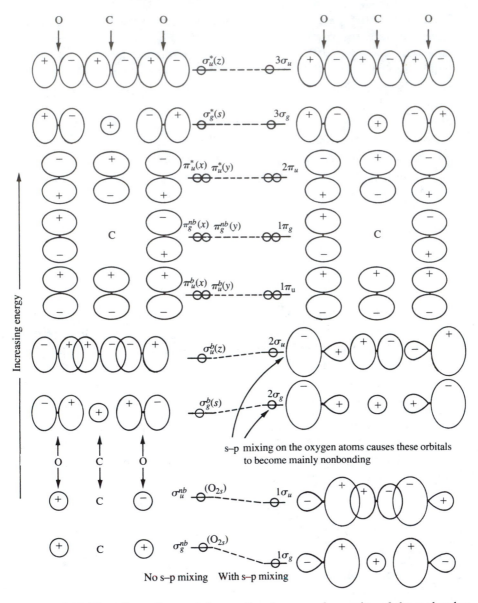

Figure 5-28 The effect of s–p mixing on the shapes and energies of the molecular orbitals for a linear symmetric molecule such as CO_2. Notice that the $2\sigma_g$ and $2\sigma_u$ molecular orbitals become relatively nonbonding as a result of s–p mixing at the oxygen atoms. No s–p mixing can occur at the carbon atom, because the carbon atom 2s and $2p_z$ orbitals do not have the same symmetry in a linear symmetric molecule.

s–p mixing for the fluorine atoms in XeF_2, where we could not do so for the oxygen atoms in CO_2. This is because the energy difference between the 2s and 2p valence orbitals of oxygen is only 16.5 eV, whereas the difference in the fluorine atom is 21.6 eV (Table 4-4). By using the energy-level scheme for CO_2 (Figure 5-26), we may arrive at the ground-state electronic configuration of XeF_2:

$$[\sigma_g^{nb}(F_{2s})]^2[\sigma_u^{nb}(F_{2s})]^2[\sigma_g^b(s)]^2[\sigma_u^b(z)]^2[\pi_u^b(x,y)]^4$$
$$[\pi_g^{nb}(x,y)]^4[\pi_u^*(x,y)]^4[\sigma_g^*(s)]^2. \quad (5\text{-}8)$$

In simplified notation this becomes

$$(1\sigma_g)^2(1\sigma_u)^2(2\sigma_g)^2(2\sigma_u)^2(1\pi_u)^4(1\pi_g)^4(2\pi_u)^4(3\sigma_g)^2$$

This electronic configuration predicts that there is only one net σ bond in XeF_2 and no net π bonds. That is, each Xe—F bond has only a half bond character. This prediction agrees with the observed Xe—F average bond energy, which is only about 30 kcal mole⁻¹. The electronic configuration just described also is in agreement with the observed photoelectron spectrum of XeF_2 presented in Figure 5-29.

The photoelectron spectrum of XeF_2 has two interesting features. The first is that the lowest ionization energy belongs to electron ejection from the π_u^* orbital and not from σ_g^* as predicted by Equation 5-8. This configuration is based on the molecular-orbital energy-level scheme for CO_2 presented in Figure 5-26. The diagram for CO_2 took no account of the use of orbitals beyond the valence ns and np orbitals. However, the nd orbitals may be important in XeF_2. For example, the ground-state electronic configuration of the xenon atom is $[Kr]4d^{10}5s^25p^65d^06s^06p^0$. Inclusion of the $5d$ orbitals

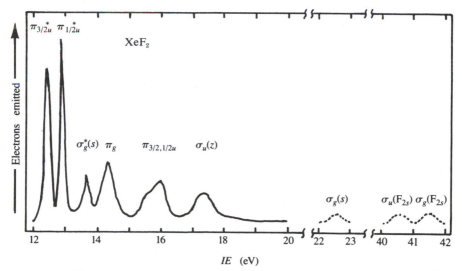

Figure 5-29 Photoelectron spectrum of XeF_2. Adapted with permission from C.R. Brundle, M.B. Robin, and G.R. Jones, *J. Chem. Phys.* 52: 3383 (1970).

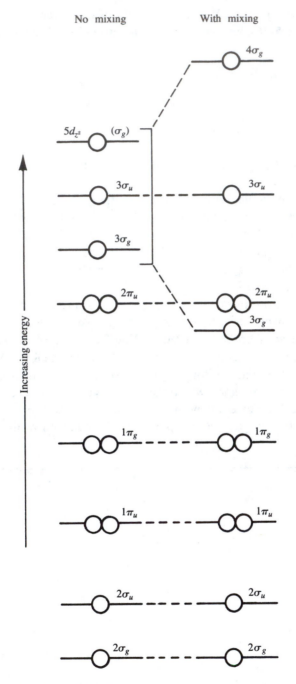

Figure 5-30 The effect of Xe $5d_{z^2}$ orbital mixing with the $\sigma_g{}^*(s)\,(=3\sigma_g)$ orbital of XeF$_2$. The exact energy of the $5d_{z^2}$ orbital relative to $3\sigma_g$ and $3\sigma_u$ is uncertain. We have represented $5d_{z^2}$ as the most energetic orbital. Orbital mixing could cause the $3\sigma_g$ orbital to become more stable than the $2\pi_u$ orbital.

will have a stabilizing influence on the $\sigma_g{}^*$ orbital, because the $5d_{z^2}$ orbital has σ_g symmetry. This stabilizing effect is illustrated in Figure 5-30. The importance of the valence nd orbitals in the bonding of noble gas compounds is controversial. In order to accommodate the five electron pairs surrounding the Xe atom in the Lewis structure of XeF_2, the hybrid-orbital method requires dsp^3 hybridization of the xenon atom (Section 3-5). Therefore, the $5d_{z^2}$ orbital is included in the bonding scheme. We have seen that molecular-orbital theory can describe the bonding in XeF_2 without including the valence nd orbitals. Unfortunately, there is no unequivocal experimental or theoretical proof as to whether nd valence orbitals do participate in the bonding for these compounds.

The second interesting feature of the XeF_2 photoelectron spectrum is that the first *two* bands are connected to ionization from the $\pi_u{}^*$ orbital (Figure 5-29). These two bands result from a coupling of the orbital angular momentum with the spin angular momentum; this commonly is called spin-orbit coupling (Section 1-15). The orbital angular momentum of a π orbital is 1, and the net spin angular momentum is $\frac{1}{2}$ in the ionized state, because there is an unpaired electron in the $\pi_u{}^*$ orbital. The electronic configuration of the $\pi_u{}^*$ orbital in the ionized state is $(\pi_u{}^*)^3$. Consequently the spin-orbit coupling can produce two states, designated $\pi_{3/2u}$ and $\pi_{1/2u}$; these are indicated in Figure 5-29. As pointed out in Section 1-14, the magnitude of spin-orbit coupling depends on the position of the elements in the periodic table. Heavy elements exhibit large spin-orbit coupling, whereas light elements exhibit negligible spin-orbit coupling. Xenon is a heavy element, so the spin-orbit coupling is prominent in the photoelectron spectrum of Figure 5-29. Because the π_g orbital contains no central-atom (xenon) orbital character (Figure 5-25), no spin-orbit coupling is observed in the band assigned to that orbital. The magnitude of the spin-orbit coupling in the $\pi_u{}^*$ orbital can be obtained by taking the energy difference between the $\pi_{3/2u}$ and $\pi_{1/2u}$ photoelectron bands (0.47 eV). Comparison of this value with the spin-orbit coupling constant (0.87 eV) in the xenon ion, Xe^+, indicates that the $\pi_u{}^*$ orbital of XeF_2 is $(0.47/0.87) \times 100 = 54$ percent, localized on the xenon atom. The remaining 46 percent is equally distributed between the two fluorine atoms.

5-7 Molecular-Orbital Theory and Molecular Topology

In Section 2-16 we discussed the fact that the central atom in a molecule usually is the least electronegative atom, provided that low formal charges are maintained and the maximum number of bonds is formed. We stated that the central atom should be able to share its electron density with other atoms, and consequently it should have a low electronegativity.

Molecular-orbital theory provides a reason for the low electronegativity of the central atom in terms of the delocalized molecular orbitals. The higher-energy occupied molecular orbitals have nodes that pass through the center of the molecule (Figures 5-3 and 5-25). The nodes reduce the electron density in the center of the molecule and build up electron density at the terminal atoms. Thus the central atoms must be able to give up electron density, and the terminal atoms must be able to accept it.

5-8 Delocalized Molecular Orbitals in Carbon Compounds

The molecular orbitals of carbon compounds can be analyzed in the same way as those of the other molecules we have discussed. In this section we consider two—ethylene and benzene.

Ethylene

Ethylene is an unsaturated, planar hydrocarbon with the following Lewis structure:

$$
\begin{array}{ccc}
\text{H} & & \text{H} \\
\diagdown & & \diagup \\
& \text{C}=\text{C} & \\
\diagup & & \diagdown \\
\text{H} & & \text{H}
\end{array}
$$

Atomic-orbital overlap for the occupied C_2H_4 molecular orbitals is shown in Figure 5-31. The molecular-orbital energy-level scheme for C_2H_4, shown in Figure 5-32, is constructed as follows. The most stable $\psi_1(\sigma_g)$ orbital contains no nodes. The next two orbitals, $\psi_2(\sigma_u)$ and $\psi_3(\sigma_u)$, contain one node each and are C—H bonding orbitals. The fourth orbital, $\psi_4(\sigma_g)$, is both C—C and C—H bonding and contains two nodes, as shown in Figure 5-31. The fifth molecular orbital, $\psi_5(\sigma_g)$, is C—H bonding and also contains two nodes. Finally, the highest occupied molecular orbital is $\psi_6(\pi_u)$, which contains only one node and is C—C π bonding.

The orbital assignment of the photoelectron bands shown in Figure 5-33 is based on quantum-mechanical calculations, a study of the vibrational structure exhibited in the photoelectron spectrum, and other spectroscopic measurements of ethylene. Notice that the highest occupied molecular orbital of ethylene is a π orbital. As a general rule, we will find that the π electrons of unsaturated compounds reside in the higher-energy occupied molecular orbitals. Although $\psi_6(\pi_u)$ has only one node, it is less stable than $\psi_5(\sigma_g)$, which contains two nodes. This can be understood by realizing that

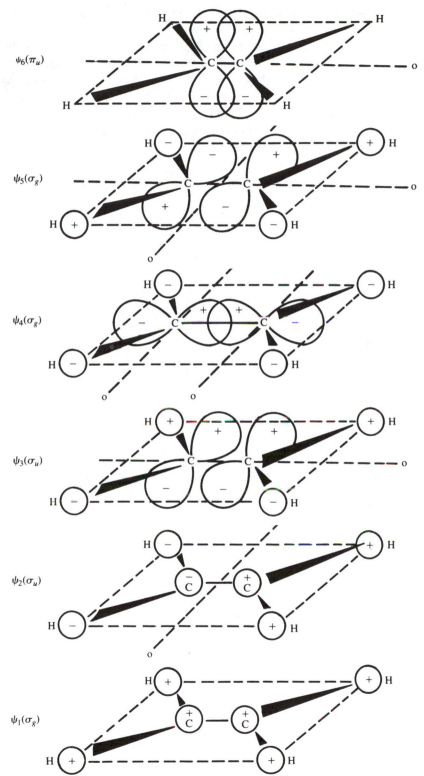

Figure 5-31 Overlap of atomic orbitals to form molecular orbitals in C_2H_4.

the π orbitals have less electron density between the nuclei than the σ orbitals. Notice also that the lowest unoccupied molecular orbital of ethylene is the antibonding π^* orbital. Unsaturated hydrocarbons generally have π^* orbitals at lower energy than σ^* orbitals. By the same token that π orbitals are less bonding than σ orbitals, so also π^* orbitals are less antibonding than σ^* orbitals. In general, π^* orbitals concentrate less electron density outside the nuclei than do σ^* orbitals.

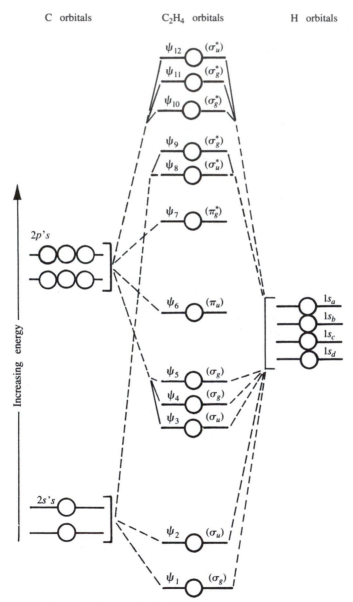

Figure 5-32 Relative orbital energies in C_2H_4.

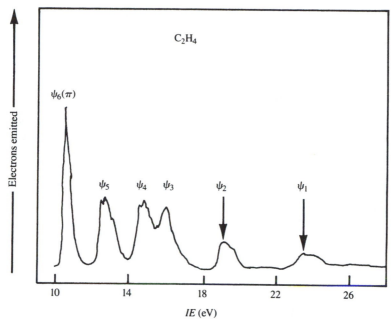

Figure 5-33 Photoelectron spectrum of C_2H_4. Adapted from C.R. Brundle, M.B. Robin, H. Basch, M. Pinsky, and A. Bond, *J. Amer. Chem. Soc.* 92: 3863 (1970), © American Chemical Society.

Unsaturated hydrocarbons such as ethylene absorb light at longer wavelengths than do saturated hydrocarbons. For example, ethylene has an ultraviolet absorption peak at 171 nm (58,500 cm^{-1}), whereas ethane does not begin to absorb strongly before 160 nm (62,500 cm^{-1}). This fact suggests that the separation between σ bonding and σ antibonding orbitals in hydrocarbons is larger than the separation between π bonding and π antibonding orbitals (Figure 5-32), as concluded earlier based on the nature of π and σ orbitals. For this reason it is common to include only π^b and π^* orbitals in simplified molecular-orbital formulations of unsaturated hydrocarbons. Spatial representations and relative energies of the π^b and π^* orbitals of C_2H_4 are shown in Figure 5-34. Using the π-orbital electronic structure of C_2H_4, we assign the absorption peak at 58,500 cm^{-1} to the electronic transition $\pi^b \rightarrow \pi^*$.

If the π-electron system of an unsaturated hydrocarbon is more extensive than that of ethylene, the energy separation between the highest occupied π^b orbital and the lowest unoccupied π^* level becomes smaller, and the energy absorption occurs at longer wavelengths. Such extensive π-electron systems are found in *conjugated polyenes,* compounds in which conventional structural formulas show alternate single and double bonds:

$$-\overset{|}{\underset{|}{C}}=\overset{}{\underset{|}{C}}-\overset{|}{\underset{|}{C}}=\overset{}{\underset{|}{C}}-\overset{|}{\underset{|}{C}}=\overset{}{\underset{|}{C}}-$$

Conjugated polyene skeleton

Polyenes having ten or more conjugated double bonds absorb visible light, hence they are colored. The pigments responsible for light perception in the human eye contain long, conjugated polyene chains, as do some vegetable pigments such as carotene, the substance which gives carrots their color.

Benzene

The structure of benzene was discussed in Section 3-6, where it was pointed out that the benzene molecule has 24 electrons in the σ orbitals and 6 electrons in the π orbitals. The Lewis structure can be represented by two

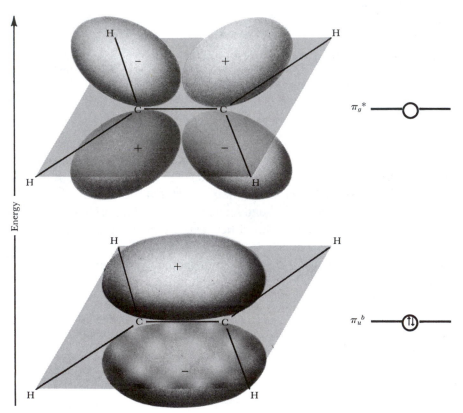

Figure 5-34 The bonding and antibonding π molecular orbitals in C_2H_4. The electronic structure of the ground state is $(\pi_u^b)^2$.

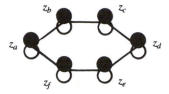

Figure 5-35 $2p(\pi)$ valence orbitals in C_6H_6.

resonance forms to indicate the equivalence of all the C—C bonds in benzene:

We could draw spatial representations for all of the σ and π molecular orbitals of benzene just as we did for ethylene. For many applications in chemistry, however, we are interested mainly in the symmetry characteristics of the higher occupied and the lower unoccupied molecular orbitals. We know that the occupied σ orbitals are generally more stable than the occupied π orbitals for most unsaturated organic molecules, so we shall describe the forms of the π orbitals only.

The general expressions for the delocalized π molecular orbitals in benzene are given by appropriate SALCs of the six $2p(\pi)$ valence orbitals.* Using the lettering system shown in Figure 5-35, we can formulate the bonding orbital of lowest energy as

$$\psi(\pi_1{}^b) = \frac{1}{\sqrt{6}}(z_a + z_b + z_c + z_d + z_e + z_f)$$

The highest-energy antibonding orbital has nodes between the nuclei:

$$\psi(\pi_6{}^*) = \frac{1}{\sqrt{6}}(z_a - z_b + z_c - z_d + z_e - z_f)$$

The other molecular orbitals have energies between $\pi_1{}^b$ and $\pi_6{}^*$:

$$\psi(\pi_2{}^b) = \frac{1}{2\sqrt{3}}(2z_a + z_b - z_c - 2z_d - z_e + z_f)$$

*The rules for constructing the benzene molecular orbitals are straightforward, but require symmetry and orthogonality principles that are not presented in this book.

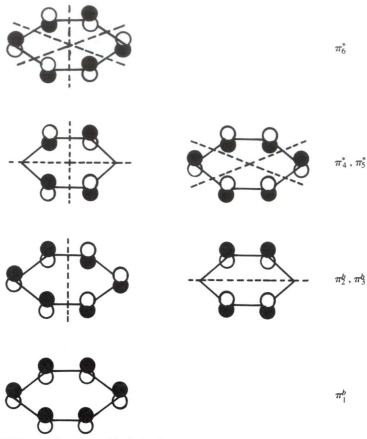

Figure 5-36 π molecular orbitals for benzene.

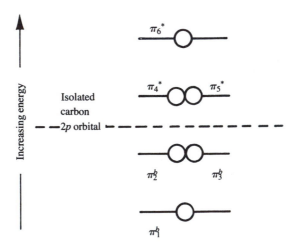

Figure 5-37 Energy-level scheme for the π molecular orbitals in benzene.

$$\psi(\pi_3{}^b) = \tfrac{1}{2}(z_b + z_c - z_e - z_f)$$

$$\psi(\pi_5{}^*) = \frac{1}{2\sqrt{3}}(2z_a - z_b - z_c + 2z_d - z_e - z_f)$$

$$\psi(\pi_4{}^*) = \tfrac{1}{2}(-z_b + z_c - z_e + z_f)$$

The π molecular orbitals for benzene are shown in Figure 5-36. A calculation of the energies of the six π molecular orbitals gives the energy-level scheme for C_6H_6 shown in Figure 5-37.

There are a total of 30 valence electrons in benzene. Of these, 24 are used in σ bonding, producing six C—C bonds and six C—H bonds, and leaving 6 electrons for the π molecular orbitals shown in Figure 5-37. In the ground state the π electrons have the configuration $(\pi_1{}^b)^2(\pi_2{}^b)^2(\pi_3{}^b)^2$, giving a total of three π bonds. It follows that each carbon–carbon bond consists of a full σ bond and a half π bond. And indeed, the carbon–carbon bond length in C_6H_6 is 1.390 Å, which is between the C—C and C=C bond lengths.

The photoelectron spectrum of benzene agrees with the energy-level scheme presented in Figure 5-37. The first ionization band has been assigned to electron ejection from the degenerate pair of molecular orbitals $\psi(\pi_2{}^b)$ and $\psi(\pi_3{}^b)$.

5-9 The Frontier-Orbital Concept

The Japanese chemist Kenichi Fukui has introduced the *frontier-orbital* concept to describe the reaction of Lewis acids and bases in terms of the molecular-orbital theory. The frontier orbitals of any molecule are the highest occupied molecular orbital (HOMO) and the lowest unoccupied molecular orbital (LUMO). Recall that a Lewis acid is an electron-pair acceptor and that a Lewis base is an electron-pair donor.

We already have used the frontier orbital theory implicitly in discussing the electronic structure and molecular shape of $H_3{}^+$. In Section 5-1 we described the formation of $H_3{}^+$ in terms of proton attack on the occupied σ^b molecular orbital of H_2 to form the triangular $H_3{}^+$ species:

$$
\begin{array}{c}
\text{H}^+ \\
\downarrow \\
\text{H——H}
\end{array}
\quad\longrightarrow\quad
\left[
\begin{array}{c}
\text{H} \\
\text{H}\ \ \text{H}
\end{array}
\right]^+
$$

In frontier orbital terminology, this reaction results in a shift of electron density from the HOMO (σ^b) of the Lewis base (H_2) to the LUMO ($1s$) of the Lewis acid (H^+).

We also used the frontier-orbital concept implicitly in presenting the electronic structure and molecular shape of diborane. In Section 3-8 we imagined forming B_2H_6 in the hypothetical reaction

$$B_2H_4{}^{2-} + 2H^+ \rightarrow B_2H_6$$

Now, $B_2H_4{}^{2-}$ is isoelectronic with C_2H_4 and therefore its HOMO is a π^b molecular orbital (see Figures 3-34 and 5-36). Since the proton is a Lewis acid, its LUMO ($1s$) can withdraw electron density from the HOMO (π^b) of the Lewis base $B_2H_4{}^{2-}$ in forming B_2H_6.

These two examples, $H_3{}^+$ and B_2H_6, provide the necessary background for a general statement that may be made regarding Lewis acid-base reactions. *Electron density is transferred from the HOMO of the Lewis base to the LUMO of the Lewis acid.* This statement does not imply that all other molecular orbitals will be unaltered in the reaction. However, it does indicate that the most important interaction between orbitals will be that involving the HOMO–LUMO pair. It is easy to see why the frontier orbitals are the most important in a chemical reaction. These orbitals are the least bound energetically, and therefore the most readily available for interaction with other molecules.

Proton Affinity and the Frontier-Orbital Concept

Let us now investigate the relationship betweem orbital energies and the energy released when a proton attacks the HOMO of a Lewis base. The *proton affinity, PA,* of an atom or molecule, B, is the energy released upon reaction of B with a proton in the gas phase:

$$B + H^+ \rightarrow BH^+ \qquad PA(B) = -\Delta H \qquad (5\text{-}9)$$

The development of *PA* may arbitrarily be divided into two hypothetical steps:

$$B + H^+ \rightarrow B^+ + H \qquad \Delta H_1 = IE(B) - IE(H) \qquad (5\text{-}10)$$
$$B^+ + H \rightarrow BH^+ \qquad \Delta H_2 = -DE(B^+ - H) \qquad (5\text{-}11)$$

In the first step, Equation 5-10, the proton attacks B and one electron is transferred from B to H^+. According to Fukui's frontier-orbital concept, the proton's initial point of attack will be the position of highest electron density in the HOMO. In the second step, Equation 5-11, the separated atoms B^+ and H unite to form BH^+ with a bond energy $DE(B^+ - H)$. In this simple analysis, the final position of proton attachment will correspond to the intial site of attack.

By combining Equations 5-9 through 5-11 we obtain

$$PA(B) = IE(H) - IE(B) + DE(B^+ - H) \tag{5-12}$$

where $IE(H) = 13.598$ eV. Equation 5-12 shows that the smaller the first ionization energy of molecule B, $IE(B)$, the larger its proton affinity will be. This is reasonable because the proton is an electrophile and can better remove electron density from B if the ionization energy of B is low.

For molecules isoelectronic with Ne, we can use the Lewis electron-dot structures to depict their reaction with the proton:

$$PA \text{ (kcal mole}^{-1})$$

$$:\ddot{\text{N}}\text{e}: + \text{H}^+ \rightarrow [:\ddot{\text{N}}\text{e}-\text{H}]^+ \qquad\qquad 48 \tag{5-13}$$

$$\text{H}-\ddot{\text{F}}: + \overset{+}{\text{H}} \rightarrow \left[\begin{array}{c} \text{H}-\ddot{\text{F}}: \\ | \\ \text{H} \end{array} \right]^+ \qquad 112 \tag{5-14}$$

$$\text{H}-\overset{..}{\underset{\text{H}}{\text{O}}}: + \text{H}^+ \rightarrow \left[\text{H}-\overset{..}{\underset{|}{\text{O}}}-\text{H} \atop \text{H} \right]^+ \qquad 164 \tag{5-15}$$

$$\text{H}-\overset{..}{\underset{|}{\text{N}}}-\text{H} + \text{H}^+ \rightarrow \left[\begin{array}{c} \text{H} \\ | \\ \text{H}-\text{N}-\text{H} \\ | \\ \text{H} \end{array} \right]^+ \qquad 201 \tag{5-16}$$

$$\begin{array}{c} \text{H} \qquad \text{H} \\ \diagdown \quad \diagup \\ \text{C} \\ \diagup \quad \diagdown \\ \text{H} \qquad \text{H} \end{array} \quad + \text{H}^+ \rightarrow \text{CH}_5^+ \qquad 126 \tag{5-17}$$

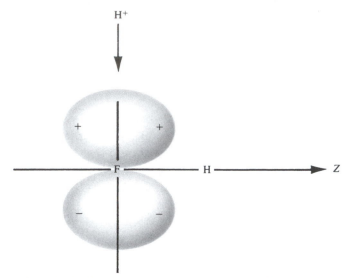

Figure 5-38 Proton attack on the highest occupied molecular orbital of HF produces an angular H_2F^+ ion.

According to the frontier-orbital concept, the proton attacks the HOMO at the position of highest electron density. Hence if we know the symmetry of the HOMO, we can predict the structure of the molecule-ion BH^+. For example, the HOMO of HF is the 1π molecular orbital localized on fluorine (Section 4-11). As shown in Figure 5-38, the proton attack on the 1π molecular orbital should produce an angular H_2F^+ molecule, a structure that agrees with the prediction based on the VSEPR method discussed in Section 2-15. Using the frontier-orbital approach one also can predict that H_3O^+ will be trigonal pyramidal and that NH_4^+ will be tetrahedral. The structure of CH_5^+ is not known, but the frontier-orbital method predicts proton attack on one of the σ_x, σ_y, or σ_z molecular orbitals, because they are all degenerate, as shown in Figure 5-20. We may predict such an attack, then, along a line bisecting an H—C—H angle, as shown in Figure 5-39. Consequently, we shall not expect a trigonal-bipyramidal molecule to be the final result.

Beyond wishing to predict the shape of protonated molecules, we are also interested in understanding the trends in PA for the isoelectronic series Ne → CH_4 (Equations 5-13 through 5-17). It is not too surprising that the PA increases through the series Ne, HF, H_2O, and NH_3, because each molecule has a lone pair of electrons which the proton can attack, and the ionization energy decreases in the order Ne > HF > H_2O > NH_3. An intriguing feature of the proton-affinity values given in Equation 5-13 through 5-17 is that the proton affinity of methane is *greater* than that of hydrogen fluoride, although HF has *three* lone electron pairs for the proton to attack,

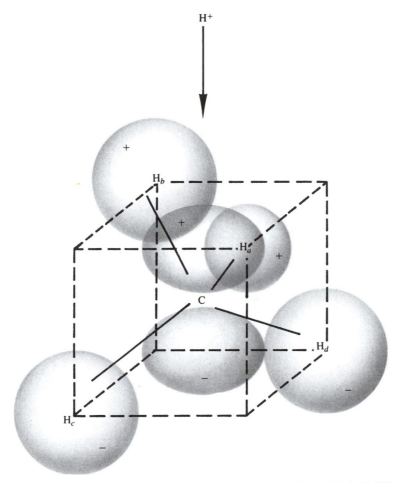

Figure 5-39 Proton attack on the highest occupied molecular orbital of CH_4. The final shape of CH_5^+ is predicted to be irregular, that is, neither square pyramidal nor trigonal bipyramidal.

whereas there are *no* lone pairs in the methane molecule. To understand this apparent anomaly we must concentrate on the frontier-orbital concept and Equation 5-12, which relates *PA* to *IE*. In Figure 5-40, *PA* is plotted versus *IE* for the isoelectronic series Ne, HF, H_2O, NH_3, and CH_4. It can readily be seen that the proton affinity of CH_4 is greater than that of HF, because the ionization energy of CH_4 is less than that of HF. In other words, as is evident from Equation 5-12, the proton does not need to attack a lone electron pair in forming BH^+. The crucial factor is the ionization energy of the highest-energy occupied molecular orbital. Although the HOMO of CH_4 is a bonding orbital (degenerate σ_x^b, σ_y^b, and σ_z^b), its ionization energy

(12.05 eV) is less than that of the lone-pair orbital (1π) of HF (16.05 eV). Consequently the proton affinity of CH_4 is greater than that of HF.

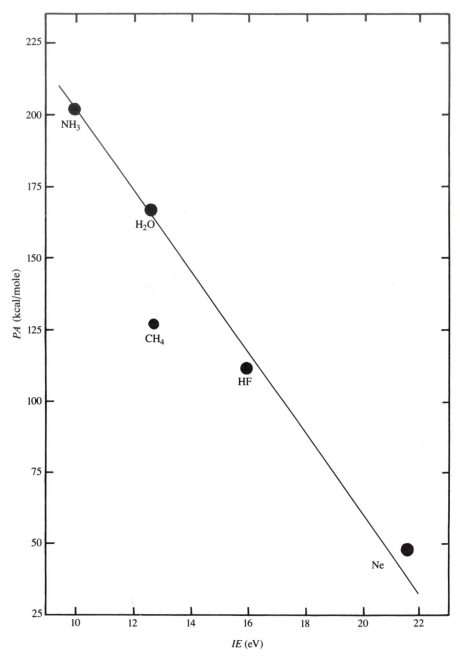

Figure 5-40 Plot of proton affinity versus ionization energy for Ne, HF, H_2O, NH_3, and CH_4.

The Frontier-Orbital Concept Applied to Reactions of Carbon Monoxide

We shall consider three reactions of carbon monoxide:

$$H^+ + CO \rightarrow HCO^+$$

$$O + CO \rightarrow CO_2$$

$$BH_3 + CO \rightarrow BH_3CO$$

In the first reaction, the CO molecule is attacked by a proton. As discussed in Section 4-11, the HOMO of CO is 3σ, predominantly localized on the carbon atom. Consequently, we predict that the proton will attack this orbital to form the linear HCO^+ ion.

Next, we consider the reaction of an oxygen atom with carbon monoxide to form carbon dioxide:

$$:\ddot{O} + :C\equiv O: \rightarrow :\ddot{O} - C\equiv O: \tag{5-18}$$

In Equation 5-18 we have taken the oxygen atom in an excited spin-paired state, $1s^2 2s^2 2p_x^2 2p_y^2 2p_z^0$. The LUMO of the oxygen atom $(2p_z)$ can attack the HOMO of the CO molecule to form the linear CO_2 molecule as shown in Figure 5-41(a). This reaction should cause a drastic stabilization in the energy of the 3σ molecular orbital of CO. In fact, the photoelectron spectra of CO (Problem 4-21) and CO_2 (Figure 5-27) show that the first σ orbital of CO_2 ionizes at about 18 eV, whereas the 3σ orbital of CO ionizes at about 14 eV. Consequently, the 3σ orbital of CO is stabilized by about 4 eV. This

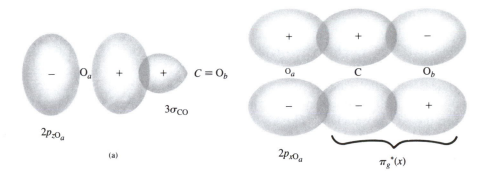

(a)

(b)

Figure 5-41 Orbital interactions in the reaction $O + CO \rightarrow CO_2$: (a) formation of the O_a—C bond by electron donation from the 3σ orbital on CO to the empty $2p_z$ orbital on the O_a atom; (b) π back bonding from the filled $2p_x$ orbital on O_a to the empty $\pi_g^*(x)$ orbital on CO. A similar interaction occurs between $2p_y$ and $\pi_g^*(y)$.

orbital stabilization indicates strong O—CO bond formation, in agreement with the large experimental bond dissociation energy:

$$CO_2 \rightarrow O + CO \qquad \Delta E = 127 \text{ kcal mole}^{-1}$$

Figure 5-41(a) depicts only the formation of the σ bond in O—CO. The p_x and p_y electrons can also become involved in the bonding:

$$:\ddot{O}_a—C\equiv O:_b \rightarrow :\ddot{O}_a=C=\ddot{O}:_b$$

We can depict this orbital interaction as shown in Figure 5-41(b). The oxygen atom p_x and p_y electrons are delocalized into the π^* orbital of CO. This delocalization causes a strengthening of the O_a—C bond but a weakening of the C—O_b bond. The final result is that instead of one single and one triple bond, we have two double bonds in the carbon dioxide molecule.

The orbital interaction shown in Figure 5-41(b) is referred to as π *back bonding* or *back donation*. This name arises because the oxygen atom *accepts* electron density in forming the σ bond but *releases* electron density in forming the π bond. Consequently, the oxygen atom behaves as both a Lewis acid and a Lewis base in the reaction $O + CO \rightarrow CO_2$.

As our final example of the application of the frontier-orbital concept to reactions of carbon monoxide, we consider borane carbonyl, BH_3CO. We can imagine the reaction of the transient BH_3 molecule and CO to form BH_3CO:

$$\begin{array}{c} H \qquad H \\ \diagdown \diagup \\ B \qquad + :C\equiv O: \rightarrow \quad H \blacktriangleright B—C\equiv O: \\ | \qquad\qquad\qquad\qquad \diagup \\ H \qquad\qquad\qquad\quad H \end{array} \qquad (5\text{-}19)$$

Upon bonding to CO, the trigonal-planar BH_3 moiety becomes trigonal pyramidal. The reaction shown in Equation 5-19 is a Lewis acid-base reaction.

The borane molecule is isoelectronic with the oxygen atom. We can consider BH_3 to be formed by the hypothetical withdrawal of three protons from the oxygen nucleus. The six valence electrons of the oxygen atom, $2s^2 2p^4$, correspond to the $(\sigma_s^b)^2(\sigma_{x,y}^b)^4$ electrons of BH_3 (Section 5-3). Consequently, there is a very close analogy between the BH_3CO molecule and the CO_2 molecule.

Equation 5-19 can be understood in terms of the frontier-orbital concept. In this approach the HOMO of CO (3σ) is considered to donate electron density to the LUMO of BH_3 ($2p_z$) to form the B—C σ bond in BH_3CO. A contribution to B—C π bonding can also occur by interaction between the HOMO of BH_3 ($\sigma_{x,y}^b$) and the LUMO of CO (2π). These interactions are illustrated in the energy-level scheme for BH_3CO in Figure 5-42. The B—C σ bond formation takes place largely in the 4σ molecular orbital and results in a transfer of electron density to the BH_3 moiety. To decrease the build-up of electron density of the BH_3 moiety, the $\sigma_{x,y}{}^b$ orbital of BH_3 donates electron density to the empty 2π orbital of CO, as is shown in the

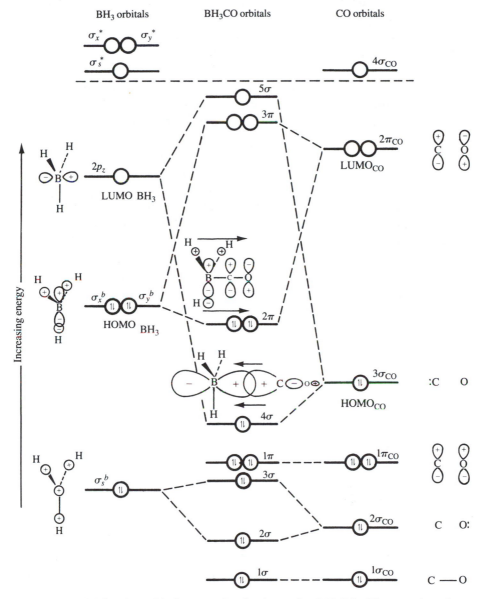

Figure 5-42 Molecular-orbital energy-level scheme for BH_3CO. Electron donation from the HOMO of CO to the empty $2p_z$ boron orbital of BH_3 takes place in the 4σ molecular orbital. Back donation occurs in the 2π molecular orbital of BH_3CO.

2π molecular orbital of BH_3CO. Because electron density is removed from the Lewis acid fragment (BH_3) this interaction results in back bonding from the BH_3 moiety to the CO moiety.

The photoelectron spectrum of BH_3CO has been measured. The assignment of the bands is in agreement with the electron configuration exhibited in Figure 5-42. The ionization energy of the highest filled σ orbital of BH_3CO (4σ) is 14.13 eV, hardly stabilized at all relative to the ionization energy of the 3σ orbital on CO (14.01 eV). The correlation diagram of Figure 5-42 indicates that the 3σ donor orbital of CO should be stabilized by the donor-acceptor interaction. From the fact that there is little change in the energy levels of CO compared to those of BH_3CO, we conclude that the B—C bond in BH_3CO is weak. Experimentally, only about 20 kcal mole^{-1} is required to dissociate BH_3CO into its component molecules:

$$BH_3CO \rightarrow BH_3 + CO \qquad \Delta E \simeq 20 \text{ kcal mole}^{-1}$$

This confirms our earlier deduction.

There is one important feature of the photoelectron spectrum of BH_3CO that is markedly different from that of CO_2. The stabilization of only 0.1 eV for the highest filled σ orbital in the reaction $BH_3 + CO \rightarrow BH_3CO$ is to be compared to the 4 eV stabilization of the σ orbital that we discussed for the reaction $O + CO \rightarrow CO_2$. These relative stabilizations are in agreement with the weak B—C bond in BH_3CO but the strong O—C bond in CO_2.

Symmetry Rules for Chemical Reactions

Since 1960, applications of the frontier-orbital concept have expanded considerably beyond the examples we have been discussing. The American chemist Ralph G. Pearson has formulated a set of rules that must be obeyed if two reactants are to undergo a facile reaction. These rules are:

1. As the reactants approach each other, electron density must flow from the HOMO of the donor to the LUMO of the acceptor.

2. The HOMO of the donor and the LUMO of the acceptor must approach each other so that they have a net positive overlap.

3. The HOMO of the donor and the LUMO of the acceptor must be relatively close in energy (within about 6 eV).

4. The net effect of HOMO $\rightarrow$ LUMO electron transfer must correspond to bonds to be made and bonds to be broken during the course of the reaction.

A reaction that obeys these four rules is said to be symmetry-*allowed*. A symmetry-allowed reaction generally will occur with a low activation energy;

a symmetry-forbidden reaction will require a high activation energy in order to take place.

A quick survey of the reactions we have discussed in this section will show that they are all allowed (e.g., $H^+ + H_2O$, $O + CO$, $BH_3 + CO$). In general, reactions of atoms are symmetry allowed.

Let us now discuss some examples of reactions that are symmetry forbidden. Consider first the isotopic exchange reaction between deuterium and hydrogen:

$$H_2 + D_2 \rightarrow 2HD$$

The HOMO for each of these molecules is $\sigma_g{}^b(s) = \dfrac{1}{\sqrt{2}}(1s_a + 1s_b)$ and the LUMO is $\sigma_u{}^* = \dfrac{1}{\sqrt{2}}(1s_a - 1s_b)$. Let us imagine the two molecules approaching in a "side-on" fashion to form a square-planar activated complex which then breaks down into $2HD$

$$
\begin{array}{ccc}
\text{H}\!-\!\text{H} & \text{H}\text{-}\text{-}\text{-}\text{H} & \text{H}\quad\text{H} \\
+ \quad \rightarrow & | \qquad | \quad \rightarrow & |\ +\ | \\
\text{D}\!-\!\text{D} & \text{D}\text{-}\text{-}\text{-}\text{D} & \text{D}\quad\text{D}
\end{array}
$$

To examine the symmetry requirements of this reaction, let us suppose that there is an electron donation from the HOMO of H_2 to the LUMO of D_2:

H_2 HOMO, $\sigma_g{}^b(s)$

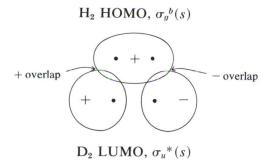

D_2 LUMO, $\sigma_u{}^*(s)$

Notice that the "+ overlap" and the "− overlap" cancel so that the net overlap is zero. Consequently, this reaction is symmetry forbidden; rule 2 is not obeyed. We therefore predict that H_2 and D_2 would not react under normal conditions of temperature and pressure.

As a second example of a symmetry-forbidden reaction, consider the reaction of hydrogen and fluorine:

$$H_2 + F_2 \rightarrow 2HF$$

There are two possibilities to examine for electron density transfer: $HOMO(H_2) \rightarrow LUMO(F_2)$ or $HOMO(F_2) \rightarrow LUMO(H_2)$. We can obtain the symmetry designation of these orbitals from our discussions in Chapter 4.

Just as for the reaction $H_2 + D_2 \rightarrow 2HD$, let us imagine that the molecules approach each other in a "side-on" fashion:

$$
\begin{array}{ccc}
\text{H} - \text{H} & \text{H---H} & \text{H \quad H} \\
+ & | \quad | & | + | \\
\text{F} - \text{F} & \text{F---F} & \text{F \quad F}
\end{array}
$$

First consider the possibility of electron transfer from $HOMO(H_2)$ to $LUMO(F_2)$.

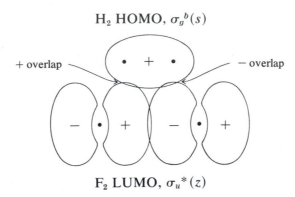

H_2 HOMO, $\sigma_g^b(s)$

F_2 LUMO, $\sigma_u^*(z)$

Notice that the "+ overlap" and the "− overlap" cancel so that the net overlap is zero. Consequently, rule 2 is not obeyed and this reaction is symmetry forbidden for electron transfer from $HOMO(H_2)$ to $LUMO(F_2)$.

Now, let us consider the possibility of electron transfer in the reverse direction, $HOMO(F_2)$ to $LUMO(H_2)$:

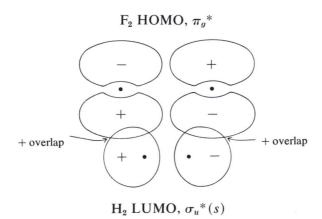

F_2 HOMO, π_g^*

H_2 LUMO, $\sigma_u^*(s)$

This interaction gives a net positive overlap. Rules 1, 2, and 3 are obeyed. Rule 4 states that the net effect of the electron transfer should correspond to forming the new bonds (H—F) and breaking the old bonds (H—H and F—F). Although this orbital interaction does correspond to forming the H—F bonds and breaking the H—H bond, we see that electron transfer from the σ_g^* orbital of F_2 will tend to *strengthen* the F—F bond. This strengthening results because we are removing electron density from an antibonding orbital of the F_2 molecule. Consequently, rule 4 is not obeyed and the reaction $H_2 + F_2 \rightarrow 2HF$ is symmetry forbidden. This reaction still could occur by a dissociation mechanism such as

$$F_2 \rightarrow 2F$$
$$F + H_2 \rightarrow HF + H$$
$$\underline{H + F \rightarrow HF}$$
$$\text{Net: } H_2 + F_2 \rightarrow 2HF$$

However, such a mechanism usually entails a high activation energy since a basic requirement is the dissociation of one of the reactant molecules.

As our final example of the application of frontier-orbital symmetry principles, consider the reaction

$$H_2 + C_2 \rightarrow HCCH$$

Recall from our discussion in Chapter 4 that the HOMO of C_2 is $\pi_u^b(x,y)$ and the LUMO is $\sigma_g^b(z)$. Imagine the reaction to occur as follows:

$$
\begin{array}{ccccc}
\text{H} \longrightarrow \text{H} & & \text{H---H} & & \\
+ & \rightarrow & \;\;\vert\;\;\;\;\vert & \rightarrow & \text{H}\!-\!\text{C}\!\equiv\!\text{C}\!-\!\text{H} \\
\text{C} \longrightarrow \text{C} & & \text{C---C} & &
\end{array}
$$

Let us determine the symmetry characteristics of the electron transfer from the HOMO of H_2 to the LUMO of C_2:

H_2 HOMO, $\sigma_g^b(s)$

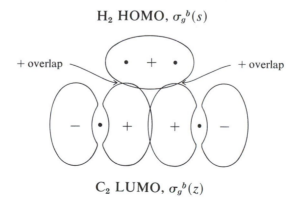

C_2 LUMO, $\sigma_g^b(z)$

Note that there is net positive overlap and that the electron transfer corresponds to breaking the H—H bond and strengthening the C—C bond, as required. The H atoms can then migrate around to the ends of the C—C molecule to form linear acetylene, H—C≡C—H. Consequently, the reaction $H_2 + C_2 \rightarrow C_2H_2$ is symmetry allowed.

Symmetry principles have been used extensively in the field of organic chemistry. The American chemists Robert B. Woodward and Roald Hoffmann have made important contributions in this area. Application of symmetry principles to reactions in organic chemistry is given in the problems at the end of the chapter.

5-10 Molecular-Orbital Theory for Transition-Metal Molecules Containing One Unsaturated Ligand

The term *ligand* applies to any atom or chemical moiety that is attached to a transition-metal atom. For the ScO molecule (Section 4-11), the oxygen atom is a ligand. The discussion in Section 4-11 applies to transition-metal molecules containing ligands that cannot accept electrons beyond those required by the ionic formulation. Such ligands are called *saturated;* typical examples are H^-, F^-, Cl^-, and O^{2-}. In ScO, for example, the ionic formulation is $Sc^{2+}O^{2-}$. Because O^{2-} has the closed-shell configuration $1s^2 2s^2 2p^6$, it can accommodate no further electrons.

Transition metals also bond to numerous unsaturated ligands. Such ligands include CO, CN^-, N_2, ethylene, and benzene. Examples of stable complexes of the first two ligands are $Cr(CO)_6$ and $Cr(CN)_6^{3-}$. In recent years, reactions have been effected between transition-metal atoms and the neutral ligands CO and N_2. At low temperatures the linear species MCO and MN_2 can be identified, M being a transition-metal atom. In these species, there is no "ionic" formulation because the ligand itself is neutral. In the Lewis electron-dot sense, CO and N_2 are said to be unsaturated because they can accept additional electrons; for example,

$$:C\equiv O: + 4e^- \rightarrow \overset{\cdot\cdot}{:}\overset{\cdot}{\underset{\cdot}{C}}\text{---}\overset{\cdot\cdot}{\underset{\cdot\cdot}{O}}:^{4-}$$

In terms of molecular-orbital theory, CO has the ground-state electronic configuration

$$1\sigma_{CO}{}^2 2\sigma_{CO}{}^2 1\pi_{CO}{}^4 3\sigma_{CO}{}^2 2\pi_{CO}{}^0 4\sigma_{CO}{}^0$$

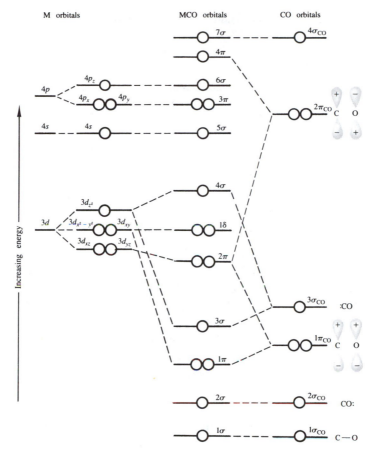

M orbitals   MCO orbitals   CO orbitals

Figure 5-43 Relative orbital energies in a linear transition-metal monocarbonyl complex, MCO. The molecular orbitals of carbon monoxide are placed at the right side of the diagram and the metal atomic orbitals at the left side. In the simplified representation presented here, the major orbital interaction occurs between the 1π, 3σ, and 2π molecular orbitals of CO and the $3d$ orbitals of the metal. The metal $4s$ and $4p$ orbitals are assumed to be so high in energy that their contribution to the bonding is negligible. The predominant character of each carbon monoxide molecular orbital is indicated at the far right of the diagram.

and the unsaturation results from the $2\pi_{CO}$ orbital, which hypothetically can accept four electrons without breaking the CO bond.*

A schematic energy-level diagram for MCO is presented in Figure 5-43. The occupied CO energy levels are placed lower in energy than the valence orbitals of the metal. This is because the *VOIE* of the metal $3d$, $4s$, and

---

*We subscript each molecular orbital of carbon monoxide with the label "CO" to distinguish it from the molecular orbitals of the MCO complex as a whole.

TABLE 5-1. VALENCE-ORBITAL IONIZATION ENERGIES (eV) FOR THE FIRST-ROW TRANSITION-METAL ATOMS

Atom	$3d^{n-1}4s \rightarrow 3d^{n-2}4s$ $3d$	$3d^{n-1}4s \rightarrow 3d^{n-1}$ $4s$	$3d^{n-1}4p \rightarrow 3d^{n-1}$ $4p$
Sc	4.7	5.7	3.2
Ti	5.6	6.1	3.3
V	6.3	6.3	3.5
Cr	7.2	6.6	3.5
Mn	7.9	6.8	3.6
Fe	8.7	7.1	3.7
Co	9.4	7.3	3.8
Ni	10.0	7.6	3.8
Cu	10.7	7.7	4.0

$4p$ orbitals varies between 3 and 10 eV, whereas the first ionization energy of CO is much higher (14.01 eV). The valence-orbital ionization energies of the first-row transition metal $3d$, $4s$, and $4p$ orbitals are presented in Table 5-1.

The atomic-orbital characteristics of the important molecular orbitals are presented in Figure 5-44. The interactions of $1\sigma_{CO}$ and $2\sigma_{CO}$ with the metal are not represented, because these orbitals are very stable and have little interaction with the metal orbitals (Figure 5-43). Likewise, the $4\sigma_{CO}$ orbital is so high in energy that it also has little interaction with the metal orbitals.

There are three major orbital interactions in the reaction of M with CO to form MCO. The most important is ligand-to-metal σ donation from the $3\sigma_{CO}$ orbital to an empty valence orbital on the metal that has the correct symmetry ($3d_{z^2}$, $4s$, or $4p_z$):

$$\overset{-}{M} \leftarrow :C\equiv\overset{+}{O}:$$

This type of interaction takes place predominantly in the 3σ molecular orbital of MCO (see Figures 5-43 and 5-44). Such ligand-to-metal σ donation is analogous to the formation of the B—C σ bond of BH_3CO (Figure 5-42).

The second type of interaction is the donation of electron density from $1\pi_{CO}$ [$= \pi^b(x,y)$] to the metal $3d_{xz,yz}(d_\pi)$ orbitals. This is another example of ligand-to-metal (L $\rightarrow$ M) π donation, which was discussed in Section 4-11 for ScO:

$$\overset{-}{M}—C\equiv\overset{+}{O}: \leftrightarrow \overset{-2}{M}=C=\overset{+2}{O}:$$

This type of interaction is important for metal atoms that have empty d_π atomic orbitals. However, since most metal carbonyls have filled d_π orbitals

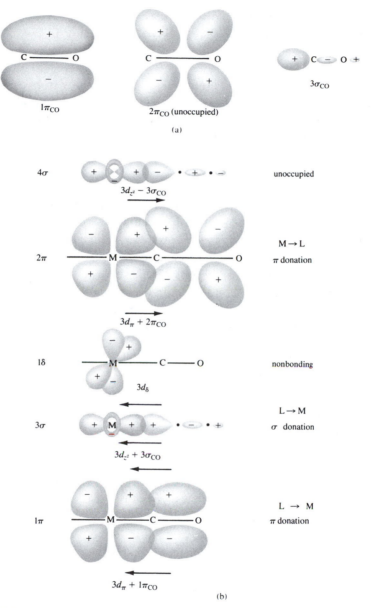

Figure 5-44 Schematic representation of orbital overlap of the $3d$ atomic orbitals of the metal atom with the $1\pi_{CO}$, $2\pi_{CO}$, and $3\sigma_{CO}$ orbitals of CO. (a) Spatial representations of the occupied $1\pi_{CO}$ and $3\sigma_{CO}$ orbitals of CO and the unoccupied antibonding $2\pi_{CO}$ orbital. (b) In the 1π molecular orbital of MCO there is electron donation from the $1\pi_{CO}$ orbital to the $3d_{xz}$ atomic orbital. A similar interaction occurs with the $3d_{yz}$ orbital, which is perpendicular to the plane of the paper. Such an interaction is termed L → M π donation. Likewise the 3σ and 2π molecular orbitals of MCO correspond to L → M σ donation and M → L π donation. The 1δ molecular orbital is nonbonding $3d_{xy}$, $_{x^2-y^2}$ and lies in a plane perpendicular to the molecular axis.

this interaction is of negligible importance; it strengthens the M—C bond at the expense of electron density on the oxygen atom. In the MCO molecule, the L → M π bonding occurs predominantly in the 1π molecular orbital (see Figures 5-43 and 5-44).

The third type of interaction removes electron density from the metal and is called *metal-to-ligand* (M → L) π donation. In this type of interaction electron density is donated from the filled metal $3d_{xz,yz}$ orbitals to the empty $2\pi_{CO}$ [$= \pi^*(x,y)$] orbital of CO:

$$\overset{-}{M}—C\equiv\overset{+}{O}: \leftrightarrow M=C=\underset{..}{O}:$$

Notice that the M → L π bonding reduces the excess negative charge on the metal brought about by the L → M σ bonding. At the same time, the M → L π bonding strengthens the M—C bond and replenishes the electron deficiency at the oxygen atom. We say that the L → M σ bonding and the M → L π bonding are *synergic* since both serve to strengthen the M—C bond. The M → L π bonding takes place in the 2π molecular orbital of MCO, as shown in Figures 5-43 and 5-44. Because M → L π bonding prevents the accumulation of excess negative charge on the metal, this type of bonding commonly is called back donation or back bonding, and is entirely analogous with the bonding interaction that we discussed in Section 5-9 for the reactions O + CO → CO$_2$ and BH$_3$ + CO → BH$_3$CO. Back donation stabilizes the 2π orbital of MCO and makes it less antibonding. (L → M) π bonding is common when the central metal ion has a positive charge and empty $3d_{xz,yz}$ orbitals; an example is ScO. (M → L) π bonding is common when the central metal has low ionic charge and filled $3d_{xz,yz}$ orbitals, as in NiCO.

The energy-level scheme in Figure 5-43 and the atomic-orbital overlap representations in Figure 5-44 utilize only the metal $3d$ orbitals in the bonding scheme. However, for a metal atom that has a filled or nearly filled $3d$ subshell, the $4s$ and $4p$ orbitals must be involved in the bonding between metal and ligand. For example, no bond could be formed in NiCO if the Ni $4s$ or $4p$ orbitals did not become involved in the bonding. The bonding effect of the 3σ orbital would be canceled by the antibonding character of the 4σ orbital, because both are occupied. Similar comments apply to the 1π and 2π orbitals. The $4s$ orbital can participate in the bonding by accepting a pair of electrons from the $3\sigma_{CO}$ orbital. It also could participate by decreasing the antibonding character of the 4σ molecular orbital. In mathematical language we are saying that

$$\psi(3\sigma) = c_1 3d + c_2 4s + c_3 3\sigma_{CO}$$

and

$$\psi(4\sigma) = c_4 3d - c_5 4s - c_6 3\sigma_{CO}$$

In the simplified representations shown in Figures 5-43 and 5-44 the $4s$ orbital is assumed to contribute only to the 5σ molecular orbital.

5-11 Photoelectron Spectroscopy of Core Electrons

Throughout most of this book, we have concentrated our attention on the valence-shell orbital energies because they generally provide information about the effects of *atomic-orbital mixing* upon molecule formation. However, the core-electron orbitals are good indicators of the *charge distribution* in a molecule. In Sections 4-9 and 4-11 we discussed the $1s$ core-electron ionization of N_2, O_2, F_2, and HF. The fluorine $1s$ binding energy was less in HF than in F_2.* We attributed this lower binding energy to the partial negative charge on the fluorine atom, which tends to destabilize its energy levels.

The gas-phase photoelectron spectra of CO and CO_2 provide a good example of this charge-distribution concept. The carbon $1s$ and oxygen $1s$ core-electron binding energies for these two molecules are presented in Figure 5-45. The experimental assignment of the bands was carried out by varying the partial pressure of the gases and examining the change in intensity of the peaks. It is clear that in the carbon $1s$ spectrum, CO_2 has the higher binding energy. In contrast, the oxygen $1s$ spectrum shows that CO_2 has the lower binding energy. These results indicate that the carbon atom in CO_2 is more positive than in CO and that the oxygen atom is more negative than in CO. This result can be understood readily on the basis of the Lewis electron-dot structures:

$$:\overset{-}{C}\equiv\overset{+}{O}: \qquad :\overset{..}{O}=C=\overset{..}{O}:$$

The variation in core-electron binding energies as a result of chemical bonding is often termed a "chemical shift."

As a second example of the chemical shift of core-electron binding energies we discuss the core photoelectron spectra of the xenon fluoride series: XeF_2, XeF_4, and XeF_6. The free xenon atom has the electronic configuration $1s^2 2s^2 2p^6 3s^2 3p^6 3d^{10} 4s^2 4p^6 4d^{10} 5s^2 5p^6$; consequently, we could examine any core-electron level between $1s$ and $4d$. For the fluorine atoms, however, only the $1s$ core electrons are available. In the experimental study that was undertaken, the binding energies of the $3d$ core electrons of xenon were examined along with fluorine $1s$ core electrons. In fact, the $3d$ orbitals

*Core-electron ionization energies are often referred to as binding energies.

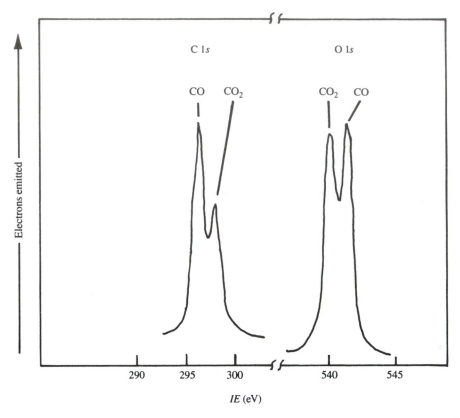

Figure 5-45 Photoelectron spectra of the carbon $1s$ and oxygen $1s$ core electrons of carbon monoxide and carbon dioxide. Adapted with permission from K. Siegbahn *et al., ESCA Applied to Free Molecules,* Amsterdam: North-Holland, 1969, p. 13.

exhibit two binding energies, a result of spin-orbit coupling. In the ionized states the $3d^9$ configuration results in the energy states $^2D_{5/2}$ and $^2D_{3/2}$. These states usually are denoted $3d_{5/2}$ and $3d_{3/2}$, respectively.

The binding energy for $3d_{5/2}$ in free atomic xenon is 676.44 eV and for fluorine $1s$ in F_2 it is 696.70 eV. Shifts of xenon $3d_{5/2}$ binding energies (in electron volts) relative to the binding energy of free atomic xenon are XeF_2, 2.87; XeF_4, 5.41, and XeF_6, 7.64. The fluorine $1s$ binding energies relative to F_2 are XeF_2, -5.48; XeF_4, -4.40; and XeF_6, -3.38. These results provide a qualitative indication that the xenon atom in XeF_n has a partial positive charge and the fluorine atoms have partial negative charges.

The investigation of molecular structure is best carried out by using the various types of molecular spectroscopy briefly discussed in Section 4-4.

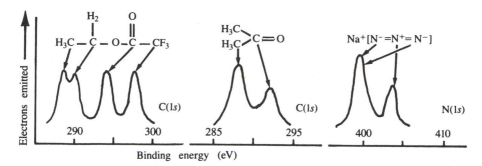

Figure 5-46 Spectra of $C(1s)$ core electrons in ethyl trifluoroacetate and acetone, and the $N(1s)$ electrons in sodium azide. Adapted with permission from W.C. Price in *Electron Spectroscopy: Theory, Techniques and Applications,* vol. 1, edited by C.R. Brundle and A.D. Baker, New York: Academic, 1977, p. 151.

However, these methods tell us little about the *electronic* energy levels in a molecule. Since we are mainly interested in electronic structure and its relationship to molecular structure, we shall discuss some instances in which photoelectron spectroscopy can provide a clue to molecular structure.

We already have seen that the valence-region photoelectron spectra of the isoelectronic sequence Ne, HF, H_2O, NH_3, and CH_4 are consistent with the known molecular shapes of the molecules H_2O, NH_3, and CH_4. For example, a bent H_2O molecule should exhibit four bands in its valence photoelectron spectrum, whereas only three bands should be observed for linear H_2O. Experimentally, four bands are observed (Figure 5-8), in agreement with the bent shape of the water molecule.

We now discuss the core photoelectron spectra of three molecules in order to illustrate the relationship between these spectra and molecular structure. In Figure 5-46, we give the spectra of carbon $1s$ electrons in ethyl trifluoroacetate and acetone, and the nitrogen $1s$ electrons in sodium azide. Notice that for ethyl trifluoroacetate there are four $C(1s)$ peaks, corresponding to the four different types of carbon atoms in the compound. Each band can be assigned to a particular type of carbon atom. For example, the band at the highest energy is assigned to $C(1s)$ ejection from the CF_3 group. This carbon atom is expected to have the highest positive charge and therefore would require the most energy to ionize the $1s$ electrons. The $C(1s)$ spectrum of acetone exhibits only two bands, as we would expect based upon its known molecular structure. Finally, the $N(1s)$ spectrum of sodium azide exhibits two bands; the higher energy band corresponds to the central nitrogen atom and the lower energy band to the terminal nitrogen atoms.

QUESTIONS AND PROBLEMS

1. In Section 5-1 the form of the most stable molecular orbital of tri-
    angular H_3^+ is given as

    $$\frac{1}{\sqrt{3}}(1s_a + 1s_b + 1s_c)$$

    (a) What are the forms of the other two molecular orbitals of H_3^+?
        (*Hint:* BH_3 also has a triangular array of H atoms. Choose the
        coordinate system we took for the H atoms of BH_3 shown in Fig-
        ure 5-9.)
    (b) Are the second and third molecular orbitals of H_3^+ bonding or
        antibonding?
    (c) Are the second and third molecular orbitals degenerate or non-
        degenerate?
    (d) Draw a molecular-orbital energy-level diagram for triangular H_3^+
        with the orbitals of H_2 on the left, that of H on the right, and the
        molecular orbitals of H_3^+ in the center.

2. Consider the molecule-ion H_3^-.
    (a) If H_3^- were triangular in shape, would it be diamagnetic or para-
        magnetic?
    (b) H^- Is a *nucleophile*. Consider the reaction $H^- + H_2 \rightarrow H_3^-$. Which
        orbital does H^- attack on the H_2 molecule?
    (c) From your answer in part (b), do you predict H_3^- to be triangular
        or linear?
    (d) Apply the molecular-orbital theory to linear H_3^-. What H—H
        bond order do you predict?

3. The species HF_2^- is linear and symmetrical, F—H—F$^-$, and has one
    of the strongest known systems of "hydrogen bonding" (Chapter 7).
    (a) Draw atomic overlap representations for each of the σ and π
        orbitals of HF_2^-.
    (b) Draw a molecular-orbital energy-level scheme for HF_2^-. Place the
        H atom $1s$ orbital on the left, the fluorine $2s$ and $2p$ atomic orbitals
        (or the appropriate SALCs) on the right, and the orbitals of HF_2^-
        in the center.
    (c) Given the energy-level scheme derived in part (b), would you expect
        HF_2^- to be diamagnetic or paramagnetic?
    (d) What is the H—F bond order in HF_2^-?

4. The species I_3^- is linear and symmetrical, I—I—I$^-$. Consider the
    bonding in I_3^- as involving *only* the $5p$ orbitals on the I atoms (i.e.,
    consider the $5s^2$ electrons as part of the "core").

(a) Draw atomic overlap representation for each of the σ and π orbitals of I_3^-.

(b) Draw a molecular-orbital energy-level scheme for I_3^-.

(c) What is the I—I bond order in I_3^-?

5. Use the molecular-orbital theory to describe the bonding in N_3. What type of orbital does the unpaired electron occupy? The electron affinity of the N_3 molecule is nearly as large as that of the fluorine atom. Explain.

6. Use the molecular-orbital theory to describe the bonding in carbonyl sulfide, OCS.

7. Consider the borohydride anion, BH_4^-, to be formed in the reaction

$$H^- + BH_3 \rightarrow BH_4^-$$

Which orbital will the H^- nucleophile attack on the BH_3 molecule?

8. Consider the borane-ammonia addition compound, BH_3NH_3. Draw a molecular-orbital energy-level diagram for BH_3NH_3 with the BH_3 orbitals on the left, the NH_3 orbitals on the right, and the orbitals of BH_3NH_3 in the center. In which molecular orbital of BH_3NH_3 does the donor-acceptor interaction take place? Do you expect the back donation to be as important in BH_3NH_3 as in BH_3CO? Why or why not?

9. Calculate the tetrahedral angle in CH_4 by inscribing the molecule in a unit cube, as shown in Figure 5-16.

10. Use the molecular-orbital theory to describe the bonding in acetylene, HCCH. Sketch a predicted photoelectron spectrum for acetylene.

11. Draw atomic-orbital overlaps of the π molecular orbitals for the linear molecules C_4 and NCCN. Assume that each of these molecules has ten electrons in the σ molecular orbitals, which you need not describe. These σ electrons account for one bond between each pair of atoms and the terminal lone pairs, :C—C—C—C: and :N—C—C—N: .

(a) Sketch the π molecular orbitals on an energy-level diagram such as was provided for C_2H_4 (Figure 5-34) and C_6H_6 (Figure 5-37).

(b) Label the molecular orbitals according to whether they are *gerade* or *ungerade*, and bonding or antibonding.

(c) Write the π electronic configuration of these molecules and indicate whether the molecules are diamagnetic or paramagnetic.

(d) Determine the terminal and central C—C bond orders in C_4, counting σ and π bonds.

(e) Determine the C—N and C—C bond orders in NCCN, counting σ and π bonds.

12. The electron affinity of the C_2 molecule is 3.1 eV, whereas that of C_3 is 1.8 eV. Apply the molecular-orbital theory to each of these molecules. Assume a linear structure for C_3. Explain why the electron affinity of C_2 is larger than that of C_3.

13. Consider the hypothetical square-planar molecule XeH_4 in the following coordinate system:

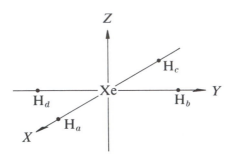

(a) Write appropriate SALCs of the hydrogen 1s atomic orbitals which will overlap with each of the following xenon orbitals that can form molecular orbitals: $5s$, $5p_x$, $5p_y$, and $5d_{x^2 - y^2}$.
(b) Draw atomic-orbital overlaps with the appropriate Xe orbital for each of these SALCs.
(c) Sketch a molecular-orbital energy-level diagram for XeH_4, showing the Xe atomic orbitals on the left, the H $1s$ orbitals on the right, and the XeH_4 orbitals in the center.
(d) Indicate which molecular orbitals are occupied, and calculate the net Xe—H bond order, based on the molecular-orbital model.

14. Formaldehyde (H_2CO) exhibits a strong electronic absorption band in the ultraviolet region, which may be assigned to a $\pi^b \rightarrow \pi^*$ transition (see the discussion of ethylene in Section 5-8). In addition, the spectrum of H_2CO [like that of all organic compounds containing a carbonyl (C=O) group] exhibits a weaker, longer wavelength peak in the region of 270–300 nm. Formulate the molecular orbitals for H_2CO and suggest a possible assignment for the long-wavelength peak.

15. Which molecule do you expect to have the higher ionization energy, ethylene or ethane? Why?

16. Do you expect benzene to absorb light of lower, or higher, energy than ethylene? Why?

17. Sketch the predicted photoelectron spectrum for the core fluorine $1s$ electrons of ClF_3, SF_4, and PF_5.

18. For a series of closely related compounds containing oxygen atoms, the proton affinity has been found to vary with the negative of the charge in oxygen $1s$ binding energy; that is,

$$\Delta(PA) \simeq -\Delta E(O\ 1s)$$

Such a series of compounds might include H_2O, CH_3OH, CH_3CH_2OH, and CF_3CH_2OH. Explain qualitatively why the proton affinity of a molecule is expected to increase as the binding energy of the oxygen $1s$ orbital decreases.

19. Consider the electrophilic attack of Cl^+ on ClF,

$$ClF + Cl^+ \rightarrow Cl_2F^+$$

Draw a molecular-orbital energy-level diagram for ClF, and consider the reaction with Cl^+ in terms of the frontier-orbital approach. Do you predict the structure of Cl_2F^+ to be $Cl—Cl—F^+$, or $Cl—F—Cl^+$?

20. Consider the formation of ethylene oxide, C_2H_4O, as a Lewis acid-base reaction between atomic oxygen (in a singlet state) and ethylene:

$$:\overset{..}{O} + C_2H_4 \rightarrow C_2H_4O$$

(a) What structure do you predict for ethylene oxide?
(b) What orbitals are involved in the σ donor-acceptor interaction?
(c) What orbitals are involved in the π back-bonding interaction?
(d) Is this reaction symmetry allowed?

21. Consider the ionic compound $Na_2N_2O_3$. An x-ray diffraction study in the solid state has shown the following structure for the anion $N_2O_3{}^{2-}$:

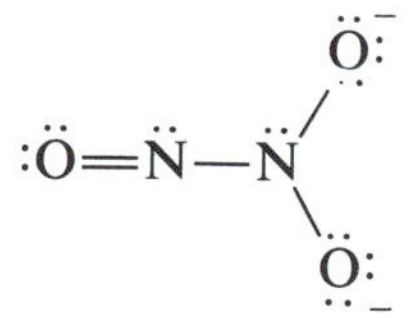

(a) Sketch the predicted photoelectron spectrum for the core nitrogen $1s$ electrons of $N_2O_3{}^{2-}$.
(b) Repeat part (a) for the oxygen $1s$ electrons.

22. In Section 5-9 we examined the symmetry rules for the $H_2 + C_2 \rightarrow HCCH$ reaction. We found that the flow of electron density from $HOMO(H_2)$ to $LUMO(C_2)$ is symmetry allowed. For this problem, apply the symmetry rules to the reverse situation: $HOMO(C_2)$ to $LUMO(H_2)$.

23. Apply the symmetry rules (Section 5-9) to the following reactions in organic chemistry:
    (a) $H_2 + C_2H_4 \rightarrow C_2H_6$
    (b) $C_2H_4 + C_2H_4 \rightarrow C_4H_8$ (cyclobutane)
    (c) $C_2H_2 + C_2H_2 \rightarrow C_4H_4$ (cyclobutadiene)

24. The N_2S_2 molecule has very nearly a perfect square structure:

$$
\begin{array}{ccc}
N & - & S \\
| & & | \\
S & - & N
\end{array}
$$

    This molecule has 22 valence electrons, of which 16 are σ electrons. (The four σ bond pairs and one σ-type lone pair on each atom account for these 16 σ electrons.)
    (a) Sketch the π molecular orbitals on an energy-level diagram such as was provided for C_2H_4 (Figure 5-34) and C_6H_6 (Figure 5-37).
    (b) Label the molecular orbitals according to whether they are *gerade* or *ungerade,* and bonding or antibonding.
    (c) Write the π electronic configuration of N_2S_2 and indicate whether it is diamagnetic or paramagnetic.
    (d) Determine the N—S bond order in N_2S_2, counting σ and π bonds.

25. Examine Figure 5-40, which exhibits a plot of PA versus IE for the isoelectronic series Ne, HF, H_2O, NH_3, and CH_4. Notice that CH_4 lies *below* the line drawn through the values for NH_3, H_2O, and HF. Use Equation 5-12 to explain why the PA of CH_4 is considerably less than that of H_2O, although their ionization energies are nearly identical. What physical explanation lies behind the fact that $PA(CH_4) < PA(H_2O)$?

26. In terms of σ donor-acceptor interactions and π back bonding, compare and contrast the bonding in BH_3CO (Section 5-9) with that in NiCO (Section 5-10).

27. The first ionization energies of the isoelectronic molecules CO and N_2 are 14.01 eV and 15.58 eV, respectively. This ionization process corresponds to electron ejection from the 3σ molecular orbital of CO and the $2\sigma_g$ molecular orbital of N_2.
    (a) Using ionization energy data, predict which of these unsaturated ligands will be a better σ donor in transition-metal molecules such as MCO and MN_2.
    (b) Using the orbital-contour diagram of the 3σ (CO) and $2\sigma_g$ (N_2) molecular orbitals, predict which of these unsaturated ligands will be a better σ donor in transition-metal molecules such as MCO and MN_2.

Suggestions for Further Reading

Alderice, D.S., Collins, G., and Foon, R., "Induced Electron Emission Processes and the Einstein Photoelectric Law," *J. Chem. Educ.* 48: 720 (1971).

Baird, N.C., "Molecular Geometry Predictions using Simple MO Theory," *J. Chem. Educ.* 55: 412 (1978).

Baker, A.D., and Betteridge, D., *Photoelectron Spectroscopy: Chemical and Analytical Aspects,* Elmsford, N.Y.: Pergamon Press, 1972.

Ballhausen, C.J., and Gray, H.B., *Molecular Orbital Theory,* Menlo Park, Calif.: Benjamin, 1964.

Beauchamp, J.L., "Ion Cyclotron Resonance Spectroscopy," in *Annual Review of Physical Chemistry,* vol. 22, edited by H. Eyring, C.J. Christensen, and H.S. Johnston, Palo Alto, Calif.: Annual Reviews, Inc., 1971, p. 527.

Bock, H., and Mollère, P.D., "Photoelectron Spectra," *J. Chem. Educ.* 51: 506 (1974).

Bradley, J.D., and Gerrans, G.C., "Frontier Molecular Orbitals," *J. Chem. Educ.* 50: 463 (1973).

Buenker, R.J., and Peyerimhoff, S.D., "Molecular Geometry and the Mulliken-Walsh MO Model. An *Ab initio* Study," *Chem. Rev.* 74: 127 (1974).

Burdett, J.K., "The Shapes of Main-Group Molecules; a Simple Semi-Quantitative Molecular Orbital Approach," *Structure and Bonding,* vol. 31, New York: Springer-Verlag, 1976, p. 67.

DeKock, R.L., and Barbachyn, M.R., "Proton Affinity, Ionization Energy, and the Nature of Frontier Orbital Electron Density," *J. Amer. Chem. Soc.* 101: 6516 (1979).

Fujimoto, H., and Fukui, K., "Molecular Orbital Theory of Chemical Reactions," in *Advances in Quantum Chemistry,* vol. 6, edited by P.-O. Löwdin, New York: Academic, 1972, p. 177.

Gimarc, B.M., *Molecular Structure and Bonding,* New York: Academic, 1978.

Gray, H.B., *Chemical Bonds,* Menlo Park, Calif.: Benjamin, 1973.

Gray, H.B., *Electrons and Chemical Bonding,* Menlo Park, Calif.: Benjamin, 1965.

Hollander, J.M., and Jolly, W.L., "X-Ray Photoelectron Spectroscopy," *Accounts of Chemical Research* 3: 193 (1970).

James, T.L., "Photoelectron Spectroscopy," *J. Chem. Educ.* 48: 712 (1971).

Karplus, M., and Porter, R.N., *Atoms and Molecules: An Introduction for Students of Physical Chemistry*, Menlo Park, Calif.: Benjamin, 1970.

Lowe, J.P., *Quantum Chemistry*, New York: Academic, 1978.

McWeeny, R., *Coulson's Valence*, New York: Oxford, 1979.

Murrell, J.N., Kettle, S.F.A., and Tedder, J.M., *The Chemical Bond*, New York: Wiley, 1978.

Pearson, R.G., "Symmetry Rules for Chemical Reactions," *Accounts of Chemical Research* 4: 152 (1971).

Pearson, R.G., *Symmetry Rules for Chemical Reactions*, New York: Wiley-Interscience, 1976.

Pimentel, G.C., and Spratley, R.D., *Understanding Chemistry*, San Francisco: Holden-Day, 1971.

Pitzer, K.S., "Multicentered Bonding," in *Physical Chemistry: An Advanced Treatise*, vol. 5, edited by H. Eyring, New York: Academic, 1970, p. 483.

Price, W.C., "Ultraviolet Photoelectron Spectroscopy: Basic Concepts and Spectra of Small Molecules," in *Electron Spectroscopy: Theory, Techniques, and Applications*, vol. 1, edited by C.R. Brundle and A.D. Baker, New York: Academic, 1977, p. 152.

Rabalais, J.W. *Principles of Ultraviolet Photoelectron Spectroscopy*, New York: Wiley-Interscience, 1976.

Siegbahn, K., "Electron Spectroscopy and Molecular Structure," *Pure and Applied Chemistry* 48: 77 (1976).

Sweigart, D.A., "Lone Pair Orbital Energies in Group VI and VII Hydrides," *J. Chem. Educ.* 50: 322 (1973).

Turner, D.W., Baker, C., Baker, A.D., and Brundle, C.R., *Molecular Photoelectron Spectroscopy*, New York: Wiley-Interscience, 1970.

Woodward, R.B., and Hoffmann, R., "The Conservation of Orbital Symmetry," *Angewandte Chemie*, International Edition 8: 781 (1969).

6 Transition-Metal Complexes

The transition metals (or d-electron elements) form a wide variety of compounds with interesting spectroscopic and magnetic properties. Many of these compounds, or complexes, as they often are called, are catalysts in important industrial chemical processes. Furthermore, several transition-metal ions are necessary components of biochemical systems. In discussing the structures of metal complexes, it is convenient to use the term *coordination number,* to refer to the number of atoms bonded directly to the central metal atom. As pointed out in Section 5-10, the groups attached to the metal are called *ligands*. Each ligand has one or more donor atoms that bond to the metal. The most common coordination numbers exhibited by metal complexes are four and six, and this chapter will deal mainly with these cases. Examples of the most important geometries for coordination numbers four, five, and six are shown in Figure 6-1.

6-1 Structures and Stabilities

Almost all of the six-coordinate complexes have octahedral structures. Both square-planar and tetrahedral geometries are prominent for four-coordinate complexes. A large number of complexes of metal ions with d^8 or d^9 electronic configurations have the square-planar structure. For example, most of the $Pd^{2+}(4d^8)$, $Pt^{2+}(5d^8)$, and $Au^{3+}(5d^8)$ complexes are square planar. Although tetrahedral complexes are formed by many metal ions of the first transition series (Sc–Zn), in heavier metal ions ($4d$ and $5d$ series) this structure appears mainly in d^0 and d^{10} configurations (*e.g.,* MoO_4^{2-}, ReO_4^-, and HgI_4^{2-}).

Recent research has shown the importance of the once rare square-pyramidal and trigonal-bipyramidal five-coordinate structures in the ground-state stereochemistry of many metal ions, particularly that of Ni^{2+}. One

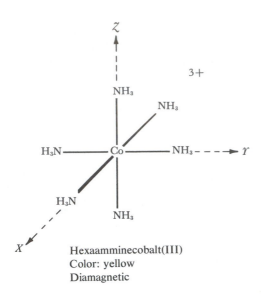

Hexaamminecobalt(III)
Color: yellow
Diamagnetic

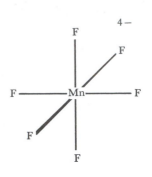

Hexafluoromanganate(II)
Color: pink
Paramagnetic (5 unpaired electrons)

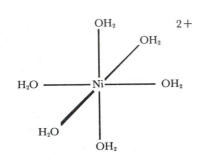

Hexaaquonickel(II)
Color: green
Paramagnetic (2 unpaired electrons)

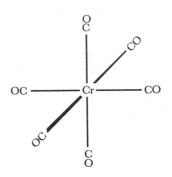

Hexacarbonylchromium(0)
Colorless
Diamagnetic

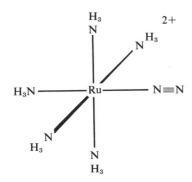

Pentaamminedinitrogenruthenium(II)
Colorless
Diamagnetic

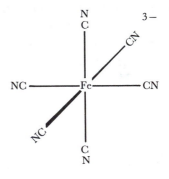

Hexacyanoferrate(III) or "ferricyanide"
Color: red
Paramagnetic (1 unpaired electron)

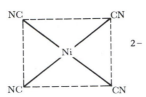

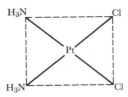

Square planar

Tetracyanonickelate(II)
Color: yellow
Diamagnetic

cis-Dichlorodiammineplatinum(II)
Color: yellow
Diamagnetic

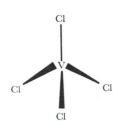

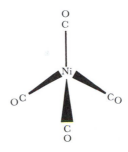

Tetrahedral

Tetrachlorovanadium(IV)
Color: red-brown
Paramagnetic (1 unpaired electron)

Tetracarbonylnickel(0)
Colorless
Diamagnetic

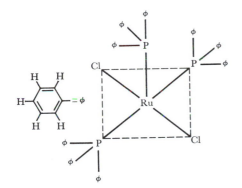

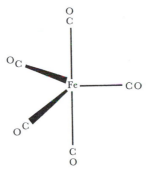

Square pyramidal

trans-Dichloro(tris)triphenyl-
phosphineruthenium(II)
Color: brown
Diamagnetic

Trigonal bipyramidal

Pentacarbonyliron(0)
Color: yellow
Diamagnetic

Figure 6-1 Several metal complexes, with systematic names, colors, and magnetic properties.

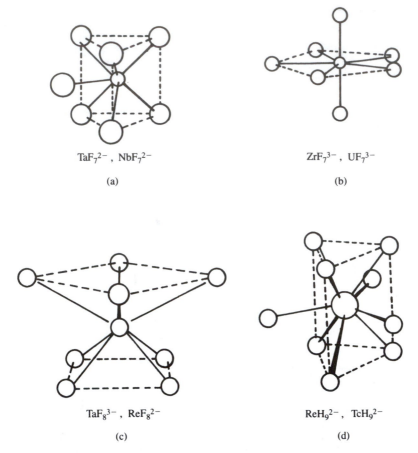

TaF$_7^{2-}$, NbF$_7^{2-}$ ZrF$_7^{3-}$, UF$_7^{3-}$

(a)   (b)

TaF$_8^{3-}$, ReF$_8^{2-}$ ReH$_9^{2-}$, TcH$_9^{2-}$

(c)   (d)

Figure 6-2 The structures of some transition-metal complexes with coordination numbers of seven, eight, and nine: (a) monocapped trigonal prism; (b) pentagonal bipyramid; (c) square antiprism; (d) tricapped trigonal prism.

such complex, the anion Ni(CN)$_5^{3-}$, exhibits both square-pyramidal and trigonal-bipyramidal structures in the solid state. Coordination numbers greater than six are also known. The structures of some of these higher coordination-number complexes are shown in Figure 6-2.

Hard and Soft Metal Ions and Ligands

The reaction between a ligand and a metal ion often is classified as a Lewis acid-base interaction, in which the ligand (base) donates an electron pair to the metal ion (acid). Ralph G. Pearson has suggested that central metal ions be designated either hard or soft, according to their ability to attach various ligands.

Hard metal ions are those that form more stable complexes with ligands containing the "hard" donor atoms N, O, and F than those containing the "softer" P, S, and Cl atoms. Hard metal ions bind ligand donor atoms in the following orders of decreasing stability: $N \gg P > As > Sb; O \gg S > Se > Te$; and $F \gg Cl > Br > I$. Hard metal ions have high positive charges, small radii, and closed-shell (Al^{3+}) or half-filled d-shell (Mn^{2+}, Fe^{3+}) configurations. For instance, Al^{3+} and Fe^{3+} are hard ions that form much more stable complexes with F^- and OH^- than with Cl^- and HS^- (or S^{2-}).

Soft metal ions form stronger complexes with ligands containing heavier donor atoms than those containing N, O, and F. Most soft metal ions have low positive charges, large radii, nonclosed-shell configurations, and are found at or near the right side of each transition series. For example, Hg^{2+} is a soft metal ion; it forms the very weak complex HgF^+ and the increasingly more stable complexes $HgCl_4^{2-}$, $HgBr_4^{2-}$, and HgI_4^{2-}. Soft metal ions also tend to complex more strongly with NH_3 than with H_2O, and more strongly with H_2O than with HF. Thus Ag^+ forms $Ag(NH_3)_2^+$ in aqueous NH_3 solution.

Metal ions with the electronic structures $4d^{10}5s^2$ and $5d^{10}6s^2$ are intermediate in behavior. When the metal ion is bonded to one ligand, it exhibits hard behavior, but when it is bonded to four to six ligands, it exhibits soft behavior. Thus PbF^+ is more stable than $PbCl^+$, but $PbCl_4^{2-}$ forms in concentrated HCl, whereas PbF_4^{2-} is not known to exist in aqueous solution. Table 6-1 classifies some common metal ions as hard, soft, or intermediate.

TABLE 6-1. CLASSIFICATION OF POSITIVE IONS (ACIDS) AS COMPLEX FORMERS

Hard	Intermediate	Soft
Be^{2+}	Tl^+	Cu^{2+}
B^{3+}	Pb^{2+}	Ag^+
Mg^{2+}	Bi^{3+}	Pt^{2+}
Al^{3+}	Sn^{2+}	Pt^{4+}
Ti^{4+}	Sb^{3+}	Pd^{2+}
Fe^{3+}		Pd^{4+}
Mn^{2+}		Ir^{3+}
Zn^{2+}		Rh^{3+}
Ga^{3+}		Hg^{2+}
Si^{4+}		

Chelation and Stability

Ligands that have two or more donor atoms situated so they can bond to the central metal ion often form unusually stable complexes. An example is the organic compound ethylenediamine, $H_2NCH_2CH_2NH_2$, which has two nitrogen atoms with ammonia-like structures, each of which is about as basic as ammonia:

Ethylenediamine (en)

We might expect one ethylenediamine molecule to be equivalent to two ammonia molecules in complexing ability. However, the equilibrium constants for complex formation (K_f) from Table 6-2 show that ethylenediamine is bound much more tightly to Ni^{2+} in aqueous solution than are two ammonia molecules.

Complexes with ligands that contain more than one point of attachment are called *chelates*, from the Greek word for crab's claw. Ligands with two points of attachment are called *bidentates*, those with three points are called *tridentates*, and so on. Some ligands exhibit both bidentate and monodentate (one point of attachment) modes of coordination, as shown for NO_3^- in Figure 6-3.

The enhanced stability of chelates, as illustrated by the comparison of $Ni(en)_3^{2+}$ and $Ni(NH_3)_6^{2+}$, often is called the *chelate effect*. Recent research has led to discovery of a wide variety of polydentate ligands that wrap

TABLE 6-2. STANDARD FREE ENERGIES (ΔG^0 VALUES) FOR THE FORMATION OF Ni^{2+} COMPLEXES WITH NH_3 AND $H_2NCH_2CH_2NH_2$(en) FROM $Ni(H_2O)_6^{2+}$ AT 298 K

Complex	$-\Delta G^0$ (kcal)	K_f
$Ni(NH_3)_2(H_2O)_4^{2+}$	6.6	1.1×10^5
$Ni(en)(H_2O)_4^{2+}$	10.6	4.5×10^7
$Ni(NH_3)_4(H_2O)_2^{2+}$	10.3	1.0×10^8
$Ni(en)_2(H_2O)_2^{2+}$	19.5	1×10^{14}
$Ni(NH_3)_6^{2+}$	11.0	5.5×10^8
$Ni(en)_3^{2+}$	25.7	4.0×10^{18}

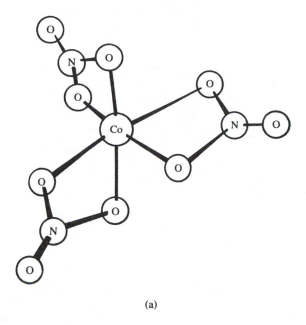

(a)

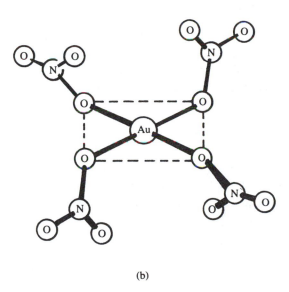

(b)

Figure 6-3 Structures of nitrate complexes: (a) bidentate coordination in $Co(NO_3)_3$; (b) unidentate coordination in $Au(NO_3)_4^-$. Adapted with permission from C.C. Addison, N. Logan, S.C. Wallwork, and C.D. Garner, *Quart. Rev. Chem. Soc.* (London) 25: 289 (1971).

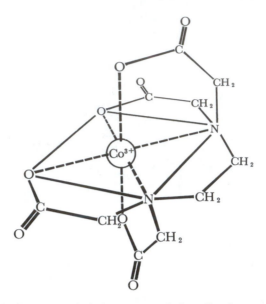

Figure 6-4 The hexadentate chelating agent ethylenediaminetetraacetate (EDTA) can occupy all six octahedral coordination positions. The Co^{3+} chelate shown is a uninegative ion and usually is abbreviated $Co(EDTA)^-$. EDTA is such a strong chelating agent that it will remove metals from enzymes, thereby inhibiting their catalytic activity completely.

around central metal ions, thereby forming extremely stable complexes because of the chelate effect. A chelate containing the important hexadentate ligand EDTA (ethylenediaminetetraacetate) is shown in Figure 6-4.

6-2 Isomerism

In Section 2-18 we discussed the various isomers of $C_2H_2Cl_2$: 1,1-dichloroethylene, *cis*-1,2-dichloroethylene, and *trans*-1,2-dichloroethylene. Each of these isomers has the same number of C–H, C–C, and C–Cl bonds but a different molecular structure. That is, the *topology* is the same although the *shape* is different for each of these isomers. Isomers that have the same topology are called *stereoisomers;* isomers that do not have the same topology are called *constitutional* isomers.

Stereoisomerism

The most common examples of stereoisomerism in transition-metal chemistry are for coordination numbers four and six. Among the square-planar four-coordinate complexes, the *cis*- and *trans*-diamminedichloroplatinum(II) molecules are an example of geometric isomerism:

cis *trans*

If we ignore the hydrogen atoms of the NH_3 ligands, each of these molecules contains a plane of symmetry. Therefore, each isomer is superimposable upon its mirror image; such stereoisomers are called *diastereoisomers*. These two geometric isomers have different chemical and physical properties. For example, the *cis*-isomer has a nonzero dipole moment, whereas the *trans*-isomer has a dipole moment of zero.

Geometric isomers also exist for six-coordinate complexes. An example is provided by the *cis*- and *trans*-tetraamminedichlorocobalt(III) cations:

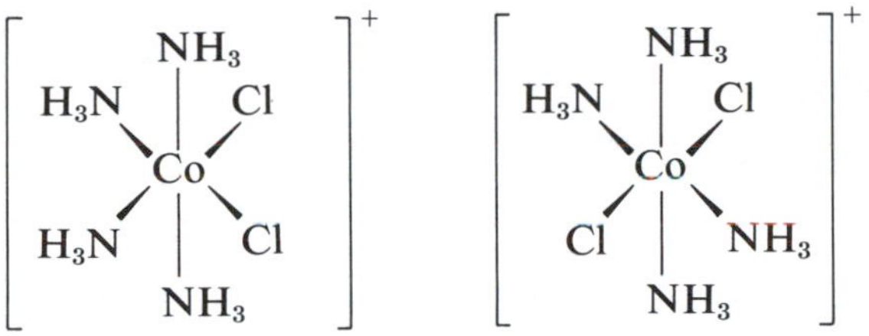

cis *trans*

Geometric isomers also are possible for six-coordinate complexes of the type MA_3B_3: (1) The ligands of each type form an equilateral triangle on one of the faces of the octahedron (the *facial* isomer) and (2) the ligands of each type lie in a symmetry plane that contains the metal atom (the *meridional* isomer). An example of this isomerism is provided by $RuCl_3(H_2O)_3$:

Facial *Meridional*

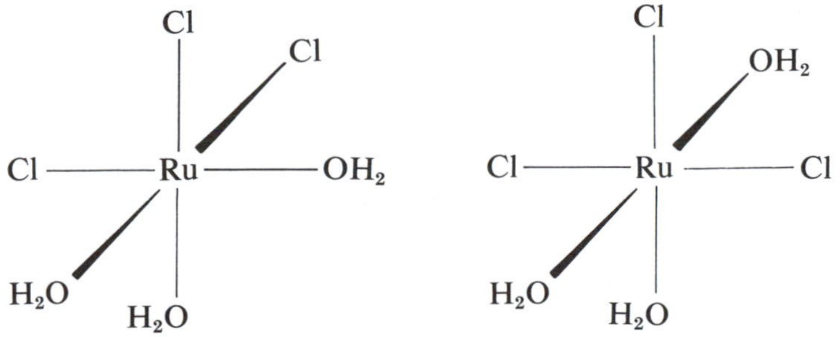

There is a more subtle type of stereoisomerism than we have discussed so far. Examples are provided by the *cis*-amminechlorobis(ethylenediamine)-cobalt(III) and tris(ethylenediamine)cobalt(III) cations:

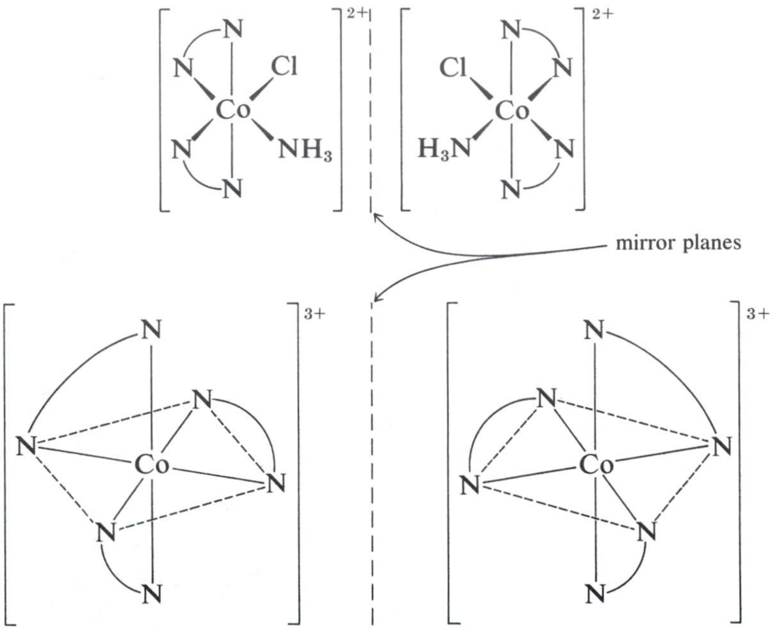

In each of these cases the isomerism arises because the molecule and its mirror image are nonsuperimposable. Such isomers are termed *enantiomers*. The lack of a plane of symmetry in the molecule results in this type of stereoisomerism. A *chiral* compound is one whose enantiomers are related to one another as the left hand is to the right. Chiral molecules have the ability to rotate the plane of plane-polarized light and are said to be *optically active*.

There is an important difference between enantiomers and geometric isomers in that the latter usually differ appreciably in their chemical and physical properties (e.g., dipole moment).

Constitutional Isomerism

Constitutional isomers differ in their atom-to-atom bonding sequence; that is, they have different molecular topologies. Therefore, they will always have distinct chemical and physical properties. We will mention four types of constitutional isomerism—*linkage, hydrate, coordination,* and *ionization* isomerism.

An example of linkage isomerism is provided by the N- and S-bonded thiocyanato complexes of $[Co(CN)_5NCS]^{3-}$:

The classical example of hydrate isomerism is the series of three hydrated chromium chloride complexes having the empirical formula $CrCl_3 \cdot 6H_2O$. The green commercial form of this compound has been shown to be $[Cr(H_2O)_4Cl_2]Cl \cdot 2H_2O$. Dissolution of this complex in water, however, causes the metal-bound chloride ions to be successively replaced by water molecules, giving first the blue-green isomer, $[Cr(H_2O)_5Cl]Cl_2 \cdot H_2O$, and then the violet isomer, $[Cr(H_2O)_6]Cl_3$.

An example of coordination isomerism is given by $[Co(NH_3)_6][Cr(CN)_6]$ and $[Cr(NH_3)_6][Co(CN)_6]$. In the first isomer the ammonia ligands are coordinated to the cobalt atom and in the second to the chromium atom. In order for coordination isomerism to be possible, the compound must exist as a salt in which both the cation and the anion consist of transition-metal complexes.

The final type of constitutional isomerism that we shall consider is ionization isomerism, of which the compounds $[Co(NH_3)_5Br]SO_4$ and $[Co(NH_3)_5SO_4]Br$ are examples. In the first compound, the bromide ligand is coordinated to the cobalt atom; in the second, it is a free anion.

6-3 Effective Atomic Number and Stability

Now that we have examined the structures and stabilities of several transition-metal complexes, we turn to a description of the chemical bonding in these chemical compounds. Just as the octet rule was useful in understanding the chemical bonding of main-group molecules, we shall find that the "effective atomic number" is useful for understanding the bonding in transition-metal complexes.

Effective Atomic Number

The maximum number of σ bonds that can be constructed from s and p valence orbitals is four. Thus four is the highest coordination number commonly encountered for central atoms with $2s$ and $2p$ valence orbitals. This simple argument explains why the octet rule is useful for molecules in which the central atom is from the second row of the periodic table (C, N, O), for instance. As an example, in CH_4 the central carbon atom is "saturated" with four σ bonds. However, if a first-row transition metal is the central atom, there are five d valence orbitals in addition to the one s and three p valence orbitals. Specifically, a first-row transition metal atom has nine valence orbitals—five $3d$ orbitals, one $4s$ orbital, and three $4p$ orbitals.

Since nine valence orbitals are available, a transition-metal atom could have as many as 18 valence electrons and 9 bonds in a transition-metal complex. However, because of the large size of most ligands it is extremely difficult to achieve a coordination number of 9. Rhenium, a large third-row transition metal, and hydrogen, a small ligand, form the complex ReH_9^{2-}, which does have the coordination number 9. The structure of this complex is shown in Figure 6-2. For purposes of electron counting, let us consider the complex ReH_9^{2-} as Re^{7+} and 9 H^- ligands. Then each hydride ligand will donate its 2 electrons to the Re^{7+} ion in forming the rhenium-hydrogen bond. Since the rhenium atom has 7 valence electrons, Re^{7+} has no valence electrons. Each H^- ligand donates 2 electrons for a total of 18 valence electrons surrounding the rhenium atom in ReH_9^{2-}.

The number of electrons surrounding the metal atom in a coordination complex is referred to as the *effective atomic number,* EAN. The fact that the EAN is 18 in many stable complexes is the basis of the "18-electron rule," which is analogous to the octet rule for main-group molecules. The simplest electron counting procedure is to consider the ligand in a form that can readily donate two electrons to the metal to form the metal-ligand bond. For example, $Co(NH_3)_6^{3+}$ is treated as Co^{3+} and 6 NH_3; MnF_6^{4-} as Mn^{2+} and 6 F^-; $Pt(NH_3)_2Cl_2$ as Pt^{2+}, 2 NH_3, and 2 Cl^-; $Ni(CO)_4$ as Ni and 4 CO.

The EAN values for several metal complexes are given in Table 6-3. The following conclusions emerge from inspection of the results:

1. Stable complexes can have far fewer than 18 electrons surrounding the metal atom. Examples from Table 6-3 are VCl_4 (9) and TaF_7^{2-} (14). Two characteristics of these complexes are that the metal comes from the left side of the transition series (few d electrons), and the ligands are highly electronegative. Consequently, the bonding in these complexes has a large ionic component.

2. The 16-electron complexes $[Ni(CN)_4^{2-}, Pt(NH_3)_2Cl_2,$ and $Au(NO_3)_4^-]$ are square planar. It is noteworthy that each of these metal atoms possesses the d^8 electronic configuration. In the last section of this chapter we shall offer an explanation as to why 16-electron complexes with d^8 metal atoms exhibit square-planar structures.

3. The types of complexes that achieve exactly 18 valence electrons around the metal involve either Co^{3+} (d^6) or the metal in a low oxidation state.

EAN VALUES FOR SEVERAL
TRANSITION-METAL COMPLEXES

Complex	Metal Electrons	Ligand-donor Electrons	EAN
VCl_4	1	8	9
TaF_7^{2-}	0	14	14
ZrF_7^{3-}	0	14	14
$Ni(CN)_4^{2-}$	8	8	16
$Pt(NH_3)_2Cl_2$	8	8	16
$Au(NO_3)_4^{-}$	8	8	16
MnF_6^{4-}	5	12	17
$Fe(CN)_6^{3-}$	5	12	17
$Co(NH_3)_6^{3+}$	6	12	18
$Cr(CO)_6$	6	12	18
$Fe(CO)_5$	8	10	18
$Ni(CO)_4$	10	8	18
$Co(NO_3)_3$	6	12	18
$Re(CO)_5NO_3$	6	12	18
$Ni(H_2O)_6^{2+}$	8	12	20
NiF_6^{4-}	8	12	20

The latter are typified by the metal carbonyls $[Cr(CO)_6, Fe(CO)_5,$ and $Ni(CO)_4]$. In fact, the 18-electron rule is obeyed so strongly for metal carbonyls that we shall see shortly how it can be used to predict the stoichiometry and structure of these compounds.

4. Some complexes exceed the 18-electron rule $[Ni(H_2O)_6^{2+}$, for instance]. The most common examples of complexes whose EANs exceed 18 are those involving Ni^{2+} (d^8), Cu^{2+} (d^9), and Zn^{2+} (d^{10}). Thus, these complexes normally involve metal atoms from the right side of the transition series that have their d orbitals filled or nearly filled. The fact that the 18-electron rule can be exceeded for the metal ions in these complexes should not be too surprising. We have seen previously that nontransition molecules can exceed the octet rule if the central atom is near the right side of the period, as is the case with PF_5, SF_6, and XeF_6.

Metal Carbonyls

Except for $V(CO)_6$, which has an EAN of only 17, most of the metal carbonyl compounds [for example, $Cr(CO)_6$, $Fe(CO)_5$, and $Ni(CO)_4$] rigorously obey the 18-electron rule. Generally speaking, metal carbonyls with low EAN numbers, such as NiCO, are as unstable as hydrocarbons that do not satisfy the octet rule. There are parallels, for example, in the CH_n and $Ni(CO)_n$ molecules:

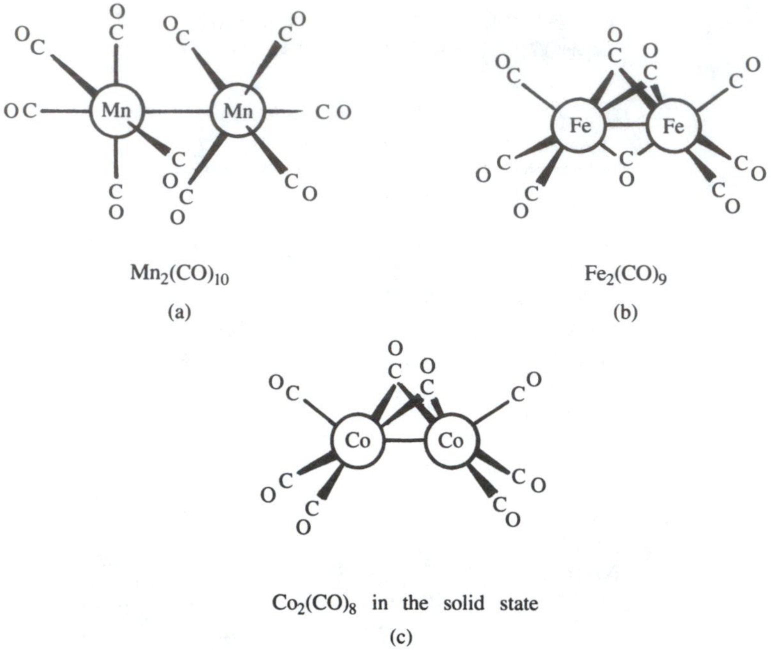

Mn$_2$(CO)$_{10}$

(a)

Fe$_2$(CO)$_9$

(b)

Co$_2$(CO)$_8$ in the solid state

(c)

Figure 6-5 Structures of Mn$_2$(CO)$_{10}$, Fe$_2$(CO)$_9$, and Co$_2$(CO)$_8$ in the solid state.

CH	CH$_2$	CH$_3$	CH$_4$
NiCO	Ni(CO)$_2$	Ni(CO)$_3$	Ni(CO)$_4$

In each case, it is the last member of the series that is stable under ordinary conditions.

The 18-electron rule is useful in formulating the structure and bonding of metal carbonyls containing more than one metal atom. Consider, for example, the structures of the binuclear (two-metal complexes) Mn$_2$(CO)$_{10}$, Fe$_2$(CO)$_9$, and Co$_2$(CO)$_8$, all shown in Figure 6-5. Notice that all three of these complexes contain metal-metal bonds and that the latter two contain bridging carbonyl groups. A bridging carbonyl group contributes one electron to each metal atom; the formation of a single M-C bond would then require that the metal atom also contribute one electron to each M–C bridging bond:

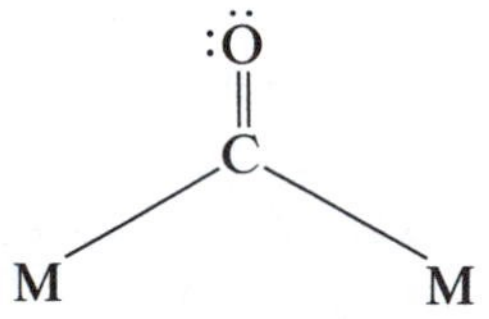

A bridging carbonyl has a formal carbon-oxygen bond order of 2, whereas a terminal carbonyl has a carbon-oxygen bond order between 3 and 2:

$$\ddot{M}-C\equiv O: \leftrightarrow M=C=\ddot{O}:$$

These bond-order predictions are consistent with vibrational spectral data (CO stretching frequencies); uncomplexed CO exhibits a vibrational band at 2143 cm^{-1}; bands due to terminal CO groups fall in the range 2000 $\pm$100 cm^{-1}; and those for bridging CO groups are in the range 1875 $\pm$75 cm^{-1}.

By using the electron-counting method described above for the molecules shown in Figure 6-5, we find that 18 electrons surround each metal atom, if one metal-metal bond is postulated in each case (Table 6-4). That is, in order to achieve an EAN of 18, each metal atom must contribute one electron to the formation of a metal-metal bond. The observed diamagnetism of each molecule, showing that all the electrons have paired spins, is experimental evidence that such metal-metal bonds do form. More convincingly, the existence of $Mn_2(CO)_{10}$ without bridging carbonyl groups is direct evidence that a metal-metal bond is present. The strengths of metal-metal single bonds fall in the range 25–55 kcal mole^{-1}.

TABLE **6-4.** EFFECTIVE ATOMIC NUMBERS FOR $Mn_2(CO)_{10}$, $Fe_2(CO)_9$, AND $Co_2(CO)_8$ IN THE SOLID STATE

Compound	Metal a Electrons	Electrons from Metal b	Electrons from Terminal CO	Electrons from Bridging CO	EAN for Metal a
$Mn_2(CO)_{10}$	7	1	10	0	18
$Fe_2(CO)_9$	8	1	6	3	18
$Co_2(CO)_8$	9	1	6	2	18

6-4 Organometallic π Complexes

Molecules that contain centrally located transition-metal atoms bonded to one or more ligands in which the donor atom is carbon are called *organometallic* complexes. Metal carbonyls are examples of organometallic complexes in which σ donor orbitals on the ligand are used to make metal-carbon σ bonds.

Another class of organometallics includes complexes in which the carbon atoms of organic groups are bonded to the metal through π-orbital networks. The first organometallic π complexes were discovered by the Danish chem-

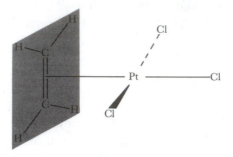

Figure 6-6 Structure of $Pt(C_2H_4)Cl_3^-$.

ist William C. Zeise in 1827. The most famous of these complexes is $K[Pt(C_2H_4)Cl_3]$, in which the organic molecule ethylene (C_2H_4) is bound to Pt^{2+}. More than one hundred years after its discovery, the structure of Zeise's salt was determined by x-ray diffraction methods. The structure of the $Pt(C_2H_4)Cl_3^-$ ion is shown in Figure 6-6. The complex can be thought of as having a square-planar structure (common to Pt^{2+}) with the ethylene bonded at one corner of the square.

The electron counting around Pt in this complex is as follows: Pt^{2+} (d^8), 2 electrons from each of the three Cl^- ligands, and 2 electrons from the π orbital of ethylene. Consequently, the total number of electrons surrounding the platinum atom is 16.

The bonding between C_2H_4 and Pt^{2+} is described conveniently by the model shown in Figure 6-7. The π molecular orbitals of C_2H_4 are used to

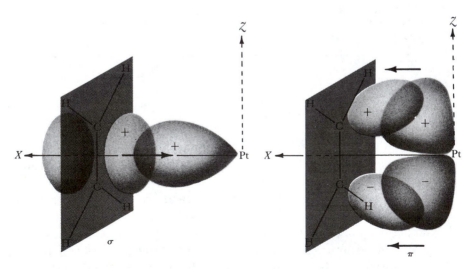

Figure 6-7 The π molecular orbitals of C_2H_4 are used to bond the molecule to Pt^{2+} in $Pt(C_2H_4)Cl_3^-$.

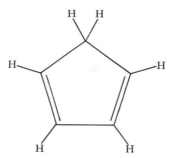

Figure 6-8 Structural formula of cyclopentadiene, C_5H_6.

bond the molecule to the central metal ion. The filled π^b molecular orbital of C_2H_4 is used to form a σ-donor bond with an available Pt^{2+} σ orbital (a combination of the $d_{x^2-y^2}$, s, and p_x orbitals in the model shown). In addition, there may be a π bond between the filled metal d_{xz} orbital and the empty π^* orbital of C_2H_4. This π back bond prevents accumulation of electron density on the metal.

Recent interest in metal π complexes began in 1948, when Samuel A. Miller and his colleagues at the British Oxygen Company discovered that the organic compound cyclopentadiene (Figure 6-8) reacted with an iron-containing catalyst. The product of this reaction was a stable, orange, crystalline substance. This work was not published until 1951, when Thomas J. Kealy and Peter L. Pauson accidentally made the same material by another method. The orange substance has the formula $(C_5H_5)_2Fe$ and is called *ferrocene.**

The structure of ferrocene is like a sandwich—the central iron atom lies between the cyclopentadienyl groups, as shown in Figure 6-9. We can consider the complex as containing the d^6 central ion Fe^{2+} and two coordinated cyclopentadienyl anions $(C_5H_5^-)$. Each $C_5H_5^-$ group furnishes six π electrons. Consequently, ferrocene obeys the 18-electron rule.

Using the 18-electron rule, chemists have reasoned (correctly) that many other sandwich complexes, in which the ligands furnish a total of 12 electrons and the metal atom furnishes 6 electrons, would be stable. A stable complex containing two molecules of benzene (six π electrons each) and a central chromium atom [$Cr(0)$, six metal valence electrons] is one noteworthy example. This sandwich complex is called *dibenzenechromium* and has the formula $(C_6H_6)_2Cr$. A schematic drawing of the structure of dibenzenechromium is given in Figure 6-10. A delocalized molecular-orbital model of the bonding in dibenzenechromium is given in Section 6-8.

---

*The IUPAC name for ferrocene is bis(η^5-cyclopentadienyl) iron. The η (Greek *eta*) stands for *hapto*, from the Greek word *haptein*, "to fasten." Thus the $C_5H_5^-$ ligand in ferrocene is said to be a *pentahapto* cyclopentadienyl group.

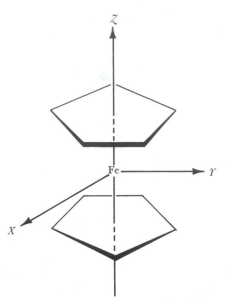

Figure 6-9 "Sandwich" structure of ferrocene, $(C_5H_5)_2Fe$.

Zeise's anion, ferrocene, and dibenzenechromium are only three examples of organometallic π complexes. Other examples are given in Figure 6-11. In the past twenty years many thousands of such complexes have been prepared and studied, and organometallic chemistry is now a major area of contemporary research. Accounts of the structures and reactions of organo-metallic complexes may be found in most inorganic chemistry texts (see Suggestions for Further Reading at the end of the chapter).

6-5 Coordination Modes of Diatomic Ligands

In Section 5-10 we discussed the bonding of simple triatomic MN_2 and MCO molecules. We pointed out that these molecules are expected to have a linear geometry. Research has shown that it is generally true that the MXY frag-

Figure 6-10 Structure of dibenzene-chromium, $(C_6H_6)_2Cr$.

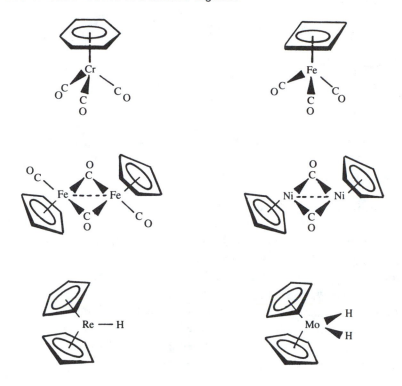

Figure 6-11 Structures of some transition-metal organometallic complexes.

ments ($XY = CN^-$, CO, or N_2) in transition-metal complexes are linear, as shown in Figure 6-1. The HOMO for these ligands is the 3σ molecular orbital. Therefore, it is not surprising that CN^-, CO, and N_2 generally adopt a linear M–X–Y structure, since the 3σ orbital achieves maximum donor ability (i.e., maximum overlap) with a metal σ orbital in such a geometric arrangement.

Not all diatomic ligands bond to transition-metal atoms in a linear "end-on" coordination mode. There are three types of structures that can be adopted by the MXY fragment in a complex. These coordination modes are

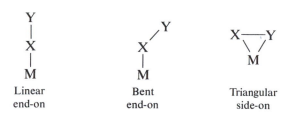

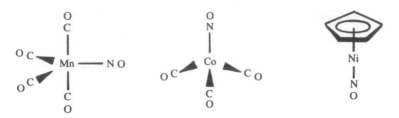

Figure 6-12 Structures of $Mn(CO)_4NO$, $Co(CO)_3NO$, and $(C_5H_5)NiNO$.

Depending on the particular complex, the nitrosyl ligand, NO, exhibits either the linear or bent end-on structure. The oxygen molecule, O_2, generally exhibits either the bent end-on or the triangular side-on mode. To complicate matters further, it is difficult to specify the electronic configuration of the ligand in MNO and MO_2 fragments, since three oxidation levels are possible in each case: NO^+, NO, NO^-; and O_2, O_2^-, O_2^{2-}.

If we treat the nitrosyl ligand as NO^+ in a given transition-metal complex, then we predict the M–N–O bond to be linear, since NO^+ is isoelectronic with CO. Examples of complexes that can be treated formally as containing NO^+ are $Mn(CO)_4NO$, $Co(CO)_3NO$, and $(C_5H_5)NiNO$, all represented in Figure 6-12. In these complexes the formal oxidation state of the metal is Mn^{1-}, Co^{1-}, and Ni^0, respectively. Note that each of these complexes obeys the 18-electron rule.

An example of a complex that exhibits a bent M–N–O structure is $[IrCl_2(NO)P\phi_3]_2$, shown in Figure 6-13. If we treat this complex as formally containing Ir^+, NO^+, and Cl^-, the electron count for the iridium atom is as follows: $Ir^+(d^8)$, four electrons from the two Cl^- anions, two electrons from NO^+, and four electrons from the two $P\phi_3$ ligands. This bonding descrip-

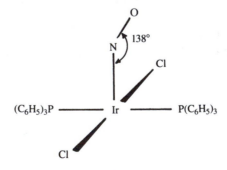

Figure 6-13 The structure of dichloronitrosylbis(triphenylphosphine) iridium.

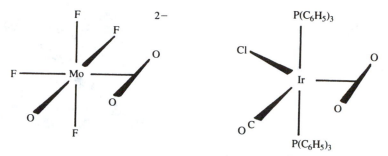

Figure 6-14 The structures of $MoOF_4(O_2)^{2-}$ and $IrCl(CO)O_2(P(C_6H_5)_3)_2$.

tion predicts a linear M–N–O structure, in contradiction to the experimental results. Another bonding model for this complex would be to consider the nitrosyl ligand as NO^-. Such a model would predict a bent M–N–O structure due to the presence of the lone pair on the nitrogen atom:

$$M—\ddot{N}\diagdown_{\underset{\cdot\cdot}{\overset{}{O}:}}$$

In this model we treat the complex as formally containing Ir^{3+}, NO^-, and Cl^-. Then the electron count surrounding the iridium atom is as follows: $Ir^{3+}(d^6)$, four electrons from the two Cl^- anions, two electrons from NO^-, and four electrons from the two $P\phi_3$ ligands. This bonding description results in only 16 electrons surrounding the iridium atom.

The description of the nitrosyl ligand as bonding either in the form NO^+ or NO^- is rather unsatisfactory. A more general approach has been adopted by Roald Hoffmann. In this model the MXY fragment is written as $[MXY]^n$, where n is the sum of the metal valence electrons and the π^* electrons of the diatomic ligand. By considering a Walsh orbital-correlation diagram in the linear, bent, and triangular forms, Hoffmann found a minimum in the orbital energies for triangular $[MXY]^4$, linear $[MXY]^6$, linear or bent $[MXY]^8$, and triangular $[MXY]^{10}$. Examples of molecules that follow these simple rules are $[MoOF_4(O_2)]^{2-}$, triangular $[MoO_2]^4$; $[Os(NH_3)_5CO]^{2+}$, linear $[OsCO]^6$; $[Os(NH_3)_5N_2]^{2+}$, linear $[OsNN]^6$; $Mn(CO)_4NO$, linear $[MnNO]^8$; $IrCl_2(NO)(P\phi_3)_2$, bent $[IrNO]^8$; $IrCl(CO)(O_2)(P\phi_3)_2$, triangular $[IrO_2]^{10}$. The structures of the two trinangular MO_2 complexes are shown in Figure 6-14. Not all MO_2 structures are triangular. In fact, oxyhemoglobin, $[FeO_2]^8$, almost certainly contains a bent FeO_2 structure.

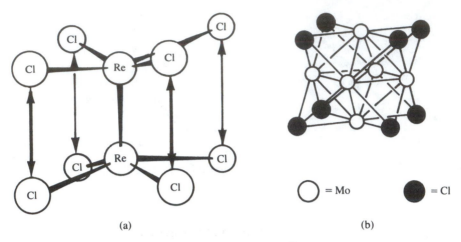

Figure 6-15 The structures of $Re_2Cl_8^{2-}$ and $Mo_6Cl_8^{4+}$.

6-6 Metal Bonds in $Re_2Cl_8^{2-}$ and $Mo_6Cl_8^{4+}$

Many complexes in addition to metal carbonyls contain metal-metal bonds. Two prototypes that have received much attention are $Re_2Cl_8^{2-}$ and $Mo_6Cl_8^{4+}$ (Figure 6-15).

The binuclear $Re_2Cl_8^{2-}$ ion exhibits two remarkable features. First, the two sets of four chlorine atoms have an *eclipsed* rotational orientation about the Re-Re axis. (Based on ligand-ligand repulsions, we could expect a *staggered* configuration.) Second, the Re-Re distance in $Re_2Cl_8^{2-}$ (2.24 Å) is much shorter than the average Re-Re distance in metallic rhenium (2.75 Å).

The d orbitals available for metal-metal bonds in $Re_2Cl_8^{2-}$ are the d_{z^2}, d_{xz}, d_{yz} and d_{xy} (the $d_{x^2-y^2}$ is used to attach the 4 Cl^- groups in each $ReCl_4^-$ unit). As shown in Section 4-10 (Figures 4-35 and 4-36), these four orbitals can combine to form four bonding molecular orbitals:

$$\sigma_g^b(z^2) = \frac{1}{\sqrt{2}}(z_a^2 + z_b^2)$$

$$\pi_u^b(xz) = \frac{1}{\sqrt{2}}(xz_a + xz_b)$$

$$\pi_u^b(yz) = \frac{1}{\sqrt{2}}(yz_a + yz_b)$$

$$\delta_g^b(xy) = \frac{1}{\sqrt{2}}(xy_a + xy_b)$$

Each Re atom in $Re_2Cl_8^{2-}$ is in the $+3$ oxidation state (d^4 configuration); thus eight electrons occupy the four bonding molecular orbitals,

$$(\sigma_g^b)^2(\pi_u^b)^4(\delta_g^b)^2$$

The result is a predicted *quadruple* rhenium-rhenium bond in $Re_2Cl_8^{2-}$. The quadruple bond obviously accounts for the short Re-Re distance in $Re_2Cl_8^{2-}$ and the δ bond accounts for the eclipsed structure of the chloride ligands. Rotation of one $ReCl_4^-$ unit relative to the other would break the δ bond (see Figure 4-35), just as rotation of one CH_2 group relative to the other CH_2 group would break the π bond in ethylene (Section 3-6).

The bonding in the $Mo_6Cl_8^{4+}$ cluster can be described in terms of an octahedral Mo_6^{12+} "core." Since a neutral molybdenum atom has 6 valence electrons, there are a total of $(6 \times 6) - 12 = 24$ electrons available for molybdenum-molybdenum bonding in the octahedral core. These 24 electrons comprise 12 electron pairs—just enough to form a single bond between each pair of molybdenum atoms. However, more realistic formulations of the bonding in $Mo_6Cl_8^{4+}$ and other polynuclear-cluster complexes require multi-center delocalized orbitals (see Suggestions for Further Reading, especially K. Wade, "The Key to Cluster Shapes").

6-7 Ligand-Field Theory for Octahedral Complexes

Most first-row metal complexes have the coordination number six and an octahedral structure. Several octahedral complexes are shown in Figure 6-1. In an octahedral structure, σ bonding uses six of the nine valence orbitals of the central atom, namely the $3d_{x^2-y^2}$, $3d_{z^2}$, $4p_x$, $4p_y$, $4p_z$, and $4s$. Notice that the $3d_{x^2-y^2}$ and $3d_{z^2}$ orbitals are directed toward the ligands, as shown in Figure 6-16. In the hybrid-orbital description, six d^2sp^3 hybrid orbitals are used to attach the six ligands, as shown in Figure 6-17 for $Co(NH_3)_6^{3+}$. This localized-orbital approach describes the bonding in $Co(NH_3)_6^{3+}$ as resulting from the donation of the sp^3 lone electron pair on each NH_3 ligand into an empty d^2sp^3 hybrid orbital on the metal atom. However, the hybrid-orbital theory is unable to explain the colors and magnetic properties of transition-metal complexes. The model of greatest utility in discussing the properties of metal complexes is based on delocalized molecular-orbital theory. Because the model emphasizes the interaction of the d valence orbitals of the metal atoms with appropriate ligand orbitals, it often is called *ligand-field theory*.

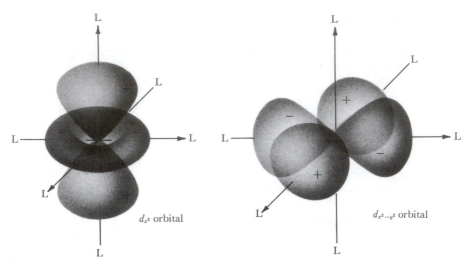

Figure 6-16 The spatial orientation of the d_{z^2} and $d_{x^2-y^2}$ orbitals in an octahedral complex. L represents a general ligand.

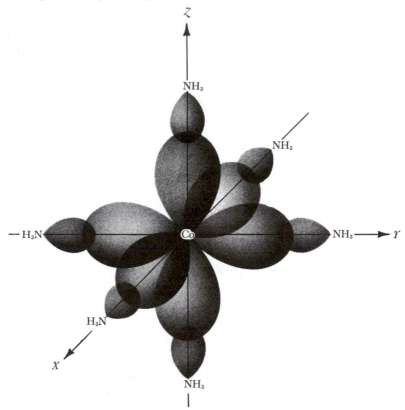

Figure 6-17 Hybrid-orbital bonding representation of $Co(NH_3)_6^{3+}$. Each NH_3 ligand is attached to the metal through a bond involving the NH_3 lone electron pair and one of the six equivalent d^2sp^3 hybrid orbitals of the central metal ion.

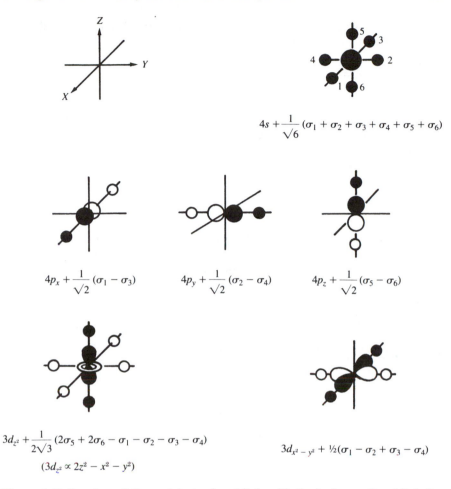

$$4s + \frac{1}{\sqrt{6}}(\sigma_1 + \sigma_2 + \sigma_3 + \sigma_4 + \sigma_5 + \sigma_6)$$

$$4p_x + \frac{1}{\sqrt{2}}(\sigma_1 - \sigma_3)$$

$$4p_y + \frac{1}{\sqrt{2}}(\sigma_2 - \sigma_4)$$

$$4p_z + \frac{1}{\sqrt{2}}(\sigma_5 - \sigma_6)$$

$$3d_{z^2} + \frac{1}{2\sqrt{3}}(2\sigma_5 + 2\sigma_6 - \sigma_1 - \sigma_2 - \sigma_3 - \sigma_4)$$

$$(3d_{z^2} \propto 2z^2 - x^2 - y^2)$$

$$3d_{x^2-y^2} + \tfrac{1}{2}(\sigma_1 - \sigma_2 + \sigma_3 - \sigma_4)$$

Figure 6-18 Overlap of the metal atomic orbitals with the hydrogen $1s$ orbitals in a hypothetical MH_6^{n+} octahedral complex.

To simplify the bonding in octahedral complexes we will consider a hypothetical MH_6^{n+} complex. For this complex we need to conider only the $3d$, $4s$, and $4p$ valence orbitals of the transition metal and the $1s$ orbital on each of the hydrogen ligand atoms.

Our first problem is to determine which SALC of hydrogen $1s$ orbitals will overlap with each of the metal valence orbitals. Due to the high symmetry of the octahedral complex, most of the SALCs can be obtained by inspection. Figure 6-18 exhibits the atomic-orbital overlaps for each of the possible molecular orbitals in MH_6^{n+}.

The metal orbitals that can form σ molecular orbitals are $3d_{x^2-y^2}$, $3d_{z^2}$, $4s$, $4p_x$, $4p_y$, and $4p_z$. Because the sign of the $4s$ orbital does not change over the boundary surface, the proper linear combination of ligand orbitals for the $4s$ orbital is

$$\frac{1}{\sqrt{6}}(\sigma_1 + \sigma_2 + \sigma_3 + \sigma_4 + \sigma_5 + \sigma_6),$$

in which σ_i $(i = 1, \ldots, 6)$ refers to the $1s$ orbital on the ith hydrogen atom. The wave function for the molecular orbital involving the metal $4s$ orbital therefore is

$$\psi(\sigma_s^b) = c_1\, 4s + c_2\left[\frac{1}{\sqrt{6}}(\sigma_1 + \sigma_2 + \sigma_3 + \sigma_4 + \sigma_5 + \sigma_6)\right]$$

We find the other molecular orbitals by matching the metal-orbital lobes with ligand σ orbitals that have the proper sign and magnitude, as shown in Figure 6-18. The wave functions for the bonding orbitals are

$$\psi(\sigma_x^b) = c_3\, 4p_x + c_4\left[\frac{1}{\sqrt{2}}(\sigma_1 - \sigma_3)\right]$$

$$\psi(\sigma_y^b) = c_3\, 4p_y + c_4\left[\frac{1}{\sqrt{2}}(\sigma_2 - \sigma_4)\right]$$

$$\psi(\sigma_z^b) = c_3\, 4p_z + c_4\left[\frac{1}{\sqrt{2}}(\sigma_5 - \sigma_6)\right]$$

$$\psi(\sigma_{x^2-y^2}^b) = c_5\, 3d_{x^2-y^2} + c_6[\tfrac{1}{2}(\sigma_1 - \sigma_2 + \sigma_3 - \sigma_4]$$

$$\psi(\sigma_{z^2}^b) = c_7\, 3d_{z^2} + c_8\left[\frac{1}{2\sqrt{3}}(2\sigma_5 + 2\sigma_6 - \sigma_1 - \sigma_2 - \sigma_3 - \sigma_4)\right]$$

Each of the SALCs of the hydrogen $1s$ orbitals can be determined readily by inspection of Figure 6-18, except the SALC that combines with the $3d_{z^2}$ orbital, which is

$$\frac{1}{2\sqrt{3}}(2\sigma_1 + 2\sigma_2 - \sigma_3 - \sigma_4 - \sigma_5 - \sigma_6)$$

Once the other five SALCs are known, the appropriate coefficient for each hydrogen $1s$ orbital can be determined by the unit-orbital and normalization conditions discussed in Sections 3-7 and 5-3. The form of this SALC also can be deduced by realizing that the actual form of the $3d_{z^2}$ orbital is $3d_{2z^2-x^2-y^2}$.

From inspection of the atomic-orbital overlaps shown in Figure 6-18, we readily conclude that the three bonding molecular orbitals involving the metal valence $4p$ orbitals will be degenerate. In octahedral symmetry the three molecular orbitals $\psi(\sigma_x^b)$, $\psi(\sigma_y^b)$, and $\psi(\sigma_z^b)$, are given the symbol t_{1u}; the corresponding antibonding molecular orbitals are designated t_{1u}^*. Although it is not obvious from Figure 6-18, the $3d_{x^2-y^2}$ and $3d_{z^2}$ orbitals are also equivalent in an octahedral complex, and $\psi(\sigma_{x^2-y^2}^b)$ and $\psi(\sigma_{z^2}^b)$ are degenerate. The bonding and antibonding molecular orbitals for this degenerate pair are labeled e_g and e_g^*, respectively, in octahedral symmetry. Therefore, including the $\psi(\sigma_s^b)$ molecular orbital (labeled a_{1g}), there are three sets of σ^b molecular orbitals in an octahedral complex: a_{1g}, t_{1u}, and e_g. Proof of the equivalence of the $3d_{x^2-y^2}$ and $3d_{z^2}$ orbitals is given in Section 6-13.

We have used all but three of the metal valence orbitals in the molecular orbitals. We are left with $3d_{xz}$, $3d_{yz}$, and $3d_{xy}$. These orbitals are situated properly for π bonding, as demonstrated in the discussion of the bonding in ScO and NiCO (Sections 4-11 and 5-10). The three d_π orbitals clearly are equivalent in an octahedral complex, and are given the symbol t_{2g}.

In Figure 6-19 we present a molecular-orbital energy-level scheme for the hypothetical MH_6^{n+} molecule. A molecule with 12 valence electrons, such as hypothetical MnH_6^+, would have all the bonding molecular orbitals occupied, and would have the resultant ground-state electronic configuration

$$(e_g)^4(a_{1g})^2(t_{1u})^6(t_{2g})^0(e_g^*)^0$$

Any valence electrons beyond these 12 must occupy the predominantly $3d$ molecular orbitals t_{2g} and e_g^*. For example, hypothetical FeH_6^+ would have the electronic configuration $\cdots (t_{2g})^1$. The two orbital sets e_g^* (mainly $3d_{z^2}$ and $3d_{x^2-y^2}$) and t_{2g} (mainly $3d_{xz}$, $3d_{yz}$, and $3d_{xy}$) are called *ligand-field levels*. As shown in Figure 6-19, the e_g^* orbitals in an octahedral complex always have higher energy than the t_{2g} orbitals. The energy separation of the e_g^* and t_{2g} levels is called the *octahedral ligand-field splitting* and is abbreviated Δ_o. Notice that the higher e_g^* orbital is of σ^* type, and the lower t_{2g} orbitals are of nonbonding π type.

Let us apply the derived energy-level scheme shown in Figure 6-19 to a known case, the metal complex $Co(NH_3)_6^{3+}$. The procedure is simply to replace the hydrogen $1s$ orbital by the lone-pair orbital on the NH_3 ligand. The 12 electrons furnished by the six NH_3 ligands occupy six bonding molecular orbitals (e_g, a_{1g}, and t_{1u}) constructed from the NH_3 lone-pair orbitals and the $3d_{z^2}$, $3d_{x^2-y^2}$, $4s$, $4p_x$, $4p_y$, and $4p_z$ metal orbitals. The bonding combinations are very stable, and need not concern us now. They are

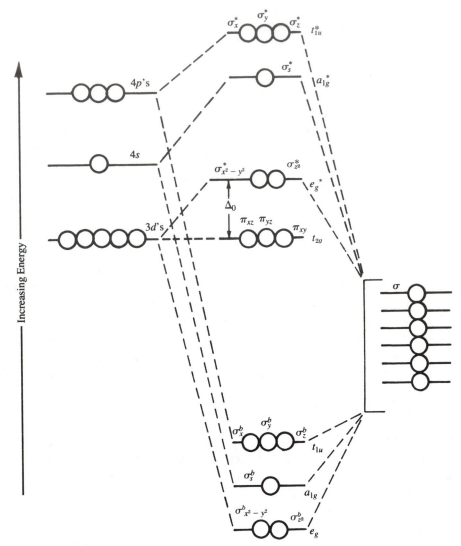

Figure 6-19 Relative orbital energies in $MH_6{}^{n+}$.

analogous to the six d^2sp^3 bonding orbitals in the hybrid-orbital description. Also, we need not consider the antibonding orbitals derived from the $4s$ and $4p$ metal atomic orbitals (a_{1g}^* and t_{1u}^*), because these orbitals are very energetic. The important orbitals for our discussion are those that make up the two ligand-field levels, t_{2g} and e_g^*, shown in Figure 6-20. Because the valence electronic structure of Co^{3+} is $3d^6$, there are six valence electrons available to place in the t_{2g} and e_g^* levels. There are two possibilities, depending on the value of Δ_0. In $Co(NH_3)_6^{3+}$, Δ_0 is sufficiently large to allow all

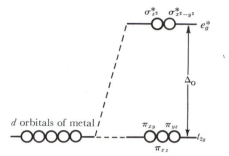

Figure 6-20 Ligand-field splitting in an octahedral complex.

six d electrons to fill the t_{2g} level, thereby giving the ground-state electronic structure $(t_{2g})^6$, with all electrons paired. However, if Δ_0 is smaller than the required electron-pairing energy, as it is in the Co^{3+} complex CoF_6^{3-}, the d electrons occupy the t_{2g} and e_g^* levels to give the maximum number of unpaired spins. The ground-state structure of CoF_6^{3-} is $(t_{2g})^4(e_g^*)^2$, with four unpaired electrons. Because of the difference in the number of unpaired electron spins in the two complexes, $Co(NH_3)_6^{3+}$ is said to be a *low-spin* complex, while CoF_6^{3-} is referred to as a *high-spin* complex.

Now let us consider the ground-state electronic configurations of octahedral complexes that contain central metal ions other than Co^{3+}. Table 6-5 gives the d^n configuration and the number of unpaired d electrons in first-row dipositive and tripositive transition-metal ions commonly observed in octahedral coordination complexes. Referring to Figure 6-20, we see that metal ions with one, two, and three valence electrons will have the respective ground-state configurations $(t_{2g})^1$, $(t_{2g})^2$, and $(t_{2g})^3$. There are two possible ground-state configurations for the metal d^4 configuration, depending on the value of Δ_0 in the complex. If Δ_0 is less than the energy required to pair two d electrons in the t_{2g} level, the fourth electron will go into the e_g^* level, thereby giving the high-spin configuration $(t_{2g})^3(e_g^*)^1$, with four unpaired electrons. Ligands that form high-spin complexes are called *weak-field* ligands.

However, if Δ_0 is larger than the required electron-pairing energy, the fourth electron will go into the lower-energy t_{2g} level and pair with one of the three electrons already present in that level. In this situation the ground state of the complex is the low-spin configuration $(t_{2g})^4$, with only two unpaired electrons. Ligands that cause large splittings and force all of the electrons to occupy the more stable t_{2g} level, giving low-spin complexes, are called *strong-field* ligands.

In filling the t_{2g} and e_g^* energy levels, the electronic configurations d^5, d^6, and d^7, as well as d^4, can exhibit either a high-spin or a low-spin ground

TABLE 6-5. UNPAIRED ELECTRONS
IN UNCOMPLEXED METAL IONS

Ion	Number of d Electrons	Number of Unpaired d Electrons
Sc^{3+}	0	0
Ti^{3+}	1	1
V^{2+}	3	3
V^{3+}	2	2
Cr^{2+}	4	4
Cr^{3+}	3	3
Mn^{2+}	5	5
Fe^{2+}	6	4
Fe^{3+}	5	5
Co^{2+}	7	3
Co^{3+}	6	4
Ni^{2+}	8	2
Cu^{2+}	9	1
Zn^{2+}	10	0

state, depending on the value of Δ_o in the complex. For a given d^n configuration the paramagnetism of a high-spin complex is larger than that of a low-spin complex. Examples of octahedral complexes with the possible $(t_{2g})^x(e_g^*)^y$ configurations are given in Table 6-6. For many first-row transition-metal complexes, the energy required to pair two d electrons is in the range 15,000–25,000 cm^{-1}. Consequently, if Δ_o is less than 15,000 cm^{-1}, a high-spin complex results; if Δ_o is more than 25,000 cm^{-1}, a low-spin complex results. The presence or absence of paramagnetism of complexes with Δ_o values between 15,000 cm^{-1} and 25,000 cm^{-1} depends upon the particular metal and ligand involved.

The first-row transition-metal ions that form the largest number of stable octahedral complexes are Cr^{3+} (d^3), Ni^{2+} (d^8), and Co^{3+} (d^6). The +3 central-ion charge for chromium and cobalt is apparently large enough for strong σ bonding, but not large enough to oxidize the ligands and destroy the complex. The orbital configurations $(t_{2g})^3$ and $(t_{2g})^6$ take maximum advantage of the low-energy t_{2g} level: $(t_{2g})^3$ corresponds to a half-filled t_{2g} shell, which requires no electron-pairing energy, and $(t_{2g})^6$ is the closed-shell structure. The Co^{3+} complexes rigorously obey the 18-electron rule since 6 electrons are contributed by the metal ion and 12 electrons by the ligands to the total bonding picture. The $(t_{2g})^6(e_g^*)^2$ configuration of octahedral Ni^{2+} complexes features a filled t_{2g} and a half-filled e_g^* level. These complexes have a total of 20 electrons surrounding the metal ion. Evidently, for metal ions at the far right side of the transition series (e.g., Ni^{2+}, Cu^{2+}, Zn^{2+}), the metal d

TABLE 6-6. ELECTRONIC CONFIGURATIONS OF
OCTAHEDRAL COMPLEXES

Electronic Configuration of the Metal Ion	Electronic Structure of the Complex	Number of Unpaired Electrons	Example
$3d^1$	$(t_{2g})^1$	1	$Ti(H_2O)_6^{3+}$
$3d^2$	$(t_{2g})^2$	2	$V(H_2O)_6^{3+}$
$3d^3$	$(t_{2g})^3$	3	$Cr(H_2O)_6^{3+}$
$3d^4$	Low-spin; $(t_{2g})^4$	2	$Mn(CN)_6^{3-}$
	High-spin; $(t_{2g})^3(e_g)^1$	4	$Cr(H_2O)_6^{2+}$
$3d^5$	Low-spin; $(t_{2g})^5$	1	$Fe(CN)_6^{3-}$
	High-spin; $(t_{2g})^3(e_g)^2$	5	$Mn(H_2O)_6^{2+}$
$3d^6$	Low-spin; $(t_{2g})^6$	0	$Co(NH_3)_6^{3+}$
	High-spin; $(t_{2g})^4(e_g)^2$	4	CoF_6^{3-}
$3d^7$	Low-spin; $(t_{2g})^6(e_g)^1$	1	$Co(NO_2)_6^{4-}$
	High-spin; $(t_{2g})^5(e_g)^2$	3	$Co(H_2O)_6^{2+}$
$3d^8$	$(t_{2g})^6(e_g)^2$	2	$Ni(NH_3)_6^{2+}$
$3d^9$	$(t_{2g})^6(e_g)^3$	1	$Cu(H_2O)_6^{2+}$

orbitals are sufficiently stable that the 18-electron rule can be exceeded, and the e_g^* orbitals can be occupied without destroying the complex.

Photoelectron Spectroscopy of Octahedral Complexes

We have seen that it is possible to explain the magnetic properties of octahedral transition-metal complexes on the basis of the t_{2g} and e_g^* ligand-field levels shown in Figures 6-19 and 6-20. To gain experimental information on the bonding e_g, a_{1g}, and t_{1u} molecular orbitals, we must resort to optical absorption spectroscopy or photoelectron spectroscopy. As we have seen in previous chapters, photoelectron spectroscopy is one of the best techniques available for determining the relative energy of occupied molecular orbitals.

Consider two octahedral complexes of tungsten, $W(CH_3)_6$ and $W(CO)_6$. The ground-state electronic configuration of the tungsten atom is $[Xe]4f^{14}5d^46s^2$. The $4f^{14}$ electrons behave as part of the core; consequently, tungsten has six valence electrons (d^4s^2). In $W(CH_3)_6$ each ligand contributes one electron to the $W-CH_3$ bond, because it has one unpaired electron that is not involved in carbon-hydrogen bonding:

$$H—\overset{\displaystyle \cdot}{C}—H$$
$$|$$
$$H$$

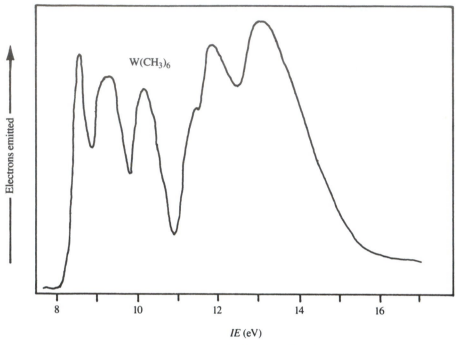

Figure 6-21 Photoelectron spectrum of $W(CH_3)_6$. The three bands between 8 eV and 11 eV are assigned to the e_g, a_{1g}, and t_{1u} orbitals. The bands that occur at greater than 11 eV are attributed to ionization from orbitals that are predominantly involved in C–H bonding within the CH_3 ligand. Adapted with permission from J.C. Green, D.R. Lloyd, L. Galyer, K. Mertis, and G. Wilkinson, *J. Chem. Soc. Dalton* 1403 (1978).

Altogether there are six electrons from the six CH_3 ligands and six electrons from the tungsten atom that are available for forming the W–CH_3 bonds. All 12 electrons occupy bonding orbitals, as in Figure 6-19, giving the ground-state electronic configuration $(e_g)^4(a_{1g})^2(t_{1u})^6$. As predicted, the experimental photoelectron spectrum of $W(CH_3)_6$ exhibits three bands between 8 and 11 eV (Figure 6-21). The exact ordering of these levels is uncertain, but that need not concern us here. The important feature is that three bands are found. The bands that occur at energies greater than 11 eV are attributed to ionization from CH_3 orbitals that involve C–H bonding.

We turn next to the bonding in $W(CO)_6$, which can be described in terms of σ donation from the lone-pair electrons on the carbon atom ($:C{\equiv}O:$) to the metal orbitals that have σ symmetry. As seen in Figure 6-18, these are the $5d_{x^2-y^2}$, $5d_{z^2}$, $6s$, $6p_x$, $6p_y$, and $6p_z$ orbitals. The CO ligands themselves furnish the 12 electrons that are necessary to fill the bonding e_g, a_{1g}, and t_{1u} molecular orbitals. In addition to these 12 electrons, we must consider the 6 valence electrons of the tungsten atom (d^4s^2). As explained for NiCO, however, the "valence state" of tungsten in $W(CO)_6$ is d^6, because all six valence electrons occupy the t_{2g} molecular orbital of $W(CO)_6$. [Carbon monoxide is a strong-field ligand so the $(t_{2g})^6$ configuration is more stable than the $(t_{2g})^4(e_g^*)^2$ configuration.] The ground-state electronic configuration of $W(CO)_6$ can then be written:

$$(e_g)^4(a_{1g})^2(t_{1u})^6(t_{2g})^6(e_g^*)^0$$

in which the t_{2g} molecular orbital is mainly W $5d$ and the e_g, a_{1g}, and t_{1u} are W-CO bonding orbitals. Because atomic tungsten has a much lower ioniza-

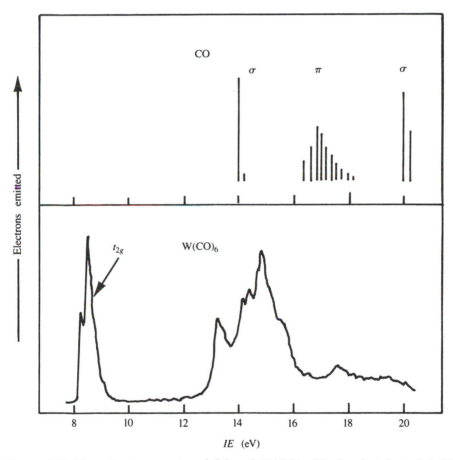

Figure 6-22 Photoelectron spectra of CO and $W(CO)_6$. The band at about 8.5 eV is assigned to the t_{2g} orbital, which is predominantly tungsten $5d$. The slight splitting of the first band is due to spin-orbit coupling. Adapted with permission from B.R. Higginson, D.R. Lloyd, P. Burroughs, D.M. Gibson, and A.F. Orchard, *J. Chem. Soc. Faraday* II: 69, 1659 (1973).

tion energy than carbon monoxide (7.98 eV, as compared to 14.01 eV), we expect the t_{2g} orbitals to be much less stable than the e_g, a_{1g}, and t_{1u} molecular orbitals.

The photoelectron spectra of $W(CO)_6$ and CO are presented in Figure 6-22. For $W(CO)_6$, one band is observed at about 8.5 eV, followed by a complex set of overlapping bands that begins at about 13 eV and extends to beyond 18 eV. The first band is assigned to ionization from the t_{2g} molecular orbitals. The slight splitting of this band is due to spin-orbit coupling, and in effect confirms that this orbital is mainly tungsten $5d$ in character. (Recall that heavy atoms exhibit large spin-orbit coupling effects; see Sections 1-14 and 5-6.) The bands between 13 and 20 eV are assigned to all of the orbitals that are predominantly M-CO and CO in character. The complexity of the $W(CO)_6$ photoelectron spectrum in the region 13–20 eV shows that the CO localized orbitals do become involved in the bonding to tungsten, but their energy position does not shift significantly.

d-d Transitions and Light Absorption

An electron in the t_{2g} level of an octahedral complex can absorb a photon in the near-infrared, visible, or ultraviolet region and make a transition to an unoccupied orbital in the more energetic e_g^* level. Electron excitations of this type are the d–d transitions that were discussed in Section 4-11. Although d–d transitions generally give rise to weak absorption bands, they are responsible for the characteristic colors of many transition-metal complexes. An example is the red-violet Ti^{3+} complex, $Ti(H_2O)_6^{3+}$, whose ground state is $(t_{2g})^1$. Excitation of the electron from the t_{2g} orbital to the e_g^* orbital occurs with light absorption in the vicinity of 500 nm (20,000 cm^{-1}). Figure 6-23 shows that maximum absorption occurs at 493 nm (20,300 cm^{-1}), with a molar extinction coefficient (ε) of 5. The value of the octahedral ligand-field splitting, Δ_o, is usually expressed in wave numbers; thus we say that for $Ti(H_2O)_6^{3+}$, $\Delta_o = 20,300$ cm^{-1}.

The colors of many other transition-metal complexes are caused by d–d transitions. The number of absorption bands depends on the molecular geometry of the complex and the d^n configuration of the central metal atom. The intensities of the d–d bands also vary with geometry. The absorption bands in octahedral complexes such as $Ti(H_2O)_6^{3+}$ are weak (ε range of 1–500) because d–d transitions of the type $t_{2g} \rightarrow e_g^*$ are orbitally forbidden. However, certain of the d–d transitions in tetrahedral complexes are fully allowed and can lead to fairly strong absorption (ε range of 200–5000). Thus, in many cases, the study of the electronic spectra of metal complexes is a powerful tool for determining molecular geometries.

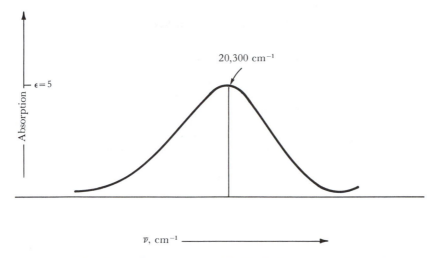

Figure 6-23 The absorption spectrum of $Ti(H_2O)_6^{3+}$ in the visible region.

A special case of interest is the high-spin d^5 configuration exhibited by the octahedral complex $Mn(H_2O)_6^{2+}$. The high-spin ground state for the five $3d$ electrons in $Mn(H_2O)_6^{2+}$ is $(t_{2g})^3(e_g^*)^2$, with five unpaired electrons. All d–d transitions from this ground state are spin-forbidden (the transitions of lowest energy are to excited states with three unpaired electrons). The color of the Mn^{2+} ion in aqueous solution is very pale pink. The color results from the extremely low intensities (ε values of approximately 0.01) of the spin-forbidden d–d absorption bands of $Mn(H_2O)_6^{2+}$, which lie in the visible region. (Weak absorption bands arising from electronic transitions that are "forbidden" according to orbital- or spin-selection rules were discussed in Section 4-4.)

Factors that Influence the Value of Δ_o

The Δ_o values for a representative selection of octahedral complexes are given in Table 6-7. The value of Δ_o depends on a number of variables, the most important ones being the nature of the ligand, the ionic charge (or oxidation number) of the central metal ion, and the principal quantum number, n, of the d valence orbitals. We discuss these variables individually in the following paragraphs.

Nature of the ligand. The order of ligands in terms of their ability to split the e_g^* and t_{2g} molecular orbitals is known as the *spectrochemical series*.

TABLE **6-7.** VALUES OF Δ_0 FOR REPRESENTATIVE
METAL COMPLEXES

Octahedral Complexes	Δ_0 (cm^{-1})	Octahedral Complexes	Δ_0 (cm^{-1})
TiF_6^{3-}	17,000	$Co(NH_3)_6^{3+}$	22,900
$Ti(H_2O)_6^{3+}$	20,300	$Co(CN)_6^{3-}$	34,500
$V(H_2O)_6^{3+}$	17,850	$Co(H_2O)_6^{2+}$	9,300
$V(H_2O)_6^{2+}$	12,400	$Ni(H_2O)_6^{2+}$	8,500
$Cr(H_2O)_6^{3+}$	17,400	$Ni(NH_3)_6^{2+}$	10,800
$Cr(NH_3)_6^{3+}$	21,600	$RhBr_6^{3-}$	21,600
$Cr(CN)_6^{3-}$	26,600	$RhCl_6^{3-}$	22,800
$Cr(CO)_6$	32,200	$Rh(NH_3)_6^{3+}$	34,100
$Fe(CN)_6^{3-}$	35,000	$Rh(CN)_6^{3-}$	44,000
$Fe(CN)_6^{4-}$	33,800	$IrCl_6^{3-}$	27,600
$Co(H_2O)_6^{3+}$	18,200	$Ir(NH_3)_6^{3+}$	40,000

The order of ligand-field splitting of some important ligands is

$$CO, CN^- > NO_2^- > NH_3 > OH_2 > OH^- > F^- > -SCN^-, Cl^- > Br^- > I^-$$

Octahedral complexes containing ligands such as CN^- and CO, which are at the strong-field end of the spectrochemical series, have Δ_0 values between 30,000 cm^{-1} and 50,000 cm^{-1}. At the other end of the series, octahedral complexes containing Br^- and I^- have relatively small Δ_0 values, in many cases less than 20,000 cm^{-1}.

We already have discussed the important types of metal-ligand bonding in transition-metal complexes. The manner in which each type affects the value of Δ_0 is illustrated in Figure 6-24. We see that a strong $(L \rightarrow M)$ σ interaction increases the energy of the e_g^* orbitals, thereby increasing the value of Δ_0. A strong $(L \rightarrow M)$ π interaction, on the other hand, destabilizes t_{2g}, thereby decreasing the value of Δ_0. A strong $(M \rightarrow L)$ π^* interaction lowers the energy of t_{2g}, thereby increasing the value of Δ_0. The spectrochemical series correlates reasonably well with the relative π-donor and π-acceptor abilities of the ligands. The π-acceptor ligands [those capable of strong $(M \rightarrow L)$ π^* bonding] cause large splittings, whereas the π-donor ligands [those capable of strong $(L \rightarrow M)$ π donation] cause small splittings. The ligands with intermediate Δ_0 values have little or no π-bonding capabilities.

The correlation of the spectrochemical series with the π-donor and π-acceptor characteristics of several important ligands may be summarized as follows:

$$CO, CN^- > NO_2^- > NH_3 > OH_2 > OH^- > F^-$$

π acceptors   non-π weak π
bonding   donors

$$> -SCN^-, Cl^- > Br^- > I^-$$

π donors

Ionic charge of the central metal ion. In complexes containing ligands that are neither strong π donors nor π acceptors, Δ_o increases with increasing ionic charge on the central metal ion. An example is the increase in Δ_o from $V(H_2O)_6^{2+}$, with $\Delta_o = 12,400$ cm^{-1}, to $V(H_2O)_6^{3+}$, with $\Delta_o = 17,850$ cm^{-1}. The larger Δ_o of $V(H_2O)_6^{3+}$ is probably due to a substantial increase in σ bonding of the H_2O ligands to the more positive V^{3+} central metal ion. The enhanced σ interaction increases the difference in energy between e_g^* and t_{2g}.

In octahedral complexes containing strong π-acceptor ligands, an increase in oxidation number of the metal ion does not seem to be accompanied by a substantial increase in Δ_o. For example, both $Fe(CN)_6^{4-}$ and $Fe(CN)_6^{3-}$ have Δ_o values of approximately 34,000 cm^{-1}. When the charge on the central metal ion is increased from $+2$ in $Fe(CN)_6^{4-}$ to $+3$ in $Fe(CN)_6^{3-}$, the t_{2g}

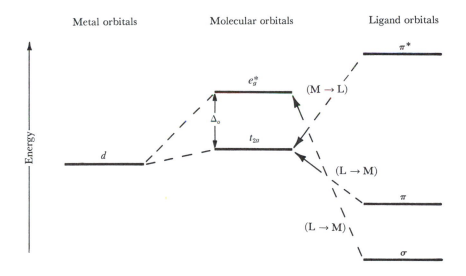

Figure 6-24 The effect on the value of Δ_o of interaction of the ligand σ, π, and π^* orbitals with the metal d orbitals.

level apparently is destabilized through a decrease in π back bonding equally as much as the energy of the e_g^* level is raised by the greater $(L \rightarrow M)$ σ bonding.

Principal quantum number of the d valence orbitals. For changes from $3d$ to $4d$ to $5d$ valence orbitals of the central metal ion in an analogous series of complexes, the value of Δ_o increases markedly. For example, the Δ_o values for $Co(NH_3)_6^{3+}$, $Rh(NH_3)_6^{3+}$, and $Ir(NH_3)_6^{3+}$ are 22,900 cm^{-1}, 34,100 cm^{-1}, and 40,000 cm^{-1}, respectively. Presumably the $4d$ and $5d$ valence orbitals of the ion are more suitable for σ bonding with the ligands than are the $3d$ orbitals, but the reason for this is not well understood. An important consequence of the much larger Δ_o values of $4d$ and $5d$ central metal ions is that low-spin ground states characterize *all* second- and third-row metal complexes. Even complexes such as $RhBr_6^{3-}$ that contain ligands at the weak-field end of the spectrochemical series, exhibit low-spin ground states.

6-8 Ligand-Field Theory for Square-Planar Complexes

As we discussed in Section 6-1, many d^8 central metal ions form square-planar complexes. Let us consider $PtCl_4^{2-}$, pictured in a reference coordinate system in Figure 6-25. The principal σ bonding involves the overlap of $3p(\sigma)Cl^-$ orbitals with the $5d_{x^2-y^2}$, $6s$, $6p_x$, and $6p_y$ metal valence orbitals. In the language of hybrid-orbital theory, the σ bonding is summarized as dsp^2 in a square-planar complex.

Of principal interest here is the ligand-field splitting of the antibonding molecular orbitals derived from the metal d valence orbitals in a square-planar complex. Examination of the overlaps of the Cl^- $3p$ orbitals with the Pt^{2+} $5d$ valence orbitals in $PtCl_4^{2-}$ (Figure 6-26) reveals that only one d

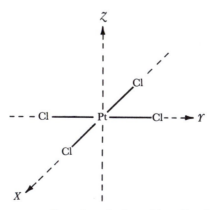

Figure 6-25 Coordinate system for a description of bonding in $PtCl_4^{2-}$.

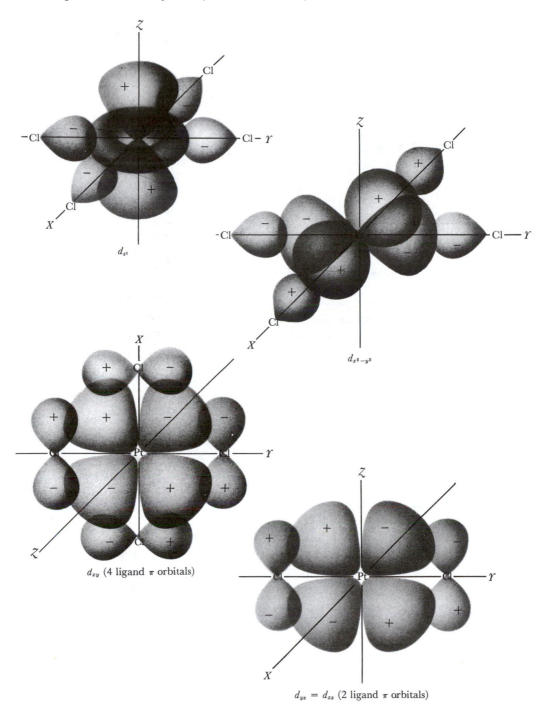

Figure 6-26 Overlap of the metal d valence orbitals with the Cl^- ligand valence ($3p$) orbitals in a square-planar complex.

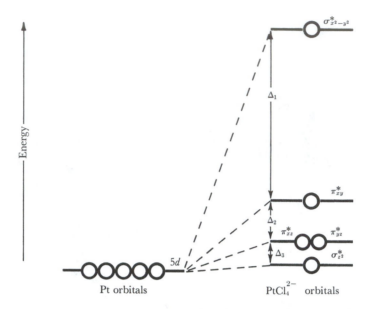

Figure 6-27 Ligand-field splitting diagram for the square-planar complex $PtCl_4^{2-}$. For certain other square-planar complexes, the $\sigma_{z^2}^*$ level may lie above π_{xz}^*, π_{yz}^* in energy.

orbital, the $d_{x^2-y^2}$ orbital, is involved in strong σ bonding in a square-planar complex. The d_{z^2} orbital interacts weakly with the four ligand σ valence orbitals because most of the d_{z^2} orbital is directed along the Z axis away from the ligands. The metal d_{xz}, d_{yz}, and d_{xy} valence orbitals are involved in π molecular orbitals with the ligands. The d_{xy} orbital interacts with $3p(\pi)$ valence orbitals on all four ligands, whereas each of the equivalent d_{xz} and d_{yz} orbitals interacts with only two ligands.

The ligand-field splitting in a square-planar complex is rather complicated, because there are four different energy levels. The ligand-field splitting diagram, which has been worked out from spectral studies of $PtCl_4^{2-}$, is shown in Figure 6-27. We may reasonably place the strongly antibonding $\sigma_{x^2-y^2}^*$ orbital highest in energy for all square-planar complexes. We may also position π_{xy}^* above $\pi_{xz}^*(\pi_{yz}^*)$ because d_{xy} interacts with all four ligands. The position of the weakly antibonding $\sigma_{z^2}^*$ level varies in square-planar complexes, depending on the nature of the ligand and the metal. As we have indicated in Figure 6-27, recent studies of $PtCl_4^{2-}$ allow us to position $\sigma_{z^2}^*$ below the $\pi_{xz}^*(\pi_{yz}^*)$ level. However, regardless of the placement of $\sigma_{z^2}^*$, the most important characteristic of the ligand-field splitting in a square-planar complex is that $\sigma_{x^2-y^2}^*$ has much higher energy than the other four orbitals, which have about the same energy.

The valence electronic configuration of the Pt^{2+} ion is $5d^8$. Because the Δ_1 splitting for $PtCl_4^{2-}$ (Figure 6-27) is much larger than the energy required to pair two electrons in this complex, the ground-state electronic structure is $(\sigma_{z^2}^*)^2(\pi_{xz,yz}^*)^4(\pi_{xy}^*)^2$. In agreement with the ligand-field model, experimental studies show that $PtCl_4^{2-}$ is diamagnetic.

We conclude from the ligand-field splitting that a particularly favorable ground-state electronic structure for square-planar complexes is the low-spin configuration $(\sigma_{z^2}^*)^2(\pi_{xz,yz}^*)^4(\pi_{xy}^*)^2$. The four relatively stable orbitals are occupied completely in this arrangement, and the high-energy $\sigma_{x^2-y^2}^*$ orbital is left vacant. The ligand-field model is consistent with the fact that complexes of the d^8 metal ions, particularly Ni^{2+}, Pd^{2+}, Pt^{2+}, and Au^{3+}, commonly exhibit square-planar geometry. All d^8 square-planar complexes are known to have low-spin ground-state configurations. Since the ligands in these four-coordinate complexes contribute a total of 8 electrons to the bonding, the d^8 square-planar complexes have only 16 electrons surrounding the metal ion.

6-9 Ligand-Field Theory for Tetrahedral Complexes

An example of a tetrahedral metal complex is VCl_4, which is shown in a coordinate system in Figure 6-28. We discussed the role of s and p valence orbitals in a tetrahedral molecule in Chapter 5. The $4s$ and $4p$ atomic orbitals

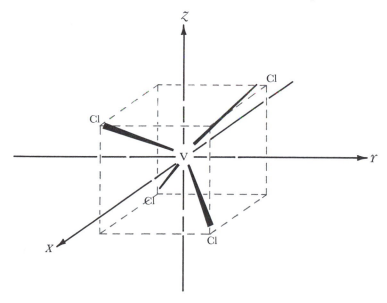

Figure 6-28 Coordinate system for tetrahedral VCl_4.

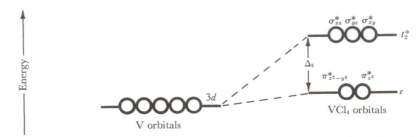

Figure 6-29 Ligand-field splitting in a tetrahedral complex. The antibonding orbitals are divided into two sets: (1) three σ^* (d) orbitals—the t_2^* set; and (2) two π (d) orbitals—the e set.

of vanadium can be used to form σ molecular orbitals. Although the overlap patterns are rather complicated, the $3d_{xz}$, $3d_{yz}$, and $3d_{xy}$ valence orbitals also are situated properly to form σ molecular orbitals. In terms of localized molecular orbitals both sd^3 and sp^3 hybrid orbitals are tetrahedrally oriented. The $3d_{x^2-y^2}$ and $3d_{z^2}$ orbitals of the central atom interact very weakly with the ligands to form π molecular orbitals.

The ligand-field splitting diagram for a tetrahedral complex such as VCl_4 is shown in Figure 6-29. The antibonding molecular orbitals derived from the $3d$ valence orbitals are divided into two sets. The orbitals formed from the $3d_{xz}$, $3d_{yz}$, and $3d_{xy}$ orbitals are of higher energy than those formed from the $3d_{z^2}$ and $3d_{x^2-y^2}$ orbitals. Thus the change from octahedral to tetrahedral geometry exactly reverses the role and the energies of the d valence orbitals of the central metal ion. In a tetrahedral complex we call the three $\sigma^*(d)$ orbitals the t_2^* set, and the two $\pi(d)$ orbitals the e set. We designate the difference in energy between t_2^* and e in a tetrahedral complex as Δ_t.

With one valence electron from $V^{4+}(3d^1)$, the ground-state structure of VCl_4 is $(e)^1$. The paramagnetism of VCl_4 is consistent with this configuration, which has one unpaired electron. Energy in the near-infrared region excites the electron in e to t_2^*, with maximum absorption at 9010 cm^{-1}. Thus for VCl_4, Δ_t is 9010 cm^{-1}. The molar extinction coefficient, ε, of the $e \rightarrow t_2^*$ band is 130, which is larger than a typical value for an octahedral complex.

Values of Δ_t for several representative tetrahedral complexes are given in Table 6-8. Tetrahedral ligand-field splitting (Δ_t) is much smaller than octahedral splitting (Δ_o). For a given metal ion and ligand, both theoretical and experimental studies have established the relationship $\Delta_t \simeq \frac{4}{9}\Delta_o$. This simple relationship is valid if the metal-ligand bond distance is nearly the same in the octahedral and tetrahedral complexes.

Molecular-orbital theory predicts that the t_2 orbitals in a tetrahedral complex will form weaker σ bonds with ligand orbitals than will the e_g octahedral orbitals, because the $3d_{xz}$, $3d_{yz}$, and $3d_{xy}$ orbitals do not "point" directly

TABLE 6-8. VALUES OF Δ_t
FOR REPRESENTATIVE
TETRAHEDRAL COMPLEXES

Tetrahedral Complexes	Δ_t (cm^{-1})
VCl_4	9010
$CoCl_4^{2-}$	3300
$CoBr_4^{2-}$	2900
CoI_4^{2-}	2700
$Co(NCS)_4^{2-}$	4700

at the ligands in tetrahedral symmetry as do the $3d_{x^2-y^2}$ and $3d_{z^2}$ orbitals in octahedral symmetry. This weakening of σ bonding results in a much less energetic t_2^* level and a relatively small Δ_t value. Because of these small Δ_t values, *all* tetrahedral transition-metal complexes have high-spin ground-state configurations.

Of the dipositive metal ions, $Co^{2+}(d^7)$ is exceptionally stable in tetra-hedral complexes. Examples are $CoCl_4^{2-}$, $Co(NCS)_4^{2-}$, and $Co(OH)_4^{2-}$.

The photoelectron spectra of the tetrahedral complexes $TiCl_4$ and VCl_4 are shown in Figure 6-30. Note that VCl_4 has an additional band at low energy, caused by ionization from the $(e)^1$ configuration. Because $TiCl_4$ has one less electron than VCl_4, it has the configuration $(e)^0$ and is diamagnetic. The complex set of overlapping bands at ionization energies greater than 11 eV are due to the Ti-Cl (or V-Cl) σ molecular orbitals and the relatively nonbonding lone-pair electrons on the chlorine atoms.

A second application of photoelectron spectroscopy to tetrahedral molecules is provided by $Ni(CO)_4$. The d^{10} electrons of nickel in its valence state occupy the ligand-field levels, $(e)^4(t_2^*)^6$. As predicted, the photoelectron spectrum of $Ni(CO)_4$ (Figure 6-31) exhibits two bands at low energy (8–11 eV). The bands at 13 eV and higher energies correspond to ionization from orbitals that are predominantly M–CO and CO in character.

6-10 Charge-Transfer Absorption Bands

Many complexes exhibit absorption bands in different positions from those associated with d–d transitions of the central metal ion. These bands are usually in the ultraviolet region, but sometimes they occur in the visible portion of the spectrum. The bands are often quite intense, because they generally involve fully allowed transitions. Schematically, the two types of electronic excitation that result in these strong absorptions are

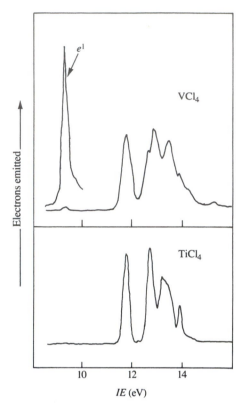

Figure 6-30 Photoelectron spectra of $TiCl_4$ and VCl_4. The first band in the spectrum of VCl_4 is due to the $(e)^1$ orbital. This orbital consists mainly of the vanadium $3d$ atomic orbital. Adapted with permission from P.A. Cox, S. Evans, A Hamnett, and A.F. Orchard, *Chem. Phys. Lett.* 7: 414 (1970).

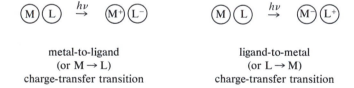

metal-to-ligand	ligand-to-metal
(or $M \to L$)	(or $L \to M$)
charge-transfer transition	charge-transfer transition

Absorptions due to $M \to L$ or $L \to M$ excitation are called charge-transfer bands because the transition involved requires the transfer of electronic charge to either the ligand ($M \to L$) or the metal ($L \to M$). Charge-transfer transitions of the type $L \to M$ were discussed in Section 4-11. The energies of charge-transfer bands are related closely to the oxidizing (or reducing) abilities of the metal and the ligands. The powerful oxidizing agent MnO_4^- exhibits strong $L \to M$ ($O^{2-} \to Mn^{7+}$) charge-transfer absorption in the

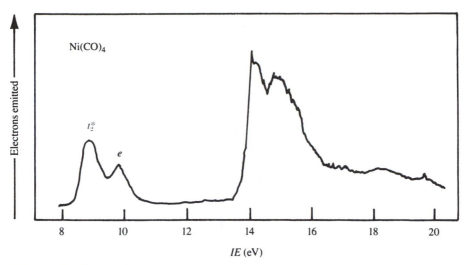

Figure 6-31 Photoelectron spectrum of $Ni(CO)_4$. The two bands between 8–11 eV are attributed to the nickel d^{10} electrons, which occupy the $(e)^4(t_2^*)^6$ orbitals in tetrahedral symmetry. Adapted with permission from I.H. Hillier, M.F. Guest, B.R. Higginson, and D.R. Lloyd, *Mol. Phys.* 27: 215 (1974).

visible region ($\bar{\nu}_{max} = 18,000$ cm^{-1}), thereby giving the permanganate ion its intense purple color. The related chromate ion, CrO_4^{2-}, is yellow, because its lowest-energy L → M ($O^{2-} \to Cr^{6+}$) band is shifted to a higher wave number ($\bar{\nu}_{max} = 26,000$ cm^{-1}). In both MnO_4^- and CrO_4^{2-} the lowest-energy L → M charge-transfer band is believed to be caused by the excitation of a nonbonding $2p$ oxygen electron to the unoccupied e ligand-field level of the d^0 tetrahedral complex.

Charge transfer of the M → L type occurs at relatively low energy if the central metal atom or ion has reducing properties and if the attached ligands have unoccupied orbitals of low enough energy to accept an electron. As one example, the complex $Cr(CO)_6$, which contains $Cr(0)$ and the strong π-acceptor ligand CO, absorbs strongly in the ultraviolet region, with $\bar{\nu}_{max} = 35,000$ cm^{-1}. This absorption has been attributed to electron excitation from the filled t_{2g} level of $Cr(0)$ to the unoccupied π^* level of CO.

6-11 Molecular-Orbital Theory for Dibenzenechromium

We choose dibenzenechromium as a prototype for the general class of organometallic π complexes of the transition metals. The structure of $Cr(C_6H_6)_2$ has been shown in Figure 6-9. We present it again in Figure 6-32,

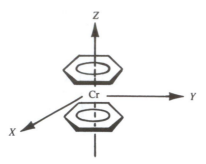

Figure 6-32 Structure of dibenzenechromium, $(C_6H_6)_2Cr$.

along with a coordinate system that will be used to discuss the molecular orbitals.

We can consider the complex as containing the d^6 central atom Cr and the two coordinated benzene rings (C_6H_6). The bonding in dibenzenechromium is commonly described in terms of a molecular-orbital model, starting with the delocalized π molecular orbitals of the C_6H_6 groups. We already have discussed the nodal characteristics and the relative energy levels of the six π molecular orbitals for a single benzene ring (Section 5-8). For two benzene rings we have a total of twelve π molecular orbitals. Schematic representations of these orbitals are shown in Figure 6-33, along with the appropriate metal orbital that has the correct symmetry to interact with the benzene π orbitals. Only the bonding combinations are shown in the figure. The forms of these bonding molecular orbitals can be approximated as follows:

$$\psi_1 = c_1 4s + c_2\left[\frac{1}{\sqrt{2}}(\pi_{1a} + \pi_{1b})\right]$$

$$\psi_2 = c_3 4p_z + c_4\left[\frac{1}{\sqrt{2}}(\pi_{1a} - \pi_{1b})\right]$$

$$\psi_3 = c_5 3d_{yz} + c_6\left[\frac{1}{\sqrt{2}}(\pi_{2a} + \pi_{2b})\right]$$

$$\psi_4 = c_5 3d_{xz} + c_6\left[\frac{1}{\sqrt{2}}(\pi_{3a} + \pi_{3b})\right]$$

$$\psi_5 = c_7 4p_y + c_8\left[\frac{1}{\sqrt{2}}(\pi_{2a} - \pi_{2b})\right]$$

$$\psi_6 = c_7 4p_x + c_8\left[\frac{1}{\sqrt{2}}(\pi_{3a} - \pi_{3b})\right]$$

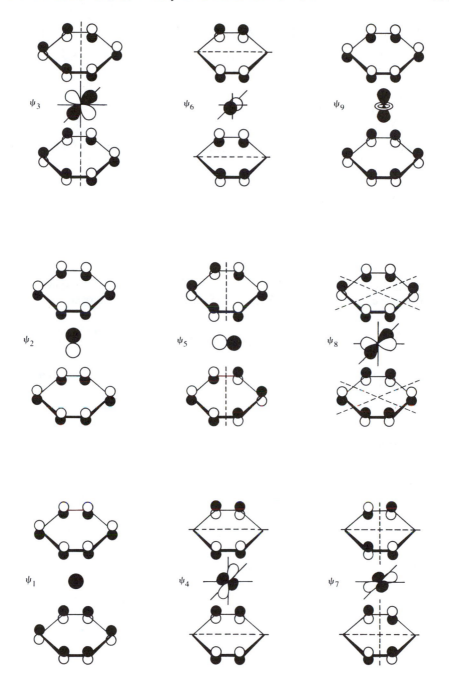

Figure 6-33 Schematic representations of the bonding molecular orbitals for dibenzenechromium. The ring lying above the chromium atom is labeled a, and the ring lying below the chromium atom is labeled b.

$$\psi_7 = c_9 3d_{xy} + c_{10}\left[\frac{1}{\sqrt{2}}(\pi_{4a} + \pi_{4b})\right]$$

$$\psi_8 = c_9 3d_{x^2-y^2} + c_{10}\left[\frac{1}{\sqrt{2}}(\pi_{5a} + \pi_{5b})\right]$$

$$\psi_9 = c_{11} 3d_{z^2} + c_{12}\left[\frac{1}{\sqrt{2}}(\pi_{1a} + \pi_{1b})\right]$$

The lowest-energy antibonding molecular orbitals are the degenerate pair ψ_{10} and ψ_{11}:

$$\psi_{10} = c_{13} 3d_{yz} - c_{14}\left[\frac{1}{\sqrt{2}}(\pi_{2a} + \pi_{2b})\right]$$

$$\psi_{11} = c_{13} 3d_{xz} - c_{14}\left[\frac{1}{\sqrt{2}}(\pi_{3a} + \pi_{3b})\right]$$

A simplified molecular-orbital energy-level diagram for dibenzenechromium is presented in Figure 6-34. It is clear that ψ_1 and ψ_2 are predominantly

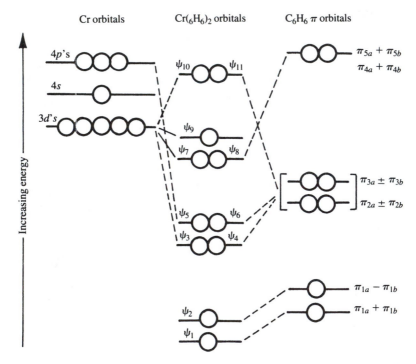

Figure 6-34 Relative orbital energies for $Cr(C_6H_6)_2$. In order to simplify the diagram, not all the correlation lines have been drawn (eg., $4s$ to Ψ_1). Also, only two antibonding orbitals are shown: ψ_{10} and ψ_{11}. The ligand-field orbitals are ψ_7, ψ_8, ψ_9, ψ_{10}, and ψ_{11}.

pure benzene orbitals; that is, $c_1 \ll c_2$ and $c_3 \ll c_4$. The major portion of the metal-ligand bonding is due to the degenerate pairs of orbitals ψ_3, ψ_4 and ψ_5, ψ_6. Although these molecular orbitals involve the metal atom $3d_{xz}$, $3d_{yz}$, $4p_x$, and $4p_y$ atomic orbitals, they are still predominantly ligand in character. The ligand-field molecular orbitals are composed of ψ_7, ψ_8, ψ_9, ψ_{10}, and ψ_{11}. That is, these molecular orbitals consist predominantly of metal $3d$ orbital character.*

Based on the molecular-orbital energy-level diagram of Figure 6-34, the electronic configuration of dibenzenechromium is

$$(\psi_1)^2 (\psi_2)^2 \ldots (\psi_9)^2$$

Consequently, the highest occupied molecular orbital is ψ_9, and the lowest unoccupied molecular orbitals are the degenerate pair ψ_{10} and ψ_{11}. The formal d^6 electronic structure of the chromium atom arises from the occupied molecular orbitals ψ_7, ψ_8, and ψ_9. Notice that dibenzenechromium obeys the 18-electron rule; all of the bonding molecular orbitals are occupied, but electrons do not populate any of the antibonding orbitals.

The photoelectron spectrum of dibenzenechromium has been studied. The energy-level diagram presented in Figure 6-34 is based on analysis of the photoelectron spectrum and on molecular-orbital calculations for $Cr(C_6H_6)_2$.

6-12 The Shapes of Transition-Metal Complexes

The VSEPR method is the simplest method available for predicting the shapes of nontransition (main-group) molecules. Essentially this method deals only with the σ bonding electrons and the lone-pair electrons around the central atom. Thus both BeH_2 and CO_2 are predicted to be linear because in each molecule the central atom is surrounded by two σ bonding electron pairs:

$$H—Be—H \qquad :\ddot{O}=C=\ddot{O}:$$

The principal difference between the electronic structure of main-group and transition-group molecules is that the lone electron pairs on the central atom of the main-group molecules occupy orbitals that have predominantly

---

*Actually, ψ_1 and ψ_9 both contain the same ligand combination, $\pi_{1a} + \pi_{1b}$. This indicates that these two molecular orbitals have the same symmetry representation. Consequently, the $3d_{z^2}$ and $4s$ orbitals could be involved in *both* of these molecular orbitals. In order to simplify the discussion, we have assigned the $4s$ atomic-orbital contribution to ψ_1 and the $3d_{z^2}$ atomic-orbital contribution to ψ_9.

s and p character, whereas in transition-group molecules the lone electron pairs occupy orbitals that have mainly d character. The main-group lone-pair electrons can always be considered as sterically "active"; we must now determine whether the transition-group lone-pairs are also sterically active. We may begin by comparing SF_6 and $Cr(CO)_6$, both octahedral molecules. Whereas SF_6 has no electron pairs on the central atom, the $Cr(CO)_6$ molecule has the $(t_{2g})^6$ electrons, which are mainly localized on the chromium atom and should be considered as d^6 lone electron pairs. Yet these $(t_{2g})^6$ electrons do not distort the molecule from its octahedral geometry. This can readily be understood by the VSEPR method, because the t_{2g} electrons are π type and therefore are not expected to distort the molecular shape. In general, however, not all of the d^n electrons will be in π orbitals. In octahedral complexes, electrons occupying the e_g^* orbitals are of σ character and may therefore be sterically active. Nevertheless, we can safely apply the VSEPR method to transition-metal complexes in cases where the d^n contribution gives a spherical electron-density function. These cases are d^0, d^5 (high-spin), and d^{10} only. In each of these situations, all of the d orbitals contain the same number of electrons: 0, 1, and 2, respectively. Consequently, the d electrons will not be capable of distorting the molecule. As explained in the discussion of $Cr(CO)_6$, the VSEPR method also is applicable if the d^n electrons are in π molecular orbitals. An example in octahedral symmetry would be any configuration t_{2g}^n for which $n = 0$–6, because t_{2g} is a π molecular orbital. In tetrahedral symmetry, the e orbitals have π character, so any electronic configuration e^n for which $n = 0$–4 should retain tetrahedral symmetry.

The Jahn-Teller Theorem

An important theorem that deals with the shapes of transition-group and main-group molecules alike is the *Jahn-Teller theorem*. This theorem says that any nonlinear molecular system in a degenerate electronic state will be unstable and will undergo some kind of distortion that will lower its symmetry and split the degenerate state.

Before attempting to apply this theorem, let us examine more carefully what is meant by a "degenerate electronic state." In practical terms, it means that there is more than one way to arrange the electrons within a degenerate energy level. Suppose that we have a doubly degenerate molecular orbital. There are five possible electron occupations of ground-state (that is, maximum-spin) configurations ranging from zero to four electrons:

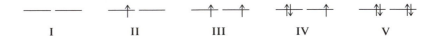

Of these five, only configurations II and IV can be written in a different way, namely

II'   IV'

Although it may appear that II is equivalent to II' and IV equivalent to IV', this is not so, as can be shown by a simple example.

Suppose that the doubly degenerate orbital previously mentioned is the e_g^* orbital of an octahedral molecule. Let us consider the case of a single e_g^* electron. If this electron is in the $d_{x^2-y^2}$ orbital, then the bonds between the metal and the four ligands in the XY plane would be weakened relative to the two ligands on the Z axis. Consequently we would have two short and four long bonds. However, if the e_g^* electron were to occupy the d_{z^2} orbital, two long and four short bonds would result. It is precisely this nonequivalence of bonding that causes the molecule to distort, with a subsequent loss of the degeneracy of the orbital.

Examples of molecules that have degenerate ground states and which therefore are subject to Jahn-Teller distortion are given in Table 6-9. Note that the Jahn-Teller effect applies to main-group molecules as well. For example, the CH_4^+ ion has the valence electronic configuration $(\sigma_s)^2(\sigma_{x,y,z})^5$ and therefore has a degenerate ground state. Detailed examination of the fine structure depicted in the photoelectron spectrum of CH_4^+ (Figure 5-21) has provided evidence that the CH_4^+ ion is distorted from tetrahedral symmetry, as predicted by the Jahn-Teller theorem.

TABLE **6-9.** MOLECULES AND IONS
SUBJECT TO JAHN-TELLER DISTORTION

Molecule	Electronic Configuration
CH_4^+	$\dots (\sigma_{x,y,z})^5$
VCl_4	$\dots (e)^1$
$V(CO)_6$	$\dots (t_{2g})^1$
ReF_6	$\dots (t_{2g})^1$
$C_6H_6^+$	$\dots (\pi_{2,3}^b)^3$
$Ti(H_2O)_6^{3+}$	$\dots (t_{2g})^1$
$Cr(H_2O)_6^{2+}$	$\dots (t_{2g})^3(e_g^*)^1$
$Cu(NH_3)_6^{2+}$	$\dots (t_{2g})^6(e_g^*)^3$

The Angular Overlap Model

Although the Jahn-Teller theorem provides a framework from which we can predict distortion of high-symmetry molecules, it does not tell us precisely

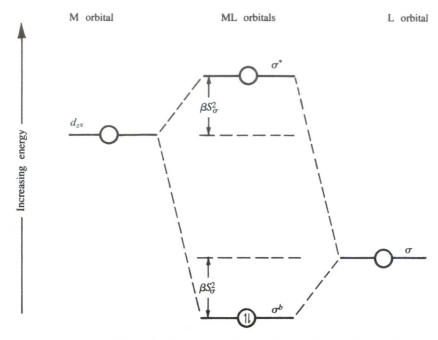

Figure 6-35 Bonding interaction between the $3d_{z^2}$ orbital and a σ orbital on the ligand. According to the angular-overlap model, the bonding molecular orbital is stabilized by $\beta S_\sigma{}^2$ and the antibonding molecular orbital is destabilized by $\beta S_\sigma{}^2$. β is an orbital interaction parameter and S_σ is the standard two-atom overlap integral shown in Figure 6-36. In a diatomic molecule, the other d orbitals exhibit π and δ symmetry, and therefore they are not included here. In our application of the angular-overlap model we are considering the σ orbitals only.

in which way a molecule will distort. Also, many molecules exhibit distortion even when distortion is not predicted by the Jahn-Teller theorem. For this reason we shall use a different approach to understanding the shapes of transition-group molecules. The approach we shall employ is called the *angular-overlap model* (AOM). According to this model, the *angular* geometry of a molecule will be determined by maximizing the bonding between the transition-element orbitals and the ligand-atom orbitals. In using AOM to predict molecular shape for transition-group molecules, it has been found that only the d orbitals on the transition-element atom and the highest occupied σ orbital on each ligand atom need to be considered.

The basis of AOM is that the interaction between a metal orbital d_j ($j = xz, xy, yz, x^2 - y^2$, or z^2), and a ligand σ orbital on atom a can be approximated as βS_{ja}^2. The parameter β is a measure of the strength of interaction between the orbitals, and S_{ja} is the overlap integral between the metal orbital d_j and the ligand orbital σ_a. The bonding orbital will be stabilized by βS_{ja}^2 and the corresponding antibonding orbital destabilized by βS_{ja}^2. This

Figure 6-36 Standard two-atom σ overlap between a d_{z^2} orbital and a ligand σ orbital. The overlap integral is designated S_σ.

result follows from a simplified form of quantum-mechanical perturbation theory. The result of σ orbital interaction for a diatomic transition-group molecule in terms of AOM is shown in Figure 6-35. The molecular axis is chosen to be the Z axis, and the standard two-atom overlap integral, S_σ is defined as the overlap of the d_{z^2} orbital with the σ orbital on the ligand atom (Figure 6-36). If we assume that the bonding σ orbital is occupied as shown in Figure 6-35, we conclude that the total amount of σ stabilization, $\Sigma(\sigma)$, is $2\beta S_\sigma^2$, because each electron produces a stabilization of βS_σ^2.

For a more complex molecule, ML_N, than the ML_a diatomic molecule just discussed, we need to determine the σ stabilization energy provided by the N ligand atoms ($a = 1, 2, \ldots, N$) on each d_j atomic orbital. This can be done by carrying out a summation over the N ligand atoms:

$$e_j = \sum_{a=1}^{N} \beta S_{ja}^2$$

To determine the total σ stabilization energy we must take into account the occupation number n_j of each orbital and sum over all five d_j orbitals:

$$\Sigma(\sigma) = \sum_{j=1}^{5} n_j e_j = \sum_{j=1}^{5} \sum_{a=1}^{N} n_j \beta S_{ja}^2 \tag{6-1}$$

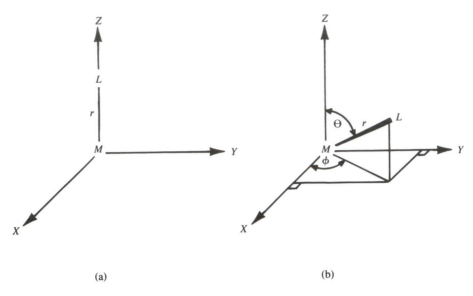

(a)   (b)

Figure 6-37 Metal-ligand coordinate system: (a) Metal and ligand both lie on the Z axis, so the standard two-atom overlap is achieved. (b) The ligand has been shifted to a new position designated by the angles θ and ϕ, as defined in the spherical polar coordinate system.

Each overlap integral S_{ja} must be expressed in terms of the standard two-atom overlap integral S_σ.

In many complexes the antibonding orbitals are also occupied. The destabilization contributed by the occupied antibonding orbitals will cancel the stabilization of the corresponding bonding orbitals. For such complexes, Equation 6-1 must be modified to show the canceling effect of the antibonding electrons. If n_j represents the number of electrons in the bonding orbitals and m_j the number in the antibonding orbitals we obtain

$$\sum (\sigma) = \sum_{j=1}^{5} \sum_{a=1}^{N} (n_j - m_j)\beta S_{ja}^2 \tag{6-2}$$

We could equally well have written Equation 6-2 as

$$\sum (\sigma) = \sum_{j=1}^{5} \sum_{a=1}^{N} h_j \beta S_{ja}^2 \tag{6-3}$$

in which $h_j = n_j - m_j$ corresponds to the number of "holes" or vacancies in the antibonding j orbital.

Although we shall not prove it here, it can readily be shown that the total stabilization of the bonding σ orbitals is $N\beta S_\sigma^2$, in which N is the number of ligands. If each of the bonding orbitals contains two electrons, then the total σ electronic stabilization is $2N\beta S_\sigma^2$. Analogous comments apply to the antibonding orbitals. That is, the total destabilization of the σ antibonding orbitals is $N\beta S_\sigma^2$, and for double occupation of each antibonding orbital the resulting destabilization is $2N\beta S_\sigma^2$.

To utilize AOM to determine the most stable molecular shape for a transition-metal complex ML_N, one must apply Equation 6-2 or 6-3 to the various possible geometries and choose the structure with the largest stabilization energy. Since β is a constant, it will not affect our determination of angular geometry. Therefore our only problem is to express the overlap integral S_σ, wherein the ligand σ orbital overlaps directly with the d_{z^2} orbital, as shown in Figure 6-36. Figure 6-37 shows the general case of a single ligand lying along the Z axis and then shifted to some other position (maintaining the same M-L bond length) characterized by an angle θ with the Z axis and ϕ with the X axis in the spherical polar-coordinate system. The effect of these *angular* changes (θ, ϕ) on the overlap integral S_{ja} is a function of trigonometry. For the overlap of a single ligand σ function and a central atom p orbital the results are particularly simple, for example, $S(p_z, \sigma) = \cos\theta\, S_\sigma$. The overlap of a single ligand σ function with each of the d orbitals is more complicated. The results are

$$S(d_{z^2}, \sigma) = [(1 + 3\cos 2\theta)/4]S_\sigma \tag{6-4}$$

$$S(d_{yz}, \sigma) = [(\sqrt{3}/2)\sin\phi \sin 2\theta]S_\sigma \tag{6-5}$$

$$S(d_{xz}, \sigma) = [(\sqrt{3}/2)\cos\phi \sin 2\theta]S_\sigma \tag{6-6}$$

$$S(d_{xy}, \sigma) = [(\sqrt{3}/4)\sin 2\phi(1 - \cos 2\theta)]S_\sigma \tag{6-7}$$

$$S(d_{x^2-y^2}, \sigma) = [(\sqrt{3}/4)\cos 2\phi(1 - \cos 2\theta)]S_\sigma \tag{6-8}$$

Although these expressions appear rather formidable, they are simple to evaluate in practice. In Table 6-10 we present the values of S_{ja}^2 for several commonly found ligand positions.

Application of AOM to Four-Coordinate Complexes

Let us now apply the angular-overlap model to a series of transition-metal complexes. Consider ML_4, for which three "high-symmetry" structures come readily to mind: tetrahedral, square planar, and the octahedral "*cis-divacant.*" The latter structure is an idealized sawhorse (SF$_4$ shape, Figure 2-9) in which all bond angles are either 90° or 180°. These structures are

TABLE 6-10. VALUES OF S_{ja}^2 FOR SEVERAL LIGAND POSITIONS IN UNITS OF S_σ^2

Ligand Positions:

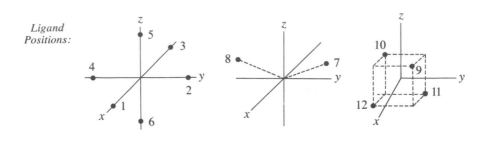

Ligand Position	Metal atomic orbitals				
	z^2	x^2-y^2	xz	yz	xy
1	$\frac{1}{4}$	$\frac{3}{4}$	0	0	0
2	$\frac{1}{4}$	$\frac{3}{4}$	0	0	0
3	$\frac{1}{4}$	$\frac{3}{4}$	0	0	0
4	$\frac{1}{4}$	$\frac{3}{4}$	0	0	0
5	1	0	0	0	0
6	1	0	0	0	0
7	$\frac{1}{4}$	$\frac{3}{16}$	0	0	$\frac{9}{16}$
8	$\frac{1}{4}$	$\frac{3}{16}$	0	0	$\frac{9}{16}$
9	0	0	$\frac{1}{3}$	$\frac{1}{3}$	$\frac{1}{3}$
10	0	0	$\frac{1}{3}$	$\frac{1}{3}$	$\frac{1}{3}$
11	0	0	$\frac{1}{3}$	$\frac{1}{3}$	$\frac{1}{3}$
12	0	0	$\frac{1}{3}$	$\frac{1}{3}$	$\frac{1}{3}$

Structure	Atoms
Linear	5 and 6
Trigonal planar	1, 7, and 8
Square planar	1–4
Tetrahedron	9–12
Trigonal bipyramidal	1, 5, 6, 7, and 8
Square pyramid	1–5
Octahedron	1–6

depicted in Figure 6-38. Recall that in main-group chemistry there are molecules that exhibit each of theses shapes. For example, CF_4 is tetrahedral, XeF_4 is square planar, and SF_4 has a shape closely related to the octahedral *cis*-divacant structure. Among transition-group molecules with four ligands, we have the following examples: $Ni(CO)_4$, $CoCl_4^{2-}$, $TiCl_4$, $Ni(CN)_4^{2-}$, $CuCl_4^{2-}$, and $Cr(CO)_4$. Of these molecules, the first three exhibit tetrahedral geometry, $Ni(CN)_4^{2-}$ and $CuCl_4^{2-}$ are square planar, and $Cr(CO)_4$ is octahedral *cis*-divacant.

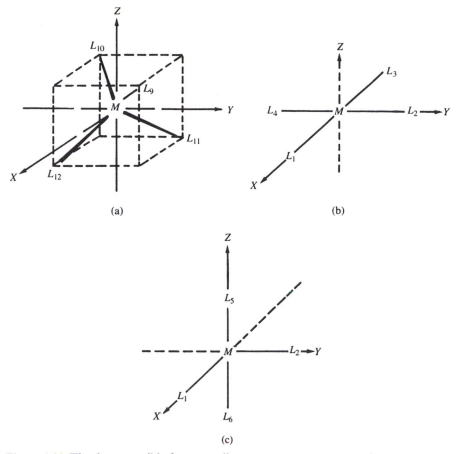

Figure 6-38 The three possible four-coordinate structures: (a) tetrahedral; (b) square-planar; and (c) octahedral "*cis*-divacant." The numbering of the ligands corresponds to that shown in Table 6-10.

We shall now calculate the relative orbital stabilization energies for the three types of four-coordinate structures. Utilizing the results of Table 6-10, we find that for the tetrahedral structure:*

$$e_{xy} = e_{xz} = e_{yz} = (4/3)\beta S_\sigma^2 \qquad (6\text{-}9)$$

$$e_{x^2-y^2} = e_{z^2} = 0 \qquad (6\text{-}10)$$

Notice that the d_{xy}, d_{xz}, and d_{yz} orbitals form a triply degenerate set, which we previously labeled t_2^* in tetrahedral symmetry. Furthermore, the d_{z^2} and $d_{x^2-y^2}$ orbitals have no σ bonding character.

*The results shown in Equations 6-9 and 6-10 are also derived in Section 6-14 by using Equations 6-4 through 6-8.

TABLE **6-11.** VALUES OF $\Sigma(\sigma)$ FOR DIFFERENT
FOUR-COORDINATE STRUCTURES (UNITS OF βS_σ^2)

Electronic Configuration	Tetrahedral	Square Planar	Octahedral cis-divacant
22222	0	0	0
22221	1.33	3.0	2.5
22220	2.67	6.0	5.0
22210	4.0	7.0	6.5
22211	2.67	4.0	4.0
22200	5.33	8.0	8.0
22111	4.0	4.0	4.0
22110	5.33	7.0	6.5
22100	6.67	8.0	8.0
22000	8.0	8.0	8.0

Similar angular-overlap calculations can be carried out for the other two four-coordinate structures presented in Figure 6-38; namely, the square-planar and the octahedral *cis*-divacant structures. The results for the square-planar structure are

$$e_{x^2-y^2} = 3\beta S_\sigma^2 \tag{6-11}$$

$$e_{z^2} = \beta S_\sigma^2 \tag{6-12}$$

$$e_{xz} = e_{yz} = e_{xy} = 0 \tag{6-13}$$

For the octahedral *cis*-divacant structure the results are

$$e_{x^2-y^2} = (3/2)\beta S_\sigma^2 \tag{6-14}$$

$$e_{z^2} = (5/2)\beta S_\sigma^2 \tag{6-15}$$

$$e_{xz} = e_{yz} = e_{xy} = 0 \tag{6-16}$$

The relative orbital energies are depicted in Figure 6-39 for all three four-coordinate structures. Only the energies of the σ^* molecular orbitals are shown, because the σ^b orbitals are normally fully occupied. In applying Equation 6-3 we need concern ourselves only with the number of "holes" in the σ^* orbitals.

In Table 6-11 we give the results obtained from Equation 6-3 for the various d electronic configurations. For the d-orbital occupations a shorthand notation is employed: for example, 22200 refers to the double occupation of the first three orbitals and no electrons in the latter two orbitals.

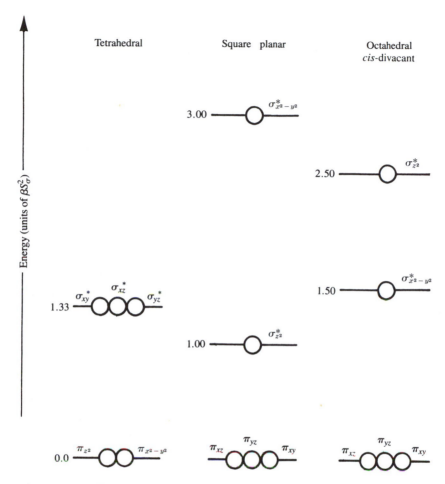

Figure 6-39 Relative energy levels of the σ molecular orbitals in the tetrahedral, square-planar, and octahedral *cis*-divacant four-coordinate structure, according to the angular-overlap model.

In Table 6-11 we calculate only the σ stabilization energy, $\Sigma(\sigma)$, and therefore there is no need to sum over π orbitals. With reference to Table 6-11, let us carry out the calculation for the 22220 (d^8 low-spin) electronic configuration for each of the three four-coordinate structures. As shown in Figure 6-39, the empty orbital must be $d_{x^2-y^2}$ in the square-planar structures, d_{z^2} in the octahedral *cis*-divacant structure, and one of the d_{xy}, d_{xz}, or d_{yz} orbitals in the tetrahedral structure. Using the results of Equation 6-9 through 6-16 and the orbital occupation numbers as deduced from Figure 6-39 we obtain:

1. For square planar:

$$\sum (\sigma) = (h_{x^2-y^2})(e_{x^2-y^2}) = (2)(3\beta S_\sigma^2) = 6\beta S_\sigma^2$$

2. For octahedral *cis*-divacant:

$$\sum (\sigma) = (h_{z^2})(e_{z^2}) = (2)(\tfrac{5}{2}\beta S_\sigma^2) = 5\beta S_\sigma^2$$

3. For tetrahedral:*

$$\sum (\sigma) = (h_{xy})(e_{xy}) = (2)(\tfrac{4}{3}\beta S_\sigma^2) = \tfrac{8}{3}\beta S_\sigma^2 = 2.67\beta S_\sigma^2$$

It can readily be seen that the largest stabilization energy for d^8 low-spin is provided by the square-planar structure. The square-planar structure also is predicted for d^9 systems. These calculated results are in agreement with the observed structure of $CuCl_4^{2-}$ (d^9) and $Ni(CN)_4^{2-}$ (d^8 low-spin). However, because the stabilization energy for d^9 ($h_j = 1$) is only one-half of that for low-spin d^8 ($h_j = 2$) there are cases in which the $CuCl_4^{2-}$ ion is distorted by packing forces in the solid state.

Further examination of the results in Table 6-11 shows that for the d electronic configurations 22222 (d^{10}), 22111 (d^7 high-spin), and 00000 (d^0) the calculated stabilization energy is the same for the square-planar, tetrahedral, and octahedral *cis*-divacant structures. In these cases the geometry is determined by ligand-ligand repulsions, which favor the tetrahedral geometry. This conclusion agrees with the observed tetrahedral geometry of $Ni(CO)_4$, (d^{10}), $CoCl_4^{2-}$ (d^7), and $TiCl_4$ (d^0). Examination of the electronic configurations for these tetrahedral molecules reveals that in each case the t_2^* orbitals are either empty, half-filled, or completely filled:

$Ni(CO)_4$	$(e)^4(t_2^*)^6$
$CoCl_4^{2-}$	$(e)^4(t_2^*)^3$
$TiCl_4$	$(e)^0(t_2^*)^0$

The use of Equation 6-3 for the d^6 (22200) $Cr(CO)_4$ molecule results in a calculated stabilization of $8\beta S_\sigma^2$ for both the square-planar and octahedral *cis*-divacant structures. In this instance we must extend the AOM approach one step beyond the S^2 terms to include the S^4 terms. When we do so we find the *cis*-divacant structure to be more stable, which is in agreement with

---

*Notice that for the tetrahedral structure we have arbitrarily chosen $h_{xy} = 2$. The calculated result is the same for any other equivalent combinations such as $h_{xy} = h_{xz} = 1$ or $h_{xz} = 2$.

experimental results for $Cr(CO)_4$. The orbital interactions for $Cr(CO)_4$ must be considerably more important than ligand-ligand repulsions in determining the correct molecular shape. The *cis*-divacant structure exhibits five ligand-ligand interactions at 90°, compared to only four such interactions in the square-planar structure.

The application of AOM to four-coordinate structures illustrates several rules which are useful in predicting the angular geometries of transition-metal complexes. These rules are:

1. Neglect any electrons in the π orbitals. The *angular* geometry is determined by the occupation numbers of the σ orbitals.

2. If the electron occupation numbers of degenerate σ orbitals are symmetrical [for example, $(t_2^*)^3$ in tetrahedral or $(e_g^*)^2$ in octahedral], then the predicted VSEPR geometry will be observed.

3. If a hole exists in the highest-energy orbital (22220, 22221, 22210, 22110), then the structure will be that of maximum overlap with the lobes of the $d_{x^2-y^2}$ orbital.

4. If two holes exist symmetrically in the two highest-energy orbitals (22200, 22211, 22100), then the structure will be that based on an octahedron containing the maximum number of *cis* ligands.

Application of AOM to Octahedral and Square-Planar Complexes

As a second application of AOM to transition-metal complexes, we are interested in the reaction

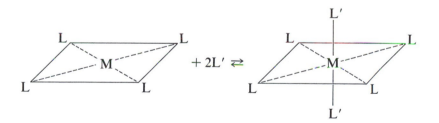

We shall pay particular attention to the relative M–L and M–L' bond lengths in the cases where L = L'. We are not seeking to determine the most stable *angular geometry,* as was the case of the four-coordinate structures previously discussed. Rather, we are interested in understanding the variation in *coordination number.* Why do some electronic configurations favor the four-coordinate square-planar structure (e.g., d^8 low-spin), while others favor the six-coordinate octahedral structure (e.g., d^6 low-spin)? In order to answer such a question, we must calculate the orbital stabilization energies for each of these structures. For the square-planar structure we utilize the results presented in Equations 6-11 to 6-13. For the octahedral structure, the results obtained from Table 6-10 are

$$e_{z^2} = e_{x^2-y^2} = 3\beta S_\sigma^2$$

$$e_{xy} = e_{xz} = e_{yz} = 0$$

The relative orbital energies for the square-planar and octahedral structures are presented in Figure 6-40.

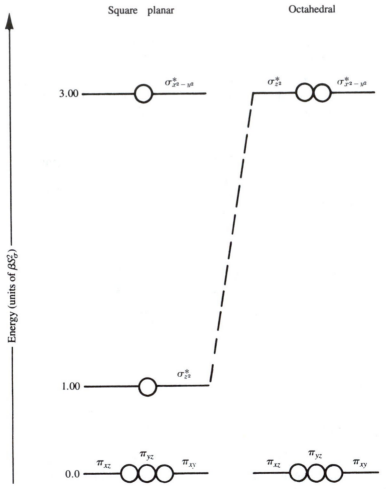

Figure 6-40 Results of angular-overlap calculations for octahedral and square-planar symmetries.

Let us apply the results given in Figure 6-40 to the specific case of d^8 and d^9 complexes. Because the topmost orbital has a relative energy of $3\beta S_{\sigma}^{2}$ for both octahedral and square-planar complexes, the $\Sigma(\sigma)$ factor is identical for both symmetries:

$$\Sigma(\sigma) = h_j(3\beta S_{\sigma}^{2})$$

in which $h_j = 1$ for d^9 and $h_j = 2$ for the d^8 electronic configuration.

The interesting aspect of the preceding example is that, in the calculated energy, there is no difference between the square-planar and octahedral complexes for low-spin d^8 and d^9 systems. This means that if two additional ligands were added to the square-planar structure, they would have to be weakly bonded, because the metal $(n + 1)s$ and $(n + 1)p$ orbitals would be required to form the new bonds. We have been assuming all along that, although these orbitals may contribute to the bonding, their effect will be smaller than that of the nd orbitals.

There is a wide variety of experimental data that provides support for the preceding calculated results. Some of it has been interpreted in terms of the Jahn-Teller effect. As we shall see here, all of the data can be explained in terms of AOM. Table 6-12 lists bond lengths for several Cu^{2+} (d^9) systems.

TABLE 6-12. SOME EXAMPLES OF BOND LENGTHS (Å) IN Cu^{2+} (d^9) SYSTEMS

Compound	Short Distances	Long Distances
CuF_2	4 F at 1.93	2 F at 2.27
K_2CuF_4	2 F at 1.95	4 F at 2.08
$CuCl_2$	4 Cl at 2.30	2 Cl at 2.95
$CsCuCl_3$	4 Cl at 2.30	2 Cl at 2.65
$CuBr_2$	4 Br at 2.40	2 Br at 3.18

It shows that there are always four bonds of one length and two bonds of another, usually longer, but in some cases shorter, than the axial bond lengths. The difference in equatorial and axial bond lengths is surprisingly large — as much as 0.65 Å. For a given ligand the equatorial bond lengths are nearly constant; a wide variation is observed in the axial bond lengths. Because the bond formation of the axial ligand is a low-energy process, the Cu^{2+} axial metal-ligand bond length should be determined predominantly by crystal forces, and therefore should show wide variation, as is observed. The equatorial bond lengths, however, should be determined by orbital interaction and should vary little from one crystal to another for a given ligand, as is also observed.

6-13 Equivalency of the d_{z^2} and $d_{x^2-y^2}$ Orbitals in an Octahedral Complex

We can demonstrate the equivalency of the d_{z^2} and $d_{x^2-y^2}$ orbitals by calculating the total overlap of the $d_{x^2-y^2}$ and d_{z^2} orbitals with their respective normalized ligand-orbital combinations. The total overlap in each case, $S_{x^2-y^2}$ and S_{z^2}, will be expressed in terms of the standard two-atom overlap between d_{z^2} and a ligand σ orbital, as shown in Figure 6-36. This overlap is called S_σ. From Table 1-2 we see that the angular functions for $d_{x^2-y^2}$ and d_{z^2} are

$$d_{z^2} = c(3z^2 - r^2)$$

and

$$d_{x^2-y^2} = \sqrt{3}c(x^2 - y^2)$$

in which $c = \sqrt{5}/(4\sqrt{\pi}r^2)$. The normalized combinations of ligand orbitals are

$$d_{z^2}: \ (1/2\sqrt{3})(2z_5 + 2z_6 - z_1 - z_2 - z_3 - z_4)$$

and

$$d_{x^2-y^2}: \ \tfrac{1}{2}(z_1 - z_2 + z_3 - z_4)$$

We first evaluate $S_{x^2-y^2}$:

$$S_{x^2-y^2} = \int \sqrt{3}c(x^2 - y^2)\tfrac{1}{2}(z_1 - z_2 + z_3 - z_4)\ d\tau$$

We may transform this integral into the standard two-atom overlap integral S_σ by rotating the metal coordinate system to coincide in turn with the coordinate systems of ligands ①, ②, ③, and ④. Using the coordinates shown in Figure 6-41 we obtain the following transformations:

M to ①	M to ②	M to ③	M to ④
$z \rightarrow y$	$z \rightarrow x$	$z \rightarrow -x$	$z \rightarrow -y$
$x \rightarrow -z$	$x \rightarrow y$	$x \rightarrow z$	$x \rightarrow -x$
$y \rightarrow x$	$y \rightarrow -z$	$y \rightarrow -y$	$y \rightarrow z$

Thus we have

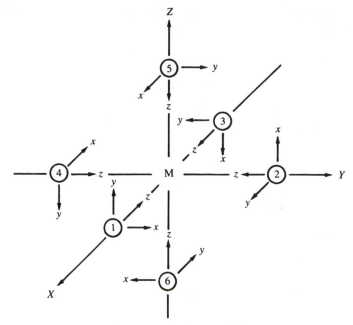

Figure 6-41 Coordinate system for an octahedral complex.

$$(\sqrt{3}/2)c(x^2-y^2)z_1 \to (\sqrt{3}/2)c(z^2-x^2)\sigma$$
$$-(\sqrt{3}/2)c(x^2-y^2)z_2 \to -(\sqrt{3}/2)c(y^2-z^2)\sigma$$
$$(\sqrt{3}/2)c(x^2-y^2)z_3 \to (\sqrt{3}/2)c(z^2-y^2)\sigma$$
$$-(\sqrt{3}/2)c(x^2-y^2)z_4 \to -(\sqrt{3}/2)c(x^2-z^2)\sigma$$

Adding the four transformed terms, we have

$$S_{x^2-y^2} = \int \sqrt{3}c(2z^2-x^2-y^2)\sigma \, d\tau = \int \sqrt{3}c(3z^2-r^2)\sigma \, d\tau = \sqrt{3}S_\sigma$$

Next we evaluate S_{z^2}:

$$S_{z^2} = \int c(3z^2-r^2)\left(\frac{1}{2\sqrt{3}}\right)(2z_5+2z_6-z_1-z_2-z_3-z_4) \, d\tau$$

The integrals involving z_5 and z_6 are simply two-atom overlaps, as shown in Figure 6-41. Thus we have

$$\int c(3z^2-r^2)\left(\frac{1}{2\sqrt{3}}\right)(2z_5+2z_6) \, d\tau = \left(\frac{2}{\sqrt{3}}\right)S_\sigma \qquad (6\text{-}17)$$

The integral involving z_1, z_2, z_3, and z_4 is transformed into S_σ using the transformation table that was used for $S_{x^2-y^2}$. Thus

$$-c(3z^2 - r^2)z_1 \rightarrow -c(3y^2 - r^2)\sigma$$

$$-c(3z^2 - r^2)z_2 \rightarrow -c(3x^2 - r^2)\sigma$$

$$-c(3z^2 - r^2)z_3 \rightarrow -c(3x^2 - r^2)\sigma$$

$$-c(3z^2 - r^2)z_4 \rightarrow -c(3y^2 - r^2)\sigma$$

Totaling the four transformed terms, we find

$$\int c(3z^2 - r^2)\left(\frac{1}{2\sqrt{3}}\right)(-z_1 - z_2 - z_3 - z_4) \, d\tau$$

$$= (1/2\sqrt{3}) \int -c(6x^2 + 6y^2 - 4r^2) \, \sigma \, d\tau$$

$$= (1/\sqrt{3}) \int c(3z^2 - r^2)\sigma \, d\tau = (1/\sqrt{3})S_\sigma \tag{6-18}$$

Finally, combining the results of Equations 6-17 and 6-18, we obtain

$$S_{z^2} = (2/\sqrt{3})S_\sigma + (1/\sqrt{3})S_\sigma = \sqrt{3}S_\sigma$$

Then

$$S_{z^2} = S_{x^2-y^2} = \sqrt{3}S_\sigma. \tag{6-19}$$

Thus the total overlap of $d_{x^2-y^2}$ and d_{z^2} with properly normalized ligand-orbital combinations is the same, and it follows that the two orbitals are equivalent in an octahedral complex.

6-14 Determining Overlap Integral for *d* Orbitals in Tetrahedral Symmetry

We wish to evaluate the overlap integrals between each of the five *d* orbitals on the central metal atom and each of the ligand atoms in tetrahedral symmetry.

To make these calculations, we use the coordinate system of Figure 6-38(a) and employ the AOM Equations 6-4 through 6-8 to express the overlap integrals in terms of the standard two-atom overlap shown in Figure 6-36.

First we shall set up a table designating the values of θ and ϕ for each of the four ligands (9), (10), (11), and (12), along with the various sine and cosine functions that are required.

Ligand	θ	$\cos 2\theta$	$\sin 2\theta$	ϕ	$\cos \phi$	$\sin \phi$	$\cos 2\phi$	$\sin 2\phi$
(9)	54.74	$-\frac{1}{3}$	$\frac{2}{3}\sqrt{2}$	45	$1/\sqrt{2}$	$1/\sqrt{2}$	0	1
(10)	54.74	$-\frac{1}{3}$	$\frac{2}{3}\sqrt{2}$	225	$-1/\sqrt{2}$	$-1/\sqrt{2}$	0	1
(11)	125.26	$-\frac{1}{3}$	$-\frac{2}{3}\sqrt{2}$	315	$1/\sqrt{2}$	$-1/\sqrt{2}$	0	-1
(12)	125.26	$-\frac{1}{3}$	$-\frac{2}{3}\sqrt{2}$	135	$-1/\sqrt{2}$	$1/\sqrt{2}$	0	-1

We now calculate each of the appropriate overlap integrals in turn.

1. For the overlap of any ligand σ function with the d_{z^2} orbital we must utilize Equation 6-4, which we repeat here:

$$S(d_{z^2}, \sigma_a) = [(1 + 3 \cos 2\theta)/4]S_\sigma$$

Since $\cos 2\theta = -\frac{1}{3}$ for each of the ligand σ functions we obtain

$$S(d_{z^2}, \sigma_a) = 0, \quad \text{for } a = 9, 10, 11, \text{ and } 12$$

2. For the d_{yz} orbital we have from Equation 6-5

$$S(d_{yz}, \sigma_a) = [(\sqrt{3}/2) \sin \phi \sin 2\theta]S_\sigma$$

Therefore,

$$S(d_{yz}, \sigma_9) = \left(\frac{\sqrt{3}}{2}\right)\left(\frac{1}{\sqrt{2}}\right)\left(\frac{2\sqrt{2}}{3}\right)S_\sigma = \left(\frac{\sqrt{3}}{3}\right)S_\sigma$$

$$S(d_{yz}, \sigma_{10}) = \left(\frac{\sqrt{3}}{2}\right)\left(-\frac{1}{\sqrt{2}}\right)\left(\frac{2\sqrt{2}}{3}\right)S_\sigma = \left(-\frac{\sqrt{3}}{3}\right)S_\sigma$$

$$S(d_{yz}, \sigma_{11}) = \left(\frac{\sqrt{3}}{2}\right)\left(-\frac{1}{\sqrt{2}}\right)\left(-\frac{2\sqrt{2}}{3}\right)S_\sigma = \left(\frac{\sqrt{3}}{3}\right)S_\sigma$$

$$S(d_{yz}, \sigma_{12}) = \left(\frac{\sqrt{3}}{2}\right)\left(\frac{1}{\sqrt{2}}\right)\left(-\frac{2\sqrt{2}}{3}\right)S_\sigma = \left(-\frac{\sqrt{3}}{3}\right)S_\sigma$$

In summary,

$$S^2(d_{yz}, \sigma_a) = \frac{1}{3}S_\sigma^2 \quad \text{for } a = 9, 10, 11, \text{ and } 12$$

3. For the d_{xz} orbital we have from Equation 6-6

$$S(d_{xz}, \sigma_a) = [(\sqrt{3}/2) \cos \phi \sin 2\theta]S_\sigma$$

Therefore,

$$S(d_{xz}, \sigma_9) = \left(\frac{\sqrt{3}}{2}\right)\left(\frac{1}{\sqrt{2}}\right)\left(\frac{2\sqrt{2}}{3}\right)S_\sigma = \left(\frac{\sqrt{3}}{3}\right)S_\sigma$$

$$S(d_{xz}, \sigma_{10}) = \left(\frac{\sqrt{3}}{2}\right)\left(-\frac{1}{\sqrt{2}}\right)\left(\frac{2\sqrt{2}}{3}\right)S_\sigma = \left(-\frac{\sqrt{3}}{3}\right)S_\sigma$$

$$S(d_{xz}, \sigma_{11}) = \left(\frac{\sqrt{3}}{2}\right)\left(\frac{1}{\sqrt{2}}\right)\left(-\frac{2\sqrt{2}}{3}\right)S_\sigma = \left(-\frac{\sqrt{3}}{3}\right)S_\sigma$$

$$S(d_{xz}, \sigma_{12}) = \left(\frac{\sqrt{3}}{2}\right)\left(-\frac{1}{\sqrt{2}}\right)\left(-\frac{2\sqrt{2}}{3}\right)S_\sigma = \left(\frac{\sqrt{3}}{3}\right)S_\sigma$$

In summary,

$$S^2(d_{xz}, \sigma_a) = \tfrac{1}{3}S_\sigma^2, \qquad \text{for } a = 9, 10, 11, \text{ and } 12$$

4. For the d_{xy} orbital we have from Equation 6-7

$$S(d_{xy}, \sigma_a) = [(\sqrt{3}/4) \sin 2\phi(1 - \cos 2\theta)]S_\sigma$$

Therefore,

$$S(d_{xy}, \sigma_9) = (\sqrt{3}/4)(1)(4/3)S_\sigma = (\sqrt{3}/3)S_\sigma$$
$$S(d_{xy}, \sigma_{10}) = (\sqrt{3}/4)(1)(4/3)S_\sigma = (\sqrt{3}/3)S_\sigma$$
$$S(d_{xy}, \sigma_{11}) = (\sqrt{3}/4)(-1)(4/3)S_\sigma = (-\sqrt{3}/3)S_\sigma$$
$$S(d_{xy}, \sigma_{12}) = (\sqrt{3}/4)(-1)(4/3)S_\sigma = (-\sqrt{3}/3)S_\sigma$$

In summary,

$$S^2(d_{xy}, \sigma_a) = \tfrac{1}{3}S_\sigma^2, \qquad \text{for } a = 9, 10, 11, \text{ and } 12$$

5. For the $d_{x^2-y^2}$ orbital we have from Equation 6-8

$$S(d_{x^2-y^2}, \sigma_a) = [(\sqrt{3}/4) \cos 2\phi(1 - \cos 2\theta)]S_\sigma$$

Since $\cos 2\phi = 0$ for each of the ligand σ functions, we obtain

$$S(d_{x^2-y^2}, \sigma_a) = 0, \qquad \text{for } a = 9, 10, 11, \text{ and } 12$$

In short, we see that the five d orbitals split in tetrahedral symmetry into a group of doubly and triply degenerate orbitals:

$$e_{z^2} = e_{x^2-y^2} = 0$$

$$e_{xz} = e_{yz} = e_{xy} = (4/3)\beta S_\sigma^2$$

Since each ligand provides a stabilization of $\frac{1}{3}\beta S_\sigma^2$, the total stabilization due to the four ligands on the d_{xz}, d_{yz}, and d_{xy} orbitals is $(\frac{4}{3})\beta S_\sigma^2$.

QUESTIONS AND PROBLEMS

1. Formulate a reasonable structure for the dichromate anion, $Cr_2O_7^{2-}$.

2. Calculate the effective atomic number for each of the transition metal organometallic complexes shown in Figure 6-11.

3. Compare and contrast the bonding between Pt(II) and ethylene in Zeise's anion, $Pt(C_2H_4)Cl_3^-$, with that in ethylene oxide (see Problem 20 in Chapter 5).

4. Formulate a structure for each of the following compounds so that the 18-electron rule is obeyed: (a) $V_2(CO)_{12}$, (b) $V_2(CO)_8$, (c) $Cr_2(CO)_4(C_5H_5)_2$.

5. The structure of $Co_2(CO)_8$ (Figure 6-5) shows that in the solid state there are two bridging carbonyl groups. Studies of $Co_2(CO)_8$ in solution have shown that the compound also exists in an unbridged form. Formulate a structure for the unbridged form in which the 18-electron rule is obeyed. Is a metal-metal bond required?

6. Apply the 18-electron rule to the following combinations of metals and ligands to predict possible complex stoichiometries (consider neutral, cationic, anionic, and dimeric species):
   (a) Fe, CO, NO   (b) Cr, C_5H_5, Cl, NO   (c) Co, $P(C_6H_5)_3$, N_2, H
   (d) Fe, PF_3, H   (e) Ni, PF_3, CO   (f) Mn, C_5H_5, C_6H_6

7. Ruthenium forms two unusual binuclear complexes: $Ru_2OCl_{10}^{4-}$ and $Ru_2(NH_3)_{10}N_2^{4+}$. In each of these complexes the metal has a coordination number of six in which the chloride and ammonia ligands occupy terminal positions. The metal linkage is *linear* Ru—O—Ru and Ru—N≡N—Ru, respectively. Each of these complexes is diamagnetic. Formulate molecular orbitals for the linear Ru—O—Ru and Ru—N≡N—Ru fragments. Which of these orbitals are occupied in the ground state? Treat the Ru—Cl and Ru—NH$_3$ bonds as localized electron pairs.

8. Which complex in each of the following pairs would you expect to be more stable:
    (a) $PtCl_4^{2-}$ or PtI_4^{2-} (b) $Fe(H_2O)_6^{3+}$ or $Fe(NH_3)_6^{3+}$
    (c) $FeCl_4^-$ or FeI_4^- (d) $ZnCl_4^{2-}$ or $ZnBr_4^{2-}$
    (e) $HgCl_4^{2-}$ or $HgBr_4^{2-}$ (f) $Ag(H_2O)_2^+$ or $Ag(NH_3)_2^+$
    (g) BF_4^- or BCl_4^- (h) $Cu(NH_3)_4^{2+}$ or $Cu(H_2O)_4^{2+}$

9. Predict whether Br^- will displace the halide ions if added to an aqueous solution of each of the following complex ions:
    (a) FeF_6^{3-} (b) $PdCl_4^{2-}$ (c) HgI_4^{2-}
    (d) BF_4^- (e) PtI_4^{2-} (f) $PtCl_4^{2-}$

10. What is a chelate? If ethylenediamine is a bidentate chelating group, and ethylenediaminetetraacetate (EDTA) is a hexadentate chelating group, how would triethylenetriamine and diethylenetriamine be described?

11. Is the following statement true or false? "Stereoisomers have the same molecular topologies, whereas constitutional isomers have different molecular topologies."

12. Give the structural formulas for all the possible isomers of each of the following. Indicate which are stereoisomers and which are constitutional isomers.
    (a) $Co(en)_2Cl_2^+$ (b) $Co(en)_2(NH_3)Cl^{2+}$
    (c) $Co(en)(NH_3)_2Cl_2^+$ (d) $Co(en)_2(Cl)NO_3^+$

13. Give the structural formula for each of the coordination isomers that could be formed starting with $[Cu(NH_3)_4][PtCl_4]$.

14. All octahedral complexes of V^{3+} have the same number of unpaired electrons, no matter what the nature of the ligand. Why is this so?

15. How does the molecular-orbital theory account for the order of ligands in the spectrochemical series?

16. Explain why $Co(CN)_6^{3-}$ is extremely stable but $Co(CN)_6^{4-}$ is not.

17. Explain the fact that most complexes of Zn^{2+} are colorless.

18. Explain why octahedrally coordinated Mn^{3+} is very unstable, whereas octahedral complexes of Cr^{3+} are extremely stable.

19. Carbon monoxide is a strong-field ligand that stabilizes transition metals in unusually low oxidation states. For example, $V(CO)_6$ and

$V(CO)_6^-$ are both stable complexes. What are the ground-state electronic configurations of these two complexes in the ligand-field levels t_{2g} and e_g^*? Which member of the series $V(CO)_6$, $Cr(CO)_6$, and $Mn(CO)_6$ would you expect to be most stable? Which would be least stable? Why?

20. A complex widely used in studying reactions of octahedral complexes is $Co(NH_3)_5Cl^{2+}$. Actually, this complex is octahedral only in an approximate sense, because NH_3 and Cl^- have different ligand-field strengths. How would you modify the ligand-field energy levels of $Co(NH_3)_6^{3+}$ in formulating a splitting diagram for $Co(NH_3)_5Cl^{2+}$? (For convenience, place the Cl^- ligand along the Z axis of a Cartesian coordinate system.) Apply your theory to explain the fact that $Co(NH_3)_6^{3+}$ is yellow (absorbs light at 430 nm), whereas $Co(NH_3)_5Cl^{2+}$ is purple (absorbs light at 530 nm). Which d orbital receives the electron in the excitation, giving rise to the 530 nm band in $Co(NH_3)_5Cl^{2+}$? Why?

21. Electronically excited molecules often emit light just as do excited atoms. However, excited molecules also may use their excess energy to break chemical bonds and thereby undergo chemical reactions that might not occur otherwise. An example involves the metal complex $W(CO)_6$. The molecule is unreactive in its ground state, but upon irradiation by light of 300 nm wavelength the following reaction occurs:

$$W(CO)_6 \rightarrow W(CO)_5 + CO$$

From this information, estimate an upper limit, in kcal mole^{-1}, for the W–CO bond energy. As the reaction proceeds the concentration of carbon monoxide increases. Is there a possibility that the secondary reaction

$$CO \xrightarrow{\lambda\,=\,300\ nm} C + O$$

can occur? Why or why not?

22. (a) Give the ligand-field electronic configuration of the ground state of $W(CO)_6$. Is the complex diamagnetic or paramagnetic? The first electronic absorption band occurs at about 31,000 cm^{-1}. Assign this band to an electronic transition in the complex. Is the photochemical dissociation of $W(CO)_6$ to $W(CO)_5 + CO$, described in Problem 21, reasonable? Explain.
(b) The $W(CO)_5$ molecule has a square-pyramidal structure. Assume a reference coordinate system in which only one CO ligand is along

the Z axis and predict the ligand-field splitting for this complex. $W(CO)_5$ absorbs light strongly at 25,000 cm^{-1}. Assign the band to an electronic transition in $W(CO)_5$ and explain why the absorption is at lower energy than in $W(CO)_6$.

23. Predict which complex in each of the following pairs will have the lower-energy d–d transition:
    (a) $Co(NH_3)_5F^{2+}$ or $Co(NH_3)_5I^{2+}$
    (b) $Co(NH_3)_5Cl^{2+}$ or $Co(NH_3)_5NO_2^{2+}$
    (c) $Pt(NH_3)_4^{2+}$ or $Pd(NH_3)_4^{2+}$
    (d) $Co(CN)_6^{3-}$ or $Ir(CN)_6^{3-}$
    (e) $Co(CN)_5H_2O^{2-}$ or $Co(CN)_5I^{3-}$
    (f) $V(H_2O)_6^{2+}$ or $Cr(H_2O)_6^{3+}$
    (g) $RhCl_6^{3-}$ or $Rh(CN)_6^{3-}$
    (h) $Ni(H_2O)_6^{2+}$ or $Ni(NH_3)_6^{2+}$

24. A common inorganic chemical compound, ferric ammonium sulfate, has the formula $Fe_2(SO_4)_3(NH_4)_2SO_4 \cdot 24H_2O$. Large crystals of the compound are very pale violet, due to weak absorption (ε values between 0.05 and 1) in the visible region of the spectrum. The Fe^{3+} in the compound is present as the hexaaquo complex ion, $Fe(H_2O)_6^{3+}$. Using ligand-field theory, formulate an explanation of the weak absorption bands.

25. The ferrocyanide ion, $Fe(CN)_6^{4-}$, does not exhibit an absorption band in the visible region, but ferricyanide, $Fe(CN)_6^{3-}$, absorbs strongly at approximately 25,000 cm^{-1}. What type of electronic transition is responsible for the strong absorption that makes $Fe(CN)_6^{3-}$ red?

26. Would you expect a given $M \rightarrow L$ charge-transfer transition to be at lower, or higher, energy in $Cr(CO)_6$ than in $Mn(CO)_6^+$? Why?

27. Give reasonable examples of the following: (a) d^5 high-spin octahedral complex; (b) low-spin square-planar complex; (c) d^0 tetrahedral complex; (d) d^5 low-spin octahedral complex; (e) d^3 complex.

28. Many complexes of Cu^{2+} have square-planar structures. What is the ground-state electronic configuration of a square-planar Cu^{2+} complex? How many d–d transitions of different energies can be expected?

29. The ligand-field splitting diagram for $PtCl_4^{2-}$ given in Figure 6-27 is

$$\sigma_{z^2}^* < \pi_{xz,yz}^* < \pi_{xy}^* < \sigma_{x^2-y^2}^*$$

Experimental measurements on $Pt(CN)_4^{2-}$ indicate a different order for the ligand-field orbitals:

$$\pi_{xy}^* < \pi_{xz,yz}^* < \sigma_{z^2}^* < \sigma_{x^2-y^2}^*$$

By considering the different π bonding capabilities of the Cl^- and CN^- ligands, interpret the difference in energy ordering for the ligand-field orbitals in these two complexes.

30. Using ligand-field theory predict the number of unpaired electrons in the following complexes: FeO_4^{2-}, $Mn(CN)_6^{3-}$, $NiCl_4^{2-}$ (tetrahedral), $PdCl_4^{2-}$ (square planar), $MnCl_4^{2-}$, $Co(en)_3^{2+}$, $Co(en)_3^{3+}$, $Rh(NH_3)_6^{3+}$, $CoBr_4^{2-}$, and $Pt(NH_3)_4^{2+}$.

31. One of the most toxic substances known to man is tetracarbonylnickel (0), $Ni(CO)_4$. Predict its geometrical structure, using VSEPR theory. Formulate the ground-state electronic structure of $Ni(CO)_4$ using ligand-field theory. Would you expect to observe d–d transitions in this compound? Why or why not?

32. The $CuCl_2$ molecule has been observed in the gas phase. It has a linear structure. Assume that the internuclear line is along the Z axis and that $3d_{z^2}$–$4s$ mixing is not important.
    (a) Draw a molecular-orbital energy-level diagram for $CuCl_2$.
    (b) Indicate which orbitals are predominantly Cu $3d$ in character, and hence correspond to the ligand-field orbitals.
    (c) Considering only the ligand-field orbitals, what is the ground-state electronic configuration of $CuCl_2$?
    (d) How many d–d transitions should be observed? What are the transition assignments?

33. Consider the following complex ions: MnO_4^{3-}, $Pd(CN)_4^{2-}$, NiI_4^{2-}, $Ru(NH_3)_6^{3+}$, $MoCl_6^{3-}$, $IrCl_6^{2-}$, $AuCl_4^-$, and FeF_6^{3-}. Use ligand-field theory to predict the structure and number of unpaired electrons in each ion.

34. (a) Calculate the energy in cm^{-1} of the absorption spectral line in isolated Li^{2+} corresponding to the transition $1s \to 3d$. Do the $1s \to 3s$ and $1s \to 3p$ transitions have the same energy? (b) Now assume that the Li^{2+} is placed in an octahedral ligand field. How many different electronic transitions would you expect to observe from the $1s$ orbital to the $n = 3$ orbitals in this ligand field? How many would be expected in a tetrahedral field? How many in a square-planar field?

35. Explain the fact that the lowest $L \to M$ charge-transfer band shifts from $18,000$ cm^{-1} in MnO_4^- to $26,000$ cm^{-1} in CrO_4^{2-}.

36. The yellow complex HgI_4^{2-} exhibits a strong absorption band at $\bar{v}_{max} = 27,400$ cm^{-1}. What type of electronic transition is responsible for this band?

37. Consider the protonation of tetracarbonylnickel (0), $Ni(CO)_4$:

    $$H^+ + Ni(CO)_4 \rightarrow [HNi(CO)_4]^+$$

    Using the frontier-orbital approach, predict which atom (Ni, C, or O) the proton will attack in $Ni(CO)_4$.

38. Dibenzenevanadium, $V(C_6H_6)_2$, has a sandwich structure like that of dibenzenechromium, $Cr(C_6H_6)_2$ (Figure 6-32). Formulate the ground-state electronic structures for these two organometallic π complexes (see Figures 6-33 and 6-34). How many unpaired electrons are there in each case?

39. The complexes $V(C_6H_6)_2$ and $Cr(C_6H_6)_2$ are both readily oxidized in air to their respective cations. How many unpaired electrons are there in each case?

40. The molecular-orbital energy-level scheme for ferrocene, $Fe(C_5H_5)_2$, is similar to that for $Cr(C_6H_6)_2$ (Figures 6-33 and 6-34). There is a whole series of complexes like ferrocene: $V(C_5H_5)_2$, $Cr(C_5H_5)_2$, $Mn(C_5H_5)_2$, $Co(C_5H_5)_2$, and $Ni(C_5H_5)_2$. Formulate the ground-state electronic structures of these organometallic π complexes assuming low-spin electronic configurations. How many unpaired electrons are there in each case? Experimentally, $Mn(C_5H_5)_2$ exists in the high-spin electronic configuration. How many unpaired electrons are there in this case?

41. Explain why $Co(C_5H_5)_2$ is readily oxidized to $Co(C_5H_5)_2^+$.

42. Use the angular-overlap model to show the following relationship between the ligand-field splitting in octahedral and tetrahedral complexes:

    $$\Delta_t = (4/9)\Delta_o$$

    Assume that the same ligand and metal are involved and that the metal-ligand bond length is unchanged in ML_4 and ML_6.

43. Use the angular-overlap model to calculate the ligand-field splitting in a trigonal-bipyramidal molecule such as $Fe(CO)_5$. Use the following coordinate system:

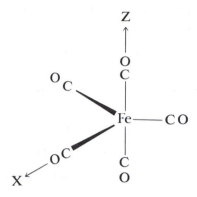

44. In the angular-overlap model, the β parameter is related to the strength of interaction between metal and ligand. Consequently, β should vary from ligand to ligand in roughly the same order as the spectrochemical series. Assume that $\beta_{CN^-} \gg \beta_{H_2O}$ and explain qualitatively in terms of the angular-overlap model why Ni^{2+} forms an octahedral complex with H_2O, $Ni(H_2O)_6^{2+}$, but a square-planar complex with CN^-, $Ni(CN)_4^{2-}$.

45. Using the Jahn-Teller theorem, predict which of the following should be distorted: $Ni(H_2O)_6^{2+}$, $CuCl_4^{2-}$, $CuCl_6^{4-}$.

46. Several stable tetrahedral complexes are formed with Co^{2+} (d^7 high-spin). However, Ni^{2+} with strong field ligands (d^8 low-spin) forms stable square-planar complexes. Interpret this result by using the $\Sigma(\sigma)$ AOM results given in Table 6-11.

47. Use the angular-overlap model (Table 6-10) to obtain the result presented in Section 6-13 by the "coordinate rotation" method

$$S_{z^2} = S_{x^2-y^2} = \sqrt{3}S_\sigma$$

48. Use Equation 6-4 to calculate the angle θ for which

$$S(d_{z^2}, \sigma) = 0$$

Suggestions for Further Reading

Basolo, F., and Johnson, R., *Coordination Chemistry,* Menlo Park, Calif.: Benjamin, 1964.

Burdett, J.K., "A New Look at Structure and Bonding in Transition Metal Complexes," *Advances in Inorganic Chemistry and Radiochemistry,* vol. 21, edited by H.J. Eméleus and A.G. Sharpe, New York: Academic, 1978, p. 113.

Coates, G.E., Green, M.L.H., Powell, P., and Wade, K., *Principles of Organo-metallic Chemistry,* London: Methuen, 1968.

Cotton, F.A., and Wilkinson, G., *Advanced Inorganic Chemistry,* 3rd ed., New York: Wiley, 1972.

Cotton, F.A., and Wilkinson, G., *Basic Inorganic Chemistry,* New York: Wiley, 1976.

Eland, J.H.D., *Photoelectron Spectroscopy,* London: Butterworth, 1974.

Gray, H.B., *Chemical Bonds,* Menlo Park, Calif.: Benjamin, 1973.

Gray, H.B., *Electrons and Chemical Bonding,* Menlo Park, Calif.: Benjamin, 1965.

Hoffmann, R., Chen, M.M.-L., and Thorn, D.L., "Qualitative Discussion of Alternate Coordination Modes of Diatomic Ligands in Transition Metal Complexes," *Inorg. Chem.* 16: 503 (1977).

Huheey, J.E., *Inorganic Chemistry: Principles of Structure and Reactivity,* 2nd ed., New York: Harper and Row, 1978.

King, R.B., *Transition Metal Organometallic Chemistry: An Introduction,* New York: Academic, 1969.

Lagowski, J.J., *Modern Inorganic Chemistry,* New York: Marcel Dekker, 1973.

Larsen, E., and La Mar, G.N., "The Angular Overlap Model," *J. Chem. Educ.* 51: 633 (1974).

Mitchell, P.R., and Parish, R.V., "The Eighteen-Electron Rule," *J. Chem. Educ.* 46: 811 (1969).

Orgel, L.E., *An Introduction to Transition Metal Chemistry,* New York: Wiley, 1966.

Pearson, R.G., "Hard and Soft Acids and Bases," *J. Chem. Educ.* 45: 581, 643 (1968).

Phillips, C.S.G., and Williams, R.L.P., *Inorganic Chemistry,* vol. 2, London: Oxford, 1966.

Purcell, K.F., and Kotz, J.C., *Inorganic Chemistry,* Philadelphia: Saunders, 1977.

Rochow, E.G., *Organometallic Chemistry,* New York: Van Nostrand Rein-hold, 1964.

Wade, K., "The Key to Cluster Shapes," *Chemistry in Britain* 11: 177 (1975).

7 Bonding in Solids and Liquids

Discrete molecules such as H_2O and CH_4 are present in many solid substances. In other solids the atoms are bonded together in an infinite array to build a "giant" molecule, or crystal. Such giant molecules can be either electrical conductors or insulators. In this chapter we examine the relationship between the properties of solid substances and the types of interatomic or intermolecular bonding interactions that are present.

7-1 Elemental Solids and Liquids

Solids that are built by weak attractive interactions between individual molecules are called *molecular solids*. At very low temperatures the noble gases (He, Ne, Ar, Kr, Xe, Rn) exist as molecular solids that are held together by weak interatomic forces. For example, argon freezes at −189 °C to make the close-packed structure shown in Figure 7-1. Examples of

3.8 Å

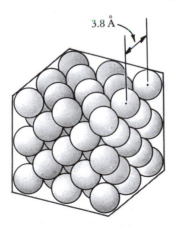

Figure 7-1 The structure of solid argon. Each sphere represents an individual Ar atom; the atoms are in cubic close packing with 3.8 Å between the atomic centers.

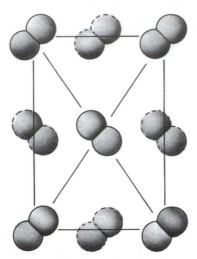

Figure 7-2 The structure of crystalline bromine, Br_2. The solid outlines indicate one layer of packed molecules, and the dashed outlines indicate a layer beneath. The molecules have been shrunk for clarity in this drawing; they are actually in close contact within a layer, and the layers are packed against one another.

elemental substances that crystallize to give molecular solids include the halogens; Br_2, for example, freezes at -7 °C to build the structure shown in Figure 7-2.

Atoms of the Group VI elements such as oxygen and sulfur (s^2p^4) have two vacancies in their valence shell and hence form two electron-pair bonds per atom. Under normal conditions of temperature and pressure, the most stable form of elemental oxygen is the diatomic molecule, whereas sulfur exists as a solid, and its two principal allotropes consist of discrete S_8 rings (Figure 7-3).* There are two other allotropes of sulfur, one of which has S_6 rings; the other contains helical chains of S atoms.

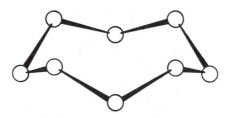

Figure 7-3 Structure of solid sulfur. The S_8 ring structure shown is the form sulfur atoms take in the two principle allotropes of crystalline sulfur—rhombic and monoclinic. Rhombohedral sulfur, a third, less stable, allotrope, consists of S_6 rings. A fourth allotrope, amorphous sulfur, contains helical chains of S atoms.

*Allotropes of an element possess different interatomic structures and have different physical and chemical properties.

TABLE 7-1. MELTING POINTS, BOILING POINTS, AND ASSOCIATED
ENTHALPIES FOR SOME ELEMENTAL MOLECULAR SOLIDS
AND LIQUIDS

Molecular Unit	mp (K)	ΔH_{fus} (kcal mole^{-1})	bp (K)	ΔH_{vap} (kcal mole^{-1})
Ne	24.55	0.080	27.07	0.431
Ar	83.78	0.28	87.29	1.56
F_2	53.54	0.06	85.02	0.78
Cl_2	172	1.53	239.0	4.88
N_2	63	0.172	77	1.333
O_2	54	0.106	90	1.63
P_4	317	0.60	357	12.4
S_8	392	0.40	717.7	2.3

Atoms of the Group V elements have three vacancies in their valence electronic configuration (s^2p^3) and thus are expected to form three electron-pair bonds per atom. The most stable form of elemental nitrogen is the diatomic molecule, whereas the allotrope white phosphorus exists as a solid containing discrete tetrahedral P_4 units (Figure 7-4).

Physical properties that provide a measure of intermolecular binding forces are melting points, boiling points, enthalpies of fusion (ΔH_{fus}), and enthalpies of vaporization (ΔH_{vap}). High melting and boiling points generally correspond to large ΔH_{fus} and ΔH_{vap}, which are an indication of large intermolecular binding forces in the solid and liquid states. For elemental molecular solids and liquids the ΔH_{fus} and ΔH_{vap} values are small and the corresponding melting points and boiling points are low, as shown in Table 7-1.

We turn next to the Group IV elements carbon and silicon, which have the valence electronic configuration s^2p^2 with only two unpaired electrons.

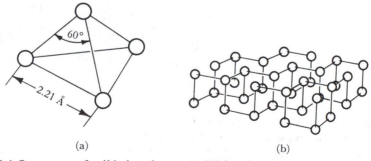

(a)   (b)

Figure 7-4 Structures of solid phosphorus: (a) White phosphorus consists of discrete P_4 molecules. (b) Black phosphorus, a more stable allotrope of the element, has an infinite network structure.

We might expect only two electron-pair bonds per atom as in the diatomic molecules :C≡C:, :Si≡Si:. However, the C_2 molecule is orbital-rich and electron deficient, since it has not achieved an octet around each atom. Therefore, each carbon atom prefers to form *four* electron-pair bonds, as illustrated by the two common allotropes, diamond and graphite (Figure 7-5). Similarly, Si_2 is electron-deficient and does not exist as an individual molecule in solid silicon. Rather, the structure of solid silicon is analogous to that of diamond [Figure 7-5(a)].

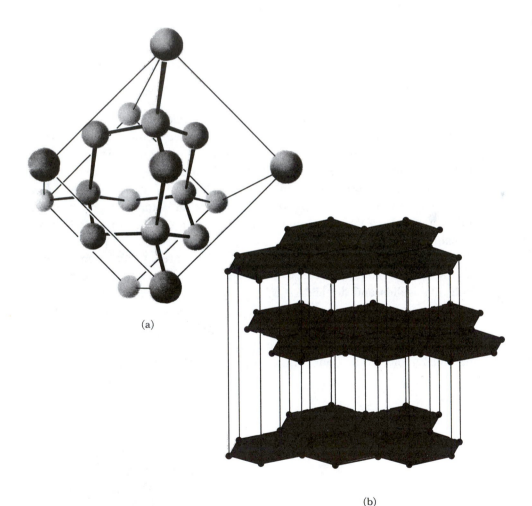

(a)

(b)

Figure 7-5 Crystalline carbon: (a) Diamond structure; the coordination number of carbon in diamond is 4. Each atom is surrounded tetrahedrally by four equidistant atoms. The C–C bond distance is 1.54 Å. (b) Graphite structure, the more stable structure of carbon. Strong carbon-carbon bonding occurs within a layer, weaker bonding occurs between layers.

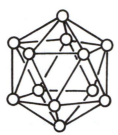

Figure 7-6 Structure of the B_{12} icosahedral unit. In the different crystalline forms of boron, these units are linked together in various ways.

Diamond and graphite are called *nonmetallic network solids,* because they consist of infinite arrays of bonded atoms; no discrete molecules can be distinguished. Thus any given piece of network solid may be considered a giant, covalently bonded molecule. Network solids generally are poor conductors of heat and electricity. Strong covalent bonds among neighboring atoms throughout the structure give these solids strength and high melting temperatures. Diamond *sublimes* (volatilizes directly to gas) instead of melting, at 3500 °C and above. Some of the hardest substances known are nonmetallic network solids.

Another measure of the strength of binding in nonmetallic network solids is the enthalpy of atomization (ΔH_{atom}), which for diamond is 170.5 kcal mole^{-1}. For solid silicon, ΔH_{atom} is 107.8 kcal mole^{-1}. Note that these values are far greater than ΔH_{vap} values for molecular solids such as formed by N_2 and O_2 (Table 7-1).

The only Group III element with nonmetallic properties is boron $(2s^2 2p^1)$. There are three principal allotropic forms of elemental boron, and all of them have network structures based on a B_{12} unit. The structure of each B_{12} unit is icosahedral (Figure 7-6). In the three allotropes the B_{12} icosahedra are linked together in different ways, but in general the bonds between individual icosahedra are weaker than those within any one icosahedron.

Atoms of the metallic elements generally have fewer valence electrons than the number of available orbitals; that is, they are electron-deficient. Consequently, these atoms tend to share their electron density with several other atoms to achieve the maximum in bonding capacity. In most metals at least eight "nearest-neighbor" atoms surround a particular atom in one of the three common structures shown in Figure 7-7. In both hexagonal and cubic close packing, each sphere touches 12 other spheres, 6 in a plane, 3 above, and 3 below. X-ray analysis reveals that two-thirds of all metals crystallize in one of these two structures. A majority of the remaining third crystallize as body-centered cubes, in which each atom has only eight nearest neighbors.

Still referring to Figure 7-7, lithium and sodium, which have the s^1 valence electronic configuration, adopt the body-centered cubic structure. Solid beryllium and magnesium (s^2 atoms) both crystallize in the hexagonal close-packed structure. The crystal structure of aluminum (s^2p^1) is cubic close packed.

The ΔH_{atom} values for Na, Mg, and Al are 26.35, 35.0 and 76.9 kcal mole^{-1}, respectively. This steady increase in ΔH_{atom} in the order $s^1 < s^2 < s^2p^1$ indicates that all the valence electrons participate in interatomic binding in these cases. The ΔH_{atom} values for the metallic elements in the first, second, and third transition series are displayed in Figure 7-8. For these series the d electrons also participate in the interatomic binding. Notice that in the second and third transition series the maximum ΔH_{atom} is achieved near the half-filled shell (d^5s^1) with a total of six valence electrons. For metals of the second and third transition series, the d electrons are used so effectively in electron-pair bonds that no electrons are left unpaired. Since unpaired d electrons are required for magnetism, the metals of the second and third transition series are not magnetic. Beyond six valence electrons the ΔH_{atom} values begin to decrease in the second and third transi-

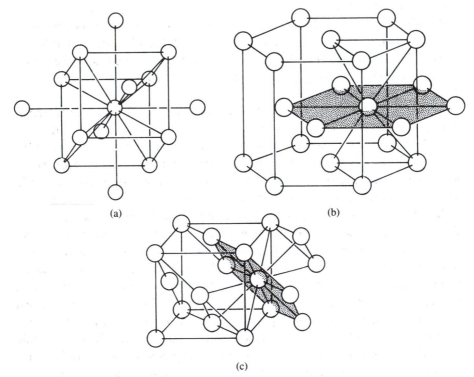

(a)   (b)

(c)

Figure 7-7 Common structures of metals: (a) Body-centered cubic (e.g., Na, V, and Ba); (b) hexagonal close packing (e.g., Mg, Ir, and Cd); (c) cubic close packing, or face-centered cubic (e.g., Al, Cu, and Au).

tion series, due to the fact that the added electrons are forced to pair up on each atom and are not available to contribute to the metallic bonding.

As seen in Figure 7-8, the ΔH_{atom} values for the first transition series exhibit a peculiar double-humped feature with a local minimum at manganese. The American chemist Leo Brewer has explained this as being due to the poor overlap of the 3d orbitals on nearest-neighbor atoms, compared to the good overlap of the 4d and 5d orbitals in the second and third transition series. For the metals starting with chromium and ending with nickel, the relative effectiveness of the 3d electron bonding causes the double-humped feature. For these metals the bonding energy that could be obtained by formation of electron-pair bonds between the nuclei is offset by the lower electron-electron repulsion that results if the electrons remain unpaired. The fact that some of the 3d electrons remain unpaired causes these metals to be magnetic.

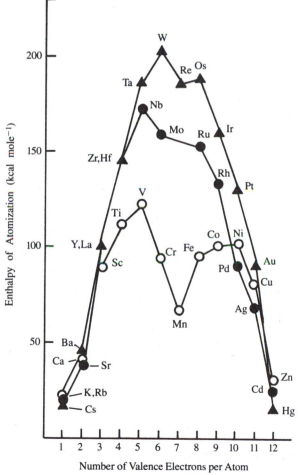

Figure 7-8 Enthalpies of atomization (ΔH_{atom}) of metals at 298 K.

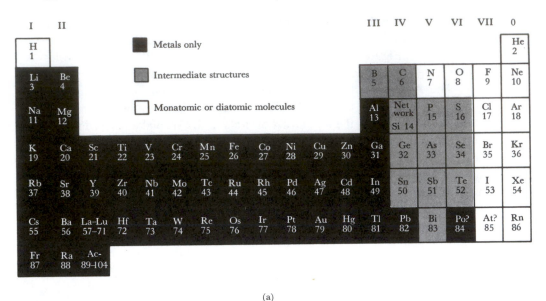

(a)

(b)

Figure 7-9 Solid structures of the elements: (a) general trends in bonding in solid elements; (b) the types of bonding in the variable zone of (a). The structures of the nonmetals are determined by their coordination numbers, which are eight less the group number except when multiple bonds are used, as in graphite, N_2, and O_2 (see Table 7-2).

TABLE 7-2. THE CORRELATION BETWEEN COORDINATION
NUMBER AND STRUCTURE IN ELEMENTAL SOLIDS

Bonding Coordination Number	Type of Solid Structure
0	Atomic solids, low melting and boiling points
1	Diatomic molecular solids, low melting and boiling points
2	Rings or chains. Solids with packed ring molecules are less metallic than those with packed chains
3	P_4 tetrahedra or sheets. Solids with packed tetrahedral molecules are less metallic than those with packed sheets
4	Three-dimensional nonmetallic networks
5	B sheet curved in on itself in a B_{12} icosahedron
6 or more	Packed metallic solids

In the periodic table shown in Figure 7-9 the elemental solids are classified as metallic, network nonmetallic, or molecular. In Table 7-2 the correlation between coordination number and structure in elemental solids is presented. Most elements crystallize in metallic structures in which each atom has a high coordination number. Included as metals are elements such as tin and bismuth, which crystallize in structures with relatively low atomic coordination numbers but which still have strong metallic properties. The gray area of the periodic table includes elements that have borderline properties. Although germanium crystallizes in a diamond-like structure in which the coordination number of each Ge atom is only four, certain of its properties resemble those of metals. This similarity to metals indicates that the valence electrons in germanium are not held as tightly as would be expected in a true nonmetallic network solid. Arsenic, antimony, and selenium exist as either molecular or metallic solids, although the so-called metallic structures have relatively low atomic coordination numbers. We know that tellurium crystallizes in a metallic structure, and it seems reasonable to predict that Te also may exist as a molecular solid. From its position in the periodic table we also predict intermediate properties for astatine, which has not been studied in detail.

7-2 Ionic Solids

Ionic solids consist of infinite arrays of positive and negative ions held together by electrostatic forces. These forces are the same as those that hold a molecule of NaCl together in the vapor phase. In solid NaCl the Na^+ and

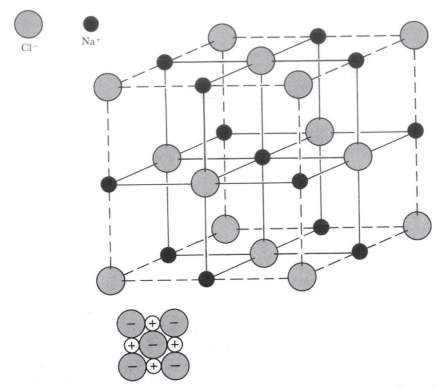

Figure 7-10 Representation of the ionic NaCl structure. The bottom figure is a representation of a cross section of the NaCl structure.

Cl⁻ ions are arranged to maximize the electrostatic attraction, as shown in Figure 7-10. The coordination number of each Na^+ ion is six, and each Cl^- ion similarly is surrounded by six Na^+ ions. Because ionic bonds are very strong, much energy is required to break down the structure in solid-to-liquid or liquid-to-gas transitions. Thus ionic compounds have high melting and boiling temperatures.

The crystal structures of several typical ionic solids are shown in Figure 7-11. Cesium chloride crystallizes in a structure in which the cation and anion each have a coordination number of eight. Zinc sulfide crystallizes in two distinct structures—the so-called zinc blende and wurtzite structures—in which the cation and anion each have a coordination number of four. Calcium fluoride crystallizes in the fluorite structure. The coordination numbers are eight for the cation (eight fluorides surround each calcium) and four for the anion. One of the crystalline forms of titanium dioxide is rutile, in which the coordination numbers are six for the cation and three for the anion.

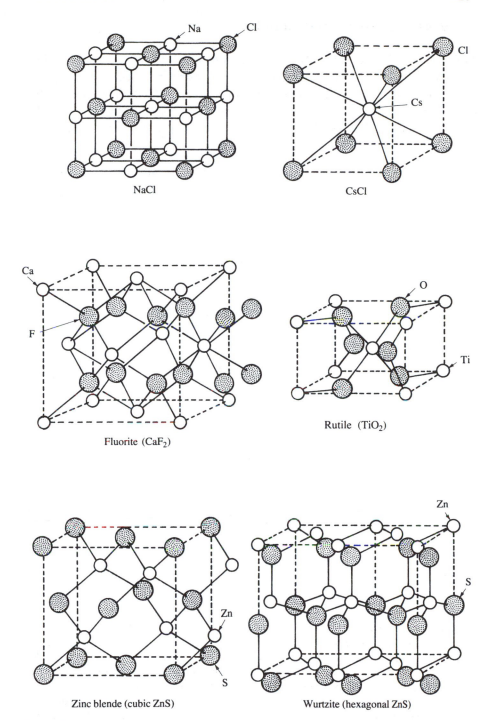

Figure 7-11 Some of the common structural types found for ionic substances.

7-3 Molecular Solids and Liquids

Molecules such as H_2, N_2, O_2, and F_2 form molecular solids because all the valence orbitals are used either for *intramolecular* bonding or are occupied with nonbonding electrons. Thus any intermolecular bonding that holds molecules together in the solid must be weak compared with the strength of the intramolecular bonding in the molecules. The weak forces that con-

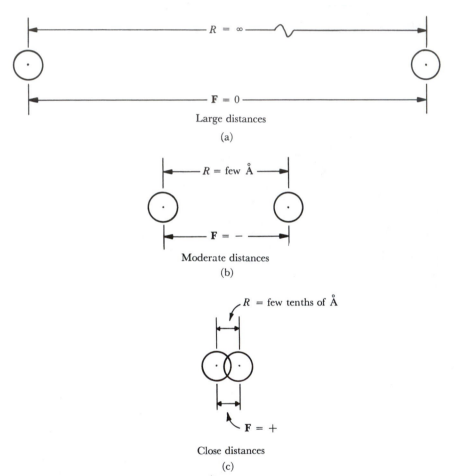

Figure 7-12 Repulsion of electrons in filled orbitals: (a) At very large distances two atoms or molecules behave toward each other as neutral species and neither repel nor attract one another. The force between them, **F**, is zero. (b) At moderate distances two atoms or molecules have not yet come close enough for repulsion to be appreciable. However, they do attract one another (see Figure 7-13) because of deformations of their electron densities. (c) At close range, when the electron density around one atom or molecule is large in the same region of space as the electron density around the other atom or molecule (i.e., when the filled orbitals overlap), coulomblike repulsion dominates, and the two molecules repel one another.

tribute to intermolecular bonding are called *van der Waals forces* after the Dutch physicist Johannes van der Waals.

Van der Waals Forces

There are two principal van der Waals forces. The most important force at short range is the repulsion between electrons in the filled orbitals of atoms on neighboring molecules. This electron-pair repulsion is illustrated in Figure 7-12. The analytical expression commonly used to describe the energy resulting from this interaction is

$$\text{van der Waals repulsion energy} = be^{-ar} \tag{7-1}$$

in which b and a are constants for two interacting atoms. Notice that this repulsion term is very small at large values of the interatomic distance, r.

The second van der Waals force is the attraction that results when electrons in the occupied orbitals of the interacting atoms synchronize their motion to avoid each other as much as possible. For example, as shown in Figure 7-13, electrons in orbitals of atoms belonging to interacting molecules can synchronize their motion to produce an instantaneous dipole-induced dipole attraction. If at any instant the left atom in Figure 7-13 has more of its electron density at the left, as shown, then the atom will be a tiny dipole with a negative left side and a positive right side. This positive side will attract electrons on the right atom in the figure and will change this atom into a dipole with similar orientation. These two atoms will attract each

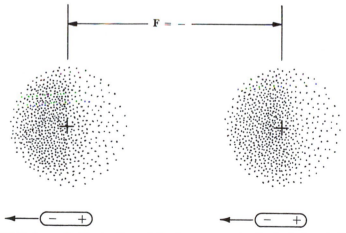

Figure 7-13 Schematic illustration of the instantaneous dipole-induced dipole interaction that gives rise to a weak attraction. For the brief instant that this figure describes, there is an attractive force, **F**, between the instantaneous dipole and the induced dipole. The effect is reciprocal; each atom induces a polarization in the other.

other because the positive end of the left atom and the negative end of the right atom are close. Similarly, fluctuation in electron density of the right atom will induce a temporary dipole, or asymmetry of electron density, in the left atom. The electron densities are fluctuating continually, yet the net effect is an extremely small but important attraction between atoms. The energy resulting from this attractive force is known as the *London energy*, after Fritz London, who derived the quantum-mechanical theory for this attraction in 1930. The London energy varies inversely with the sixth power of the separation between atoms:

$$\text{London energy} = -d/r^6 \tag{7-2}$$

in which d is a constant and r is the distance between atoms. This "inverse sixth" attractive energy decreases rapidly with increasing r, but not nearly as rapidly as the van der Waals repulsion energy. Thus at longer distances the London attraction is more important than the van der Waals repulsion, consequently a small net interatomic attraction exists.

The total potential energy of van der Waals interactions is the sum of the attractive energy of Equation 7-2 and the repulsive energy of Equation 7-1:

$$PE = be^{-ar} - d/r^6 \tag{7-3}$$

TABLE **7-3.** VAN DER WAALS ENERGY PARAMETERS

Interaction Pair	a (au)$^{-1}$*	b (kcal mole^{-1})	d [kcal mole^{-1} (au)]*
He—He	2.10	4.1×10^3	1.5×10^3
He—Ne	2.27	20.7×10^3	2.9×10^3
He—Ar	2.01	30.0×10^3	9.7×10^3
He—Kr	1.85	16.4×10^3	13.7×10^3
He—Xe	1.83	26.6×10^3	21.3×10^3
Ne—Ne	2.44	104.8×10^3	5.7×10^3
Ne—Ar	2.18	151.8×10^3	19.2×10^3
Ne—Kr	2.02	82.8×10^3	26.7×10^3
Ne—Xe	2.00	134.2×10^3	41.5×10^3
Ar—Ar	1.95	219.5×10^3	64.6×10^3
Ar—Kr	1.76	119.8×10^3	90.1×10^3
Ar—Xe	1.74	194.4×10^3	139.3×10^3
Kr—Kr	1.61	65.2×10^3	125.4×10^3
Kr—Xe	1.58	106.0×10^3	194.4×10^3
Xe—Xe	1.55	171.8×10^3	301.1×10^3

*1 au = 1 atomic unit = 0.529 Å. The value of r in Equation 7-3 must be expressed in atomic units as well.

The total van der Waals potential energy can be compared quantitatively with ordinary covalent bond energies by examining systems for which the curves of potential energy versus interatomic distance, r, are known accurately. We can calculate values for the constants a, b, and d from experimental data on the deviation of real gases from ideal gas behavior. Some of these values for interactions of noble gases are listed in Table 7-3.

The potential energy curve for van der Waals interactions between helium atoms is illustrated in Figure 7-14. At separations of more than 3.5 Å, the second term in Equation 7-3 (the London attraction) predominates. As the atoms move closer together they attract each other more, and the energy of the system decreases. However, at distances closer than 3 Å the strong electron-pair repulsion overwhelms the London attraction, and the potential energy curve in Figure 7-14 rises. A balance between attraction and repul-

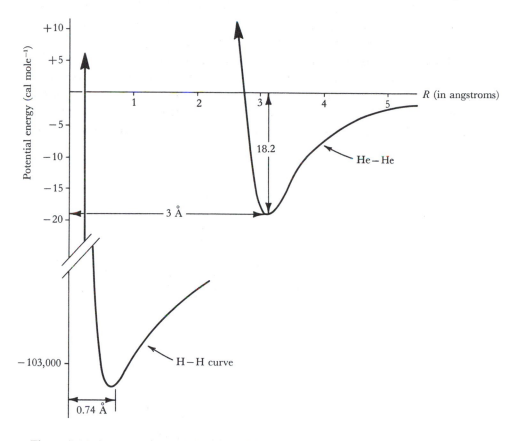

Figure 7-14 A comparison of the potential energy curves for van der Waals attraction between two He atoms and covalent bonding between two H atoms. Notice that the energy scale is in cal mole^{-1}, rather than kcal mole^{-1}. The covalent bond is more than five thousand times as stable as the van der Waals bond.

sion exists at a 3-Å separation, and the He—He "molecule" is 18.2 cal mole⁻¹ more stable than two isolated atoms. Nevertheless, the He_2 molecule does not exist because the zero-point vibrational energy exceeds this attractive energy.

Figure 7-14 also shows the marked contrast between van der Waals attraction and covalent bonding. In the H_2 molecule strong electron-proton attractions in the bonding molecular orbital cause the potential energy to decrease as the H atoms approach one another, and it is proton-proton repulsion that makes the energy increase sharply if the atoms are pushed too closely together. This proton-proton repulsion operates at smaller distances than does the electronic repulsion between the two He atoms. The H–H bond length in the H_2 molecule is 0.74 Å, whereas the equilibrium distance of van der Waals-bonded He atoms is 3 Å. Moreover, a covalent bond is much stronger than a weak van der Waals interaction. Only 18.2 cal mole⁻¹ are required to separate helium atoms at their equilibrium distance, but 103,000 cal mole⁻¹ are needed to break the covalent bond in H_2.

Molecular solids, in which only van der Waals intermolecular bonding exists, generally melt at low temperatures. This is because relatively little energy of thermal motion is needed to overcome the energy of van der Waals bonding. The liquid and solid phases of helium, which result from weak van der Waals "bonds," exist only at temperatures below 4.6 K. Even

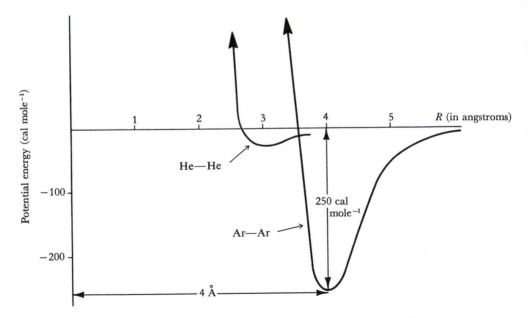

Figure 7-15 A comparison of the potential energy curves for van der Waals attraction between two Ar atoms and two He atoms. The larger Ar atoms are more tightly held, although the bond energy is still only 0.0023 that of the H–H bond in H_2.

at temperatures near absolute zero, solid helium can be produced only at high pressures (29.6 atm at 1.76 K).

Van der Waals bonds in molecular solids and liquids generally are stronger with increasing size of the atoms and molecules involved. For example, as the atomic number of the noble gases increases, the strength of the van der Waals bonding increases also, as shown by the Ar–Ar potential energy curve in Figure 7-15. The attraction between the heavier atoms is stronger, presumably because the outer electrons are held more loosely, and larger instantaneous dipoles and induced dipoles are possible. Because of this stronger van der Waals bonding, solid argon melts at −184 °C, or 89 K, which is a considerably higher temperature than solid helium.

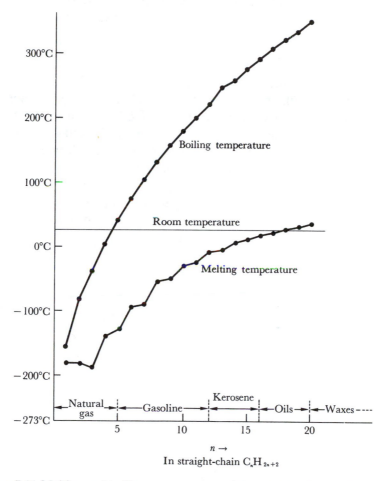

Figure 7-16 Melting and boiling temperatures of the straight-chain hydrocarbons as a function of the length of the carbon chain. More energy is required to separate two molecules of eicosane (20 carbons) than ethane (2 carbons) because of the more numerous van der Waals interactions between the two larger molecules.

An example of the effect of molecular size on melting and boiling temperatures is provided by a series of straight-chain alkanes, with formulas C_nH_{2n+2}, depicted in Figure 7-16 for $n = 1$ through $n = 20$. Part of the increase in melting and boiling temperatures with increasing molecular size and weight arises from the greater energy needed to move a heavy molecule. However, another important factor is the large surface area of a molecule such as eicosane $(C_{20}H_{42})$ compared with methane, and the greater stability that eicosane therefore can gain from intermolecular van der Waals attractions. The mass effect is similar for both melting and boiling temperatures. However, molecular surface area affects the boiling temperatures more because molecules in the liquid phase are still close enough to exert van der Waals attractions. In fact, without these attractions, which are broken during vaporization, the liquid state could not exist.

Polar Molecules and Hydrogen Bonds

Polar molecules are stabilized in a molecular solid by the attractive interaction of oppositely charged ends of the molecules. A particularly important kind of polar interaction is the *hydrogen bond*. This is a relatively weak bond—about 5 kcal mole^{-1}—between a positively charged hydrogen atom and a small, electronegative atom, usually N, O, or F. (A typical covalent or ionic bond energy is about 100 kcal mole^{-1}.)

Before discussing the effect of hydrogen bonding on the structures of molecular solids, let us examine some simple examples of hydrogen bonding in the gas phase. The structures of the gaseous dimers of HF and H_2O, and the NH_3–HF species are shown in Figure 7-17.

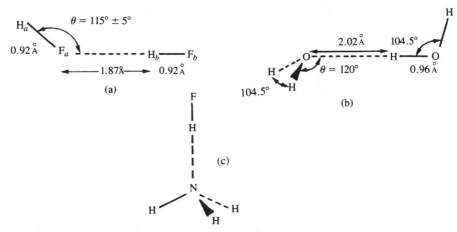

Figure 7-17 Molecular structures of the hydrogen-bonded molecules of (a) $(HF)_2$; (b) $(H_2O)_2$; and (c) NH_3–HF.

One of the simplest hydrogen bonds is represented by the gaseous dimer of HF. Its molecular structure has been determined and is shown in Figure 7-17(a). This molecular dimer exhibits several features that are common to hydrogen-bonded systems:

1. The molecular units retain their identity. The H_a–F_a and H_b–F_b bond lengths are 0.92 Å, identical within experimental error to that found for monomeric HF.

2. The F_a----H_b—F_b bond is *linear*.

3. The hydrogen atom that is bonded between the two F atoms is *asymmetrically* positioned. Only in very strong hydrogen bonds, such as FHF^-, is the hydrogen atom in a symmetrical position.

4. The angle θ is usually between 100° and 120°. For $(HF)_2$, the H_a—F_a----H_b angle is found to be 115° ±5°.

Hydrogen bonding has important consequences for the solid state structure adopted by hydrogen fluoride. The structure is found to exhibit a zig-zag chain of HF molecules, linked together by hydrogen bonds:

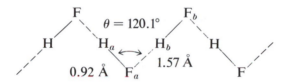

The presence of hydrogen bonds is also clear in the structure adopted by the water dimer, as shown in Figure 7-17(b). Again the hydrogen bond is observed to be linear and asymmetric.

The structure of ice provides further evidence of the importance of hydrogen bonding in molecular solids. As shown in Figure 7-18, each oxygen atom of a polar H_2O molecule is tetrahedrally coordinated to four other oxygen atoms in a structure that somewhat resembles that of diamond. Each oxygen atom is bound to its four neighbor oxygen atoms by hydrogen bonds. In two of these hydrogen bonds the central H_2O molecule supplies the hydrogen atoms; in the other two bonds the hydrogen atoms come from neighboring water molecules. Although the hydrogen bonds are relatively weak compared to typical covalent bonds, they are nonetheless important because there are so many of them in the solid and liquid phases.

Hydrogen bonding in water is responsible for many of its most important properties. Because of hydrogen bonds in both the solid and liquid phases the melting and boiling temperatures of water are unexpectedly high when

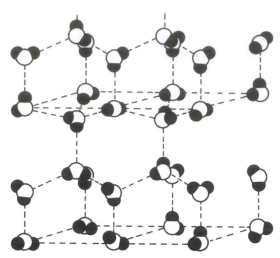

Figure 7-18 In crystalline ice each oxygen atom is hydrogen bonded to two others by means of its own hydrogen atoms, and bonded to two more oxygen atoms by means of their hydrogen atoms. The coordination is tetrahedral and the structure is similar to that of diamond.

compared with those of H_2S, H_2Se, and H_2Te, which are hydrogen compounds of elements also in Group VI of the periodic table. Solid and liquid ammonia and hydrogen fluoride show anomalous behavior similar to that of water, and for the same reason (Figure 7-19).

Since hydrogen bonding causes an open network structure in ice (Figure 7-18), ice is less dense than water at the melting temperature. Upon melting, part of the open-cage structure collapses, so that the liquid is more compact than the solid. The measured heat of fusion of ice is only 1.4 kcal mole^{-1}, whereas the energy of its hydrogen bonds is 5 kcal mole^{-1}. This contrast indicates that only about one-third of the hydrogen bonds of ice are broken when it melts. Water is not composed of isolated, unbonded molecules of H_2O; rather, it has regions or clusters of hydrogen-bonded molecules. That is, part of the hydrogen-bonded structure of the solid persists in the liquid. As the temperature is raised these clusters break up, and the volume continues to shrink. If the temperature is raised still higher, the expected thermal expansion dominates over the shrinkage caused by the collapse of the cage structures. Consequently liquid water has a minimum molar volume, or a maximum density, at 4 °C.

Because the hydrogen-bonded H_2O clusters are broken slowly as heat is added, water has a higher specific heat than any other common liquid except ammonia. Water also has an unusually high heat of fusion and heat of vaporization. All three of these properties make it possible for water to act as a large thermostat, which confines the temperature on the earth within moderate limits. Ice absorbs a large amount of heat when it melts, and water

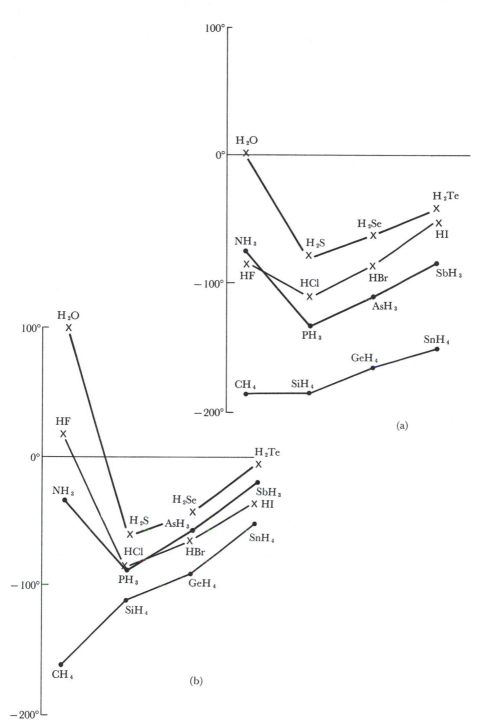

Figure 7-19 (a) Melting temperatures and (b) boiling temperatures for binary hydrogen compounds of some elements. Generally, melting and boiling temperatures increase with molecular weight within a group. The anomalous compounds HF, H_2O, and NH_3 all have hydrogen bonds between molecules in both solid and liquid states.

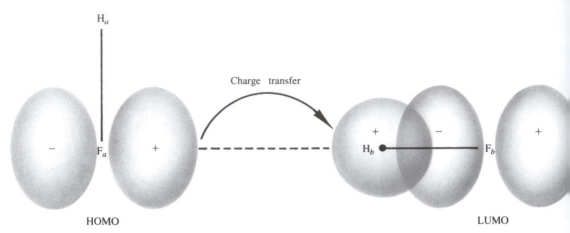

HOMO   LUMO

Figure 7-20 Representation of the charge-transfer interaction between the HOMO (1π) and LUMO (3σ) of HF in $(HF)_2$ with $\Theta = 90°$.

absorbs more heat per unit of temperature rise than almost any other substance. Correspondingly, as water cools, it gives off more heat to its surroundings than other substances do.

Due to the general importance of the hydrogen bond, many theoretical treatments have been proposed to explain its origin. Recent work on the HF dimer has indicated that there are at least four types of interactions that contribute to hydrogen bonding:

1. An *electrostatic attraction* can occur between the δ^- charge on F_a and the δ^+ charge on H_b. Alternatively, we can look on this electrostatic interaction as occurring between the negative end of the $\overrightarrow{H_a-F_a}$ dipole and the positive end of the $\overrightarrow{H_b-F_b}$ dipole. Such an interaction is referred to as a *dipole-dipole* interaction. Taken by itself, this dipole-dipole interaction would predict a linear HF dimer ($\theta = 180°$).

2. A *charge-transfer* interaction can occur between the HOMO of H_a-F_a and the LUMO of H_b-F_b. Since the HOMO of H_a-F_a is the 1π orbital and the LUMO of H_b-F_b is the 3σ orbital (Figure 7-20), this interaction alone would predict an angle of $\theta = 90°$.

3. *Electron-electron repulsion* can occur between the electron density on H_a-F_a and that on H_b-F_b. This repulsive interaction would probably be largest at $\theta = 90°$, since at this angle the HOMO electron density on H_a-F_a is closest to the electron density on H_b-F_b.

4. The electron cloud on H_a-F_a can *polarize* the electron cloud on H_b-F_b, and vice versa. This is always an attractive interaction.

Quantum-mechanical calculations on $(HF)_2$ indicate that of these four factors, the electrostatic attraction is the most important (about -6.0 kcal

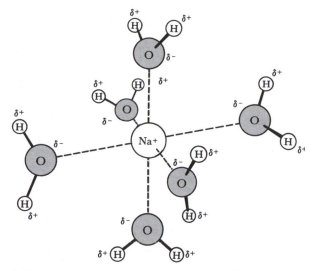

Figure 7-21 In solution, the hydrated Na$^+$ ion is surrounded by an octahedron of negative charges, but these negative charges are from the dipolar solvent molecule H$_2$O, instead of Cl$^-$.

mole^{-1}), and the charge-transfer attraction provides about half as much stabilization (-3.0 kcal mole^{-1}). The electron-electron repulsion ($+3.0$ kcal mole^{-1}) is found to nearly offset the charge-transfer attraction. Finally, the polarization is found to stabilize the dimer only slightly. Consequently, the total hydrogen bond energy in (HF)$_2$ is about 6 kcal mole^{-1}. Calculations have been performed on several hydrogen-bonded systems and it seems to be generally true that the electrostatic attraction (the dipole-dipole interaction) provides the largest contribution to the hydrogen bond.

Polar Molecules as Solvents

The polar nature of liquid water makes it an excellent solvent for ionic solids such as NaCl. Water can dissolve NaCl and separate the oppositely charged Na$^+$ and Cl$^-$ ions because the energy required to separate the ions is provided by the formation of hydrated ions (Figure 7-21). Each Na$^+$ ion in solution still has an octahedron of negative charges around it, but instead of being Cl$^-$ ions they are the negative poles of the oxygen atoms of the water molecules. The Cl$^-$ ions are also hydrated, but it is the positive end (H) of the water molecules that approach the Cl$^-$ ions. A nonpolar solvent such as gasoline, a liquid composed of hydrocarbon molecules, cannot form such *ion-dipole bonds* with Na$^+$ and Cl$^-$. Consequently NaCl and other salts are insoluble in gasoline.

Energy is required to break up a **polar** molecular solid

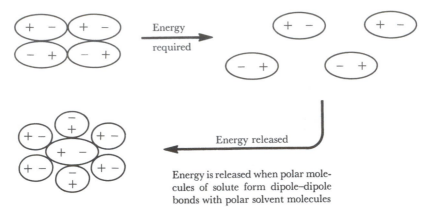

Energy is released when polar molecules of solute form dipole–dipole bonds with polar solvent molecules

Figure 7-22 When a crystalline solid composed of polar molecules dissolves, stability is lost when oppositely charged ends of neighboring molecules are removed. This loss is compensated by the stability produced by solvating the polar molecules in solution. A solvent that cannot provide such stabilization cannot dissolve the solid.

Polar solvents dissolve polar molecular solids because of dipole-dipole interactions. The energy released by the formation of dipole-dipole bonds between a polar solvent and solute molecules is sufficient to break the intermolecular forces in the molecular solids (Figure 7-22). For example, ice is soluble in liquid ammonia but not in benzene because NH_3 is a polar molecule, whereas C_6H_6 is nonpolar.

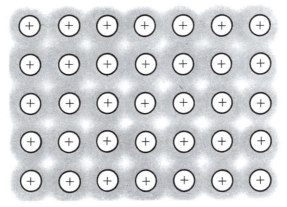

Figure 7-23 A cross section of the crystal structure of a metal that exhibits the sea of electrons. Each circled positive charge represents the nucleus and filled, non-valence electron shells of a metal atom. The shaded area surrounding the positive metal ions indicates schematically the mobile sea of electrons.

7-4 Metals

The high thermal and electrical conductivity of metals suggests that the valence electrons are relatively free to move through the crystal structure. Figure 7-23 illustrates one model in which the electrons form a "sea" of negative charges that holds the atoms tightly together. The circled positive charges represent the positively charged ions remaining when valence electrons are stripped away, thereby leaving the nuclei and the filled electron shells. Since metals generally have high melting temperatures and high densities, especially in comparison with molecular solids, the "electron sea" must strongly bind the positive ions in the crystal.

Characteristics of Metals Versus Ionic Crystals

The simple electron-sea model for metallic bonding also is consistent with two other commonly observed properties of metals: malleability and ductility. A *malleable* material can be hammered easily into sheets; a *ductile* material can be drawn into thin wires. For metals to be shaped and drawn without fracturing, the atoms in the planes of the crystal structure must be readily displaceable with respect to each other. This displacement does not result in the development of strong repulsive forces in metals because the

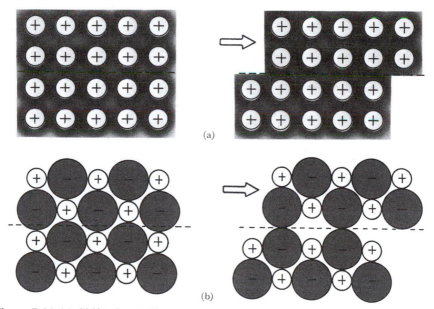

Figure 7-24 (a) Shift of metallic crystal along a plane results in no strong repulsive forces. (b) Shift of an ionic crystal along a plane results in strong repulsive forces and crystal distortion.

mobile sea of electrons provides a constant buffer, or shield, between the positive ions. This situation is in direct contrast to that of ionic crystals, in which the binding forces are due almost entirely to electrostatic attractions between oppositely charged ions. In an ionic crystal, valence electrons are bound firmly to the atomic nuclei. Displacement of layers of ions in such a crystal brings ions of like charge together and causes strong repulsions that can fracture the crystal (Figure 7-24).

Electronic Bands in Metals

The molecular-orbital theory provides an adequate model for metallic bonding. According to this model the entire block of metal is considered as a giant molecule. All of the atomic orbitals of a particular type in the crystal interact to form a set of delocalized orbitals that extend throughout the

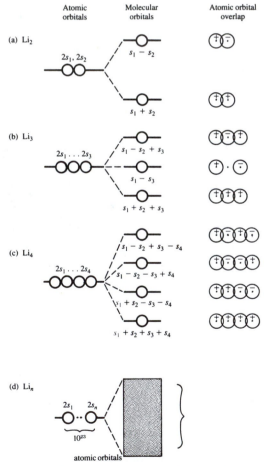

Figure 7-25 Molecular-orbital development of the band theory of metals.

entire block. For a particular crystal, assume that the number of valence orbitals is of the order of 10^{23}. To visualize the interaction of such a large number of valence orbitals, let us consider the hypothetical sequence of linear lithium molecules Li_2, Li_3, and Li_4, in which the important valence orbitals are $2s$ orbitals. Figure 7-25 shows the build-up of molecular orbitals for these three molecules. Notice that due to the delocalization of the molecular orbitals, none of the electrons is required to reside in antibonding orbitals. The spacing between the orbitals also becomes smaller. In the limit of 10^{23} equivalent atoms, the combination of atomic orbitals produces a band of closely spaced energy levels.

Figure 7-26 illustrates the three bands of energy levels formed by the $1s$, $2s$, and $2p$ orbitals of the simplest metal, lithium. The $1s$ molecular orbitals

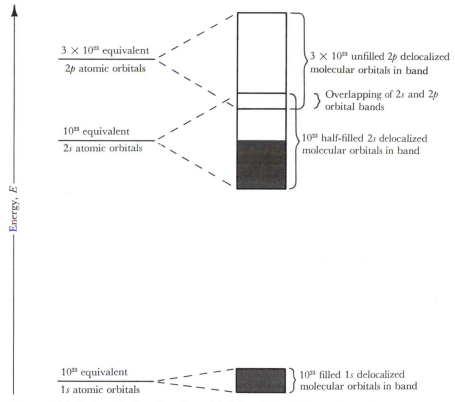

Figure 7-26 Delocalized molecular-orbital bands in lithium. The original $2s$ and $2p$ atomic orbitals are so close in energy that the molecular-orbital bands overlap. Lithium has one electron in every $2s$ atomic orbital, hence only half as many electrons as can be accommodated in the $2s$ atomic orbitals or in the delocalized molecular-orbital band. There are unfilled energy states an infinitesimal distance above the highest-energy filled state, so an infinitesimal energy is required to excite an electron and send it moving through the metal. Thus lithium is a conductor.

are filled completely because the 1s atomic orbitals in isolated lithium atoms are filled. Thus the 1s electrons make no contribution to bonding. They are part of the positive ion cores and can be eliminated from the discussion. Atomic lithium has one valence electron in a 2s orbital. If there are 10^{23} atoms in a lithium crystal, the 10^{23} 2s orbitals interact to form a band of 10^{23} delocalized orbitals. As usual, each of these orbitals can accommodate two electrons, so the capacity of the band is 2×10^{23} electrons. Lithium metal has enough electrons to fill only the lower half of the 2s band, as illustrated in Figure 7-26.

The presence of a partially filled band of delocalized orbitals accounts for bonding and electrical conduction in metals. Electrons in the lower filled orbital band move throughout the crystal in a random fashion in such a way that their motion results in no *net* separation of electrons and positive ions in the metal. For a metal to conduct an electric current, electrons must be excited to unfilled delocalized orbitals in such a way that their movement in one direction is not exactly canceled by electrons moving in the opposite direction. Such concerted electron movement occurs only when an electric potential difference is applied between two regions of a metal. Then electrons are excited to the unfilled delocalized molecular orbitals that are part of the same band (the 2s band for lithium) and just slightly higher in energy. Electrical conduction is restricted by the frequent collisions of electrons with positive ions, which have kinetic energy and thus vibrate randomly within their crystal sites. As temperature increases, vibration of the positive ions increases, and collisions with the conduction electrons are more frequent. Therefore electrical conduction in metals decreases as temperature increases.

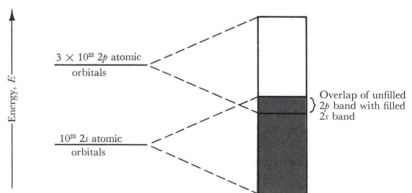

Figure 7-27 Band-filling diagram for beryllium. A Be atom has enough electrons (two) to fill its 2s orbital, so Be metal has enough electrons to fill its 2s delocalized molecular orbital band. If the 2s and 2p bands did not overlap, Be would be an insulator, because an appreciable amount of energy would be required to make electrons flow in the solid. But with the band overlap shown here, an infinitesimal amount of energy excites the electrons to the 2p band orbitals, and electrons flow.

Beryllium is a more complicated example than lithium. An isolated beryllium atom has exactly enough electrons to fill its 2s orbital. Accordingly, beryllium metal has enough electrons to fill its 2s delocalized band. If the 2p band did not overlap the 2s (Figure 7-27), beryllium would not conduct well, because an energy equal to the gap between bands would be required before electrons could move through the solid. However, the two bands do overlap and beryllium has unoccupied delocalized orbitals that are an infinitesimal distance above the most energetic filled orbitals. Consequently beryllium is a metallic conductor.

7-5 Nonmetallic Network Solids

Nonmetallic network materials such as carbon or silicon are insulators; that is, they do not conduct electrical current. There is no simple way to apply the molecular-orbital model to a discussion of the bonding in non-metallic network solids. Suffice it to say that in nonmetallic network solids it is usually possible to "count" electrons in the Lewis electron-dot sense and to show that the octet is achieved. That is, the atoms in nonmetallic network solids usually have at least as many valence electrons as the number of valence orbitals. Consequently, low coordination numbers are common and simple electron-pair bonds can be formed between each atom and its nearest neighbors. Due to the low coordination numbers, the potential energy is not constant throughout the crystal; rather, the energy is greatly lowered in the internuclear region, and the electrons are not free to move throughout the crystal as they are in metals.

In diamond, for example, each carbon atom has a coordination number of four. The hybrid-orbital model adequately describes the bonding by assigning to each carbon atom four localized tetrahedral sp^3 hybrid orbitals (Figure 7-28). The four valence electrons in each carbon atom are sufficient to fill these bonding orbitals. Thus all electrons in diamond are used for bonding, leaving none to move freely to conduct electricity.

The effect of coordination number on the electronic bands of a solid can be illustrated with respect to carbon. Calculations have shown that the electronic bands for carbon would be delocalized *if* carbon crystallized in a structure with a high coordination number such as is found for the metals (Figure 7-7). In that case the band structure would correspond to that shown in Figure 7-29(a), where one-half of the bands are occupied by electrons. (Carbon has four valence electrons and four valence orbitals, each of which can contain two electrons.) Consequently, carbon would be an electrical conductor. Experimentally it is found that carbon has a coordination number of four; this causes the extended bands shown in Figure 7-29(a) to split into

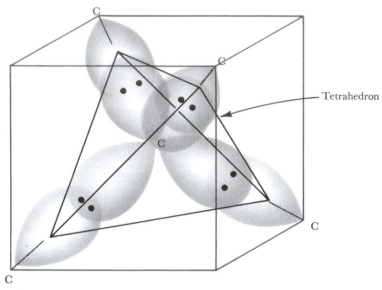

Figure 7-28 Schematic representation of the overlap of the four sp^3 hybrid orbitals of a C atom with similar orbitals from four other carbon atoms in diamond crystals.

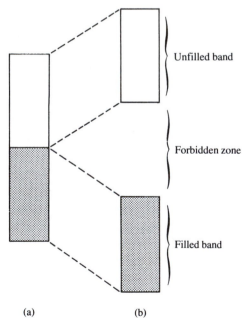

Figure 7-29 Calculated band structure of crystalline carbon: (a) assuming that carbon crystallizes in the body-centered cubic structure with each carbon atom surrounded by eight other carbon atoms; (b) assuming that carbon crystallizes in the known diamond structure wherein each carbon atom is surrounded by four other carbon atoms. In (a) carbon would behave as an electrical conductor; in (b) it would be an insulator, as observed experimentally for diamond.

two groups: a filled band corresponding to bonding orbitals and an unfilled band of antibonding orbitals. There is a "forbidden zone" between these bands [Figure 7-29(b)].

For an insulator to conduct, energy is required that is sufficient to excite electrons in the filled band across the forbidden energy zone (the band gap) into the unfilled molecular orbitals. This energy is the activation energy of the conduction process. Only high temperatures or extremely strong electrical fields will provide enough energy to an appreciable number of electrons for conduction to occur. In diamond the gap between the top of the valence band and the bottom of the conduction band is 5.2 eV, or 120 kcal mole^{-1}.

Semiconductors

The borderline between metallic and nonmetallic network structures of elements in the periodic table is not sharp (Figure 7-9); several elemental solids have properties that are intermediate between conductors and insulators. Silicon, germanium, and α-gray tin all have the diamond structure. However, the gap between filled and empty bands for these solids is much smaller than for carbon. As against 120 kcal mole^{-1} for carbon, the band gap for silicon is only 25 kcal mole^{-1}. For germanium it is 14 kcal mole^{-1}, and for α-gray tin it is 1.8 kcal mole^{-1}. The metalloids silicon and germanium are called *semiconductors*.

A semiconductor can carry a current if the relatively small energy required to excite electrons from the lower filled valence band to the upper empty conduction band is provided. Since the number of excited electrons increases as temperature increases, the conductance of the semiconductors increases with temperature. This behavior is exactly opposite to that of metals.

Conduction in materials such as silicon and germanium can be enhanced by adding small amounts of certain impurities. Although there is a forbidden energy zone in silicon, this zone can be narrowed effectively if impurities such as boron or phosphorus are added to silicon crystals. Small amounts of boron or phosphorus (a few parts per million) can be incorporated into the silicon structure when the crystal is grown. Phosphorus has five valence electrons, and thus has a free electron even after contributing four electrons to the covalent bonds of the silicon structure. This fifth electron can be moved away from a phosphorus atom by an electric field; hence phosphorus is an electron donor. Only 0.25 kcal mole^{-1} is required to free the donated electrons, thereby making a conductor out of silicon to which a small amount of phosphorus has been added. The same effect occurs in an opposite manner if boron instead of phosphorus is added to silicon. Atomic boron has one too few electrons for complete covalent bonding. Thus for each

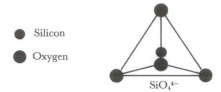

Figure 7-30 The SiO_4^{4-} tetrahedron, which is the building block of all silicate minerals. The Si atom is covalently bonded to four oxygen atoms at the corners of a tetrahedron. The covalent bonds are not shown. The lines between oxygen atoms are included only to give form to the tetrahedron.

boron atom in the silicon crystal there is a single vacancy in a bonding orbital. It is possible to excite the valence electrons of silicon into these vacant orbitals in the boron atoms, thereby causing the electrons to move through the crystal. To accomplish this conduction an electron from a silicon neighbor drops into the empty boron orbital. Then an electron that is two atoms away can fill the silicon atom's newly created vacancy. The result is a cascade effect, whereby an electron from each row of atoms moves one place to the neighboring atom. Physicists prefer to describe this phenomenon as a hole moving in the opposite direction. No matter which description is used, it is a fact that less energy is required to make a material such as silicon conduct if the crystal contains small amounts of either an electron donor such as phosphorus or an electron acceptor such as boron.

Silicates

The earth's crust—the upper 20 miles under the continents and as little as 3 miles under the ocean beds—consists mainly of silicate minerals. The mantle, a layer about 1800 miles thick that lies beneath the crust, is probably composed of dense silicates. The crust is 48 percent oxygen by weight, 26 percent silicon, 8 percent aluminum, 5 percent iron, and 11 percent calcium, sodium, potassium, and magnesium combined.

All silicates are derived from the tetrahedral orthosilicate ion, SiO_4^{4-}, shown in Figure 7-30. Each silicon atom is covalently bonded to four oxygen atoms at the corners of a tetrahedron. The SiO_4^{4-} anion occurs in simple minerals such as zircon ($ZrSiO_4$), garnet, and topaz. Two tetrahedra can share a corner oxygen atom to form a discrete $Si_2O_7^{6-}$ anion, or three tetrahedra can form a ring, shown in Figure 7-31. Benitoite, $BaTiSi_3O_9$, is the best-known example of this uncommon kind of silicate. Beryl, $Be_3Al_2Si_6O_{18}$, a common source of beryllium, has anions composed of rings of six tetrahedra with six shared oxygen atoms.

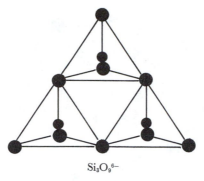

$Si_3O_9^{6-}$

Figure 7-31 A ring of three tetrahedra, with three oxygen atoms shared between pairs of tetrahedra, has the formula $Si_3O_9^{6-}$. This structure occurs as the anion in benitoite, $BaTiSi_3O_9$.

Chain structures. All of the silicates mentioned so far are made from discrete anions. A second class of silicates is made of endless strands or chains of linked tetrahedra. Some minerals have single silicate strands with the formula $(SiO_3)_n^{2n-}$. Asbestos has the double-stranded ladder structure shown in Figure 7-32. These double-stranded chains are held together by electrostatic forces between themselves and the Na^+, Fe^{2+}, and Fe^{3+} cations

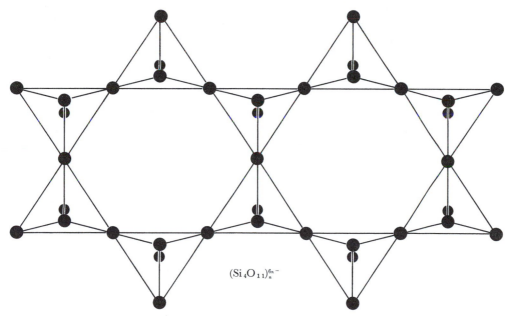

$(Si_4O_{11})_n^{6n-}$

Figure 7-32 Long double-stranded chains of silicate tetrahedra are found in fibrous minerals such as asbestos. The most common form of asbestos has the empirical formula $Na_2Fe_3^{2+}Fe_2^{3+}(Si_8O_{22})(OH)_2$.

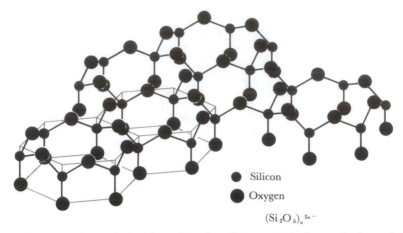

Silicon
Oxygen

$(Si_2O_5)_n^{2n-}$

Figure 7-33 In mica and the clay minerals, silicate tetrahedra each share three of their corner oxygen atoms to make endless sheets. All of the unshared oxygen atoms point down in this drawing on the same side of the sheet.

packed around them. The chains can be pulled apart with much less effort than is required to break the covalent bonds within a chain. Therefore asbestos has a stringy, fibrous texture. Aluminum ions can replace as many as one-quarter of the silicon ions in the tetrahedra. However, each replacement requires one more positive charge from another cation (such as K^+) to balance the charge on the silicate oxygen atoms. The physical properties of the silicate minerals are influenced strongly by the number of Al^{3+} ions that replace Si^{4+} ions, and by the number of extra cations which therefore are needed to balance the charge.

Sheet structure. Continuous broadening of double-stranded silicate chains produces planar sheets of silicate structures (Figure 7-33). Talc, or soapstone, has this planar structure, in which none of the Si^{4+} ions is replaced with Al^{3+}. Therefore no additional cations are required between the sheets to balance charges. The silicate sheets in talc are held together primarily by van der Waals forces. Because of these weak forces the layers slide past one another relatively easily, and produce the slippery feel characteristic of talcum powder.

Mica resembles talc, but one quarter of the Si^{4+} ions in the tetrahedra are replaced by Al^{3+}. Thus, to balance charges, an additional positive charge is required for each replacement. Mica has the layer structure shown in Figure 7-34. The layers of cations (Al^{3+} serves as a cation between layers as well as a substitute in the silicate tetrahedra) hold the silicate sheets together electrostatically with much greater strength than do the van der Waals forces in talc. Thus mica is not slippery to the touch and is not a

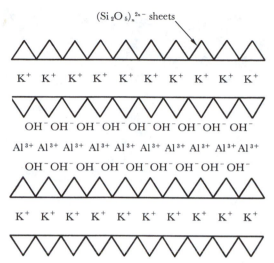

Figure 7-34 In the mica muscovite $[K_2Al_4Si_6Al_2O_{20}(OH)_4]$ anionic sheets of silicate terahedra (see Figure 7-33) alternate with layers of potassium ions and aluminum ions sandwiched between hydroxide ions. This layer structure gives mica its flaky cleavage properties.

good lubricant. However, it cleaves easily, splitting into sheets parallel to the silicate layers. Little effort is required to flake off a chip of mica, but much more strength is needed to bend the flake across the middle and break it.

The clay minerals are silicates with sheet structures like those in mica. These layer structures have enormous "inner surfaces" and often can absorb large amounts of water and other substances between the silicate layers. That is why clay soils are such useful growth media for plants, and also why clays are used as beds for metal catalysts. The common catalyst, platinum black, is finely divided platinum metal obtained by precipitation from solution. The catalytic activity of platinum black is enhanced by its large amount of exposed metal surface. The same enhancement can be achieved by precipitating a metal to be used as a catalyst (Pt, Ni, or Co) onto clays. The metal atoms coat the interior walls of the silicate sheets, but the clay structure prohibits the metal from consolidating into a useless mass.

Three-dimensional networks. The three-dimensional silicate networks, in which all four oxygen atoms of SiO_4^{4-} are shared with other Si^{4+} ions, are exemplified by quartz, $(SiO_2)_n$ (Figure 7-35). In quartz, all of the tetrahedral structures have Si^{4+} ions, but in other network minerals up to half the Si^{4+} can be replaced with Al^{3+}. These minerals include the feldspars, with a typical empirical formula $KAlSi_3O_8$. Feldspars are nearly as hard as quartz.

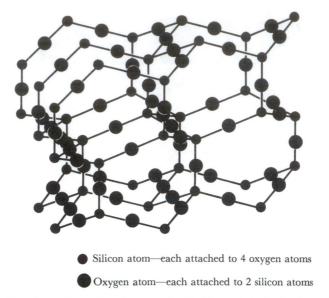

● Silicon atom—each attached to 4 oxygen atoms

● Oxygen atom—each attached to 2 silicon atoms

Figure 7-35 The three-dimensional network of silicate tetrahedra in quartz, SiO_2.

Basalt, which may be the material of the mantle of the earth, is a compact mineral related to feldspar. Granite, the primary component of the earth's crust, is a mixture of crystallites of mica, feldspar, and quartz.

Glasses are amorphous, disordered, noncrystalline aggregates with linked silicate chains of the sort depicted in Figure 7-36. Common soda-lime glass

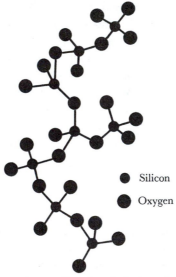

● Silicon

● Oxygen

Figure 7-36 Glasses are amorphous, disordered chains of silicate tetrahedra mixed with metal carbonates or oxides such as Na_2CO_3 or $CaCO_3$.

is made with sand (quartz), limestone ($CaCO_3$), and sodium carbonate (Na_2CO_3) or sodium sulfate (Na_2SO_4), all of which are melted together and allowed to cool. Other glasses with special properties are made by using other metal carbonates and oxides. Pyrex glass has boron as well as silicon and some aluminum in its silicate framework. Glasses are not true solids; rather, they are extremely viscous liquids.

The silicates illustrate all types of bonding in solids: covalent bonding between Si and O in the tetrahedra, van der Waals forces between silicate sheets in talc, ionic attractions between charged sheets and chains, as well as hydrogen bonds between water molecules and the silicate oxygens in clays. If we include nickel catalysts prepared on a clay support, metallic bonding is represented as well.

7-6 Lattice Energies of Ionic Solids

A quantitative measure of the stabilization of an ionic solid is the *lattice energy, U*, which is defined as the energy lowering (that is, release of energy) when one mole of a substance is formed from its constituent gas-phase ions:

$$Na^+(g) + Cl^-(g) \rightarrow NaCl(s) \qquad \Delta E = U.$$

Thus the lattice energy is always negative.

Lattice energies cannot be satisfactorily determined directly by any experimental method; they usually are determined indirectly by use of the so-called Born-Haber energy cycle. For the formation of sodium chloride from the elements sodium and chlorine, the Born-Haber cycle may be depicted as follows:

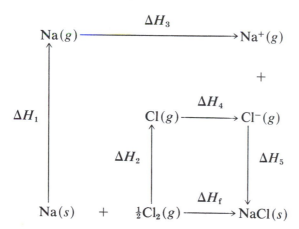

According to Hess's law

$$\Delta H_f = \Delta H_1 + \Delta H_2 + \Delta H_3 + \Delta H_4 + \Delta H_5$$

$$\Delta H_f = \Delta H_{atom}(Na) + \tfrac{1}{2}\Delta H_{atom}(Cl_2) + IE(Na) - EA(Cl) + U \quad (7\text{-}4)$$

in which ΔH_f is the enthalpy of formation of sodium chloride and the other terms have been defined previously.* If we substitute the experimental values of ΔH_f, $\Delta H_{atom}(Na)$, $\Delta H_{atom}(Cl_2)$, $IE(Na)$, and $EA(Cl)$, we obtain an experimental value for U. Writing these values in the same order as in Equation 7-4 we obtain (in kcal mole^{-1})

$$-98.6 = 26.35 + \tfrac{1}{2}(57.84) + 118.49 - 83.3 + U$$

which yields

$$U = -189.1 \text{ kcal mole}^{-1} \qquad\qquad\qquad\qquad\qquad (7\text{-}5)$$

We see that only two terms (EA and U) tend to make the enthalpy of formation negative (exothermic). Of these, the lattice energy is the more important for providing stability.

Calculation of Lattice Energies

The simplest model for the calculation of lattice energies is obtained by treating the crystal as an array of spherical ions, as in the purely ionic model. The two important forces are (a) the coulombic energy of attraction between oppositely charged ions, and (b) the repulsive energy resulting from the interpenetration of the spherically charged clouds of the ions. This repulsive energy often is referred to as the Born repulsion energy when ionic solids are discussed, but its origin is the same as the van der Waals repulsion discussed for the interaction of noble gas atoms (Equation 7-1).

Let us first evaluate the coulombic attractive energy, E_c. Referring to the structure of NaCl in Figure 7-10, we call the shortest Na-Cl distance r. Using simple trigonometry it can be shown that 1 Na$^+$ ion is surrounded by 6 Cl$^-$ ions at a distance r, 12 Na$^+$ ions at a distance $\sqrt{2}r$, 8 more distant Cl$^-$ ions at $\sqrt{3}r$, 6 Na$^+$ ions at $2r$, 24 more Cl$^-$ ions at $\sqrt{5}r$, and so on. The electrostatic interaction energy of the Na$^+$ ion with each of these surrounding ions is equal to the product of the charge on each ion, Z^+ and Z^-, divided

---

*Actually, IE, EA, and U refer to the change in internal energy at 0 K. We should convert these to the corresponding enthalpy change at 298 K. The corrections for IE and EA cancel one another and the correction for U usually is less than 1 kcal mole^{-1} and need not concern us.

TABLE 7-4. VALUES OF THE
MADELUNG CONSTANT FOR
VARIOUS STRUCTURES

Sodium chloride	1.747558
Cesium chloride	1.762670
Zinc blende (ZnS)	1.63806
Wurtzite (ZnS)	1.64132
Fluorite (CaF$_2$)	2.51939*
Rutile (TiO$_2$)	2.409†

*Use $Z^+ = 2$, $Z^- = -1$.
†Use $Z^+ = 4$, $Z^- = -2$. The exact value of the Madelung constant for the rutile structure depends upon details of the structure.

by the distance. Hence the total electrostatic energy of this positive ion can be written as

$$E_c = \frac{6e^2}{r}(Z^+)(Z^-) + \frac{12e^2}{\sqrt{2}r}(Z^+)^2 + \frac{8e^2}{\sqrt{3}r}(Z^+)(Z^-) + \frac{6e^2}{2r}(Z^+)^2$$

$$+ \frac{24e^2}{\sqrt{5}r}(Z^+)(Z^-) + \dots$$

$$= \frac{-e^2|Z|^2}{r}\left(6 - \frac{12}{\sqrt{2}} + \frac{8}{\sqrt{3}} - \frac{6}{2} + \frac{24}{\sqrt{5}} + \dots\right)$$

The sum of the geometrical terms converges to 1.747558 . . . , and is termed the *Madelung* constant, A. The Madelung constant for the commonly occurring lattices shown in Figure 7-11 have been calculated; they are given in Table 7-4. The total electrostatic energy for a mole of NaCl would be NE_c where N is Avogadro's number.

Turning next to the repulsion energy, E_R, Born and Mayer assumed it to have the same form as the van der Waals repulsion energy (Equation 7-1). That is,

$$E_R = be^{-ar} \tag{7-6}$$

The total energy for a mole of the crystal lattice containing an Avogadro's number (N) of units is

$$U = E_c + E_R = \frac{AN(Z^+)(Z^-)e^2}{r} + Nbe^{-ar} \tag{7-7}$$

The relationship between E_c and E_R to produce U is shown in Figure 7-37. In fact, b and a are related, as can be seen by realizing that at the minimum

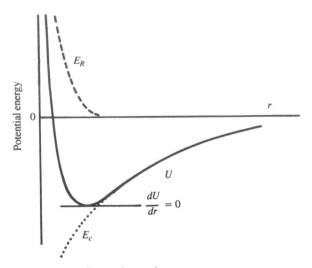

Figure 7-37 Energy curves for an ion pair.

in the energy curve $(r = r_0)$ the slope is zero:

$$\left(\frac{dU}{dr}\right)_{r=r_0} = 0 = \frac{-AN(Z^+)(Z^-)e^2}{r_0^2} - Nabe^{-ar}$$

Thus

$$b = \frac{-AN(Z^+)(Z^-)e^2e^{ar}}{r_0^2 a} \qquad (7\text{-}8)$$

Substituting Equation 7-8 into 7-7, we obtain the *Born-Mayer* equation for calculating lattice energies:

$$U_{r_0} = \frac{AN(Z^+)(Z^-)e^2}{r_0}\left(1 - \frac{1}{r_0 a}\right) \qquad (7\text{-}9)$$

in which U_{r_0} refers to the lattice energy at $r = r_0$. For alkali halide crystals (such as NaCl) the value of the constant a often is given as 2.90×10^8 cm^{-1}. Substituting this value into Equation 7-9 and evaluating at $r_0 = 2.814$ Å gives a calculated lattice energy of

$$U_{r_0} = (-75.66 \times 10^{11}\ \text{erg mole}^{-1})(2.390 \times 10^{-11}\ \text{kcal erg}^{-1})$$

$$= -180.8\ \text{kcal mole}^{-1} \qquad (7\text{-}10)$$

The conversion factor 2.390×10^{-11} kcal erg^{-1} may be calculated by using the conversion factors found in Appendix A:

$$23.060/[(6.0220 \times 10^{23})(1.6021 \times 10^{-12})]$$

The experimental value obtained from the Born-Haber cycle (Equation 7-5) is -189.1 kcal mole^{-1}. Improvements in the calculated value of the lattice energy may be obtained by correcting for the difference between enthalpy and internal energy (usually not more than 1 kcal mole^{-1}). Further corrections arise from weak van der Waals attractive forces (London energy), which are present even in ionic solids, and from the zero-point energy present in the crystalline solid. These minor corrections may be ignored. The main feature to notice is that the calculated and experimental values of the lattice energy are within 4 percent of one another. This indicates that the major contributions to the binding energy in an ionic solid such as sodium chloride are dealt with satisfactorily by the simple electrostatic model incorporating the Born repulsion energy.

QUESTIONS AND PROBLEMS

1. What types of forces hold molecules together in crystals and liquids?

2. What effect do hydrogen bonds have on the boiling temperatures of liquids? Explain, and give an example.

3. Why are nonmetallic network solids usually quite hard?

4. What physical effect is responsible for the attraction in van der Waals interactions? What is responsible for the repulsion in such interactions? Compare the origin of attraction and repulsion in van der Waals interactions with that in ionic and covalent bonds.

5. How do we determine an experimental value for the van der Waals radius of hydrogen? From a theoretical viewpoint, what determines the van der Waals radius?

6. Draw a sketch of the way in which the repulsion part of the van der Waals interaction (Equation 7-1) varies with distance r between atomic centers. Draw a similar sketch for the attraction terms (Equation 7-2). Add these two curves, approximately, and satisfy yourself that a potential curve such as the one of Figure 7-14 is the result.

7. If van der Waals bonds are extremely weak, why are they discussed at all?

8. According to the delocalized molecular-orbital theory of metals, in what sense do we say that the entire piece of metal is a large molecule?

9. Why would beryllium be an insulator if the $2s$ and $2p$ molecular-orbital bands did not overlap?

10. What is the structural difference between metals and insulators?

11. What effect do small amounts of boron or phosphorus have on the conducting properties of silicon?

12. How do hydrogen bonds participate in the structure of ice? What effects do they have on its properties?

13. How do we know that some hydrogen bonding in water persists in the liquid phase?

14. Provide a structural explanation for the fact that quartz is hard, asbestos fibrous and stringy, and mica platelike.

15. Why are clays useful in industrial catalysis?

16. It requires 5.2 eV, or 120 kcal mole^{-1}, to excite electrons in a diamond crystal from the valence band to the conduction band. What frequency of light is needed to bring about this excitation? What wavelength? What wave number? What part of the electromagnetic spectrum does this light correspond to?

17. Using data in this chapter, repeat problem 16 for the semiconductors silicon and germanium.

18. Explain the trend in the melting temperatures of the following tetrahedral molecules: CF_4, 90 K; CCl_4, 250 K; CBr_4, 350 K; CI_4, 440 K.

19. Construct the potential energy curve for the Kr-Kr van der Waals interaction. How strong is the Kr-Kr van der Waals bond? Estimate the Kr-Kr bond distance in solid krypton.

20. The molecule RbBr is held together primarily by an ionic bond. The distance between Rb$^+$ and Br$^-$ in the molecule is 2.945 Å. The closed electron shells of Rb$^+$ and Br$^-$ both have the configuration of the noble gas Kr. From the energy curve constructed in problem 19, estimate the

van der Waals energy between Rb^+ and Br^-, assuming that the energy is the same as for a pair of Kr atoms separated by a distance of 2.945 Å. Tell whether the repulsive part or the attractive part of the interaction dominates. How important is the van der Waals energy compared to the overall bond energy of 91 kcal mole^{-1} in RbBr? Examine the Kr-Kr van der Waals energy for distances of 2 Å and 1 Å and then explain what prevents Rb^+ and Br^- ions from approaching each other too closely in an ionic solid.

21. What type of solid will BF_3 and NF_3 molecules build? What kinds of intermolecular interactions are likely to be important in each solid? Which compound should have the higher melting temperature?

22. The internuclear distances in gas-phase ionic molecules are considerably smaller than those in the corresponding crystals. For example, the internuclear distance in $NaCl(g)$ is 2.3606 Å, whereas in $NaCl(s)$ the shortest Na-Cl distance is 2.814 Å. Explain why this should be so when the binding energy is larger in the crystal.

23. In calculating the lattice energies of ionic solids the repulsion energy, E_R, was expressed by Born and Landé as

$$E_R = B/r^n$$

instead of as the van der Waals expression found in Equation 7-6.
(a) Using $E_R = B/r^n$, derive an expression for U corresponding to the derivation found in the text (Equations 7-7 through 7-9).
(b) Calculate the lattice energy of $NaCl(s)$ employing the expression found for U in part (a). (The value of n often is chosen to be 8 for NaCl.) The shortest Na-Cl distance in $NaCl(s)$ is 2.814 Å.
(c) Compare the value calculated for U in part (b) to that obtained using the van der Waals repulsion expression in calculating the total lattice energy (Equation 7-10).

24. Solid carbon dioxide behaves like a molecular solid (easily compressible, sublimes at 195 K); solid silicon dioxide (quartz, Figure 7-35) is a nonmetallic network solid (very hard, melting point 1883 K). This difference in behavior is surprising at first, since CO_2 and SiO_2 are isovalent molecules. Explain this difference in behavior in terms of the relative abilities of C and Si to form π bonds with O in the molecular species CO_2 and SiO_2.

25. Magnesium oxide crystallizes in the same lattice structure as sodium chloride. The shortest Mg-O distance is 2.10 Å.
(a) Use the value $a = 2.90 \times 10^8$ cm^{-1} and calculate the lattice energy

of magnesium oxide according to Equation 7-9.

(b) The *second* electron affinity of oxygen atoms cannot be measured directly in the gas phase:

$$O^{2-} \rightarrow O^- + e^- \qquad \Delta E = EA(O^{2-})$$

Use the Born-Haber cycle (Equation 7-4), the following data, and the lattice energy calculated in part (a) to obtain a value for the second electron affinity of atomic oxygen.

	ΔH
$O^-(g) \rightarrow O(g) + e^-$	1.47 eV
$O_2(g) \rightarrow 2O(g)$	119.12 kcal mole^{-1}
$Mg(s) \rightarrow Mg(g)$	35.0 kcal mole^{-1}
$Mg(g) \rightarrow Mg^+(g) + e^-$	7.646 eV
$Mg^+(g) \rightarrow Mg^{2+}(g) + e^-$	15.035 eV
$Mg(s) + \frac{1}{2}O_2(g) \rightarrow MgO(s)$	−143.7 kcal mole^{-1}

26. Compare the structures for diamond and graphite shown in Figure 7-5.
    (a) What type of model (metallic, nonmetallic covalent, or van der Waals) best describes the bonding within a layer of the graphite structure?
    (b) What type of model best describes the bonding between layers in the graphite structure?
    (c) Explain why graphite, unlike diamond, is very soft, whereas, like diamond, it has a very high melting point.
    (d) Draw a Lewis electron-dot structure for a fragment of one graphite layer.
    (e) Graphite is a relatively good electrical conductor. Use the Lewis electron-dot structure obtained in part (d) to explain the conductivity of graphite. Which type of electrons (σ or π) do you suspect are mobile and therefore able to conduct electrical current?

27. Magnesium oxide and sodium fluoride are isoelectronic and have the same crystal structure as NaCl. Explain the observation that MgO is nearly twice as hard as NaF and has a much higher melting point than NaF (2800 °C compared to 993 °C).

28. Replacement of the zinc and sulfur atoms by carbon atoms in the wurtzite structure of zinc sulfide (Figure 7-11) results in the diamond structure (Figure 7-5). Optical and electrical measurements on samples of ZnS show that it has a band gap of approximately 3.6 eV. Discuss the nature of ZnS in terms of the covalent network model, the ionic model,

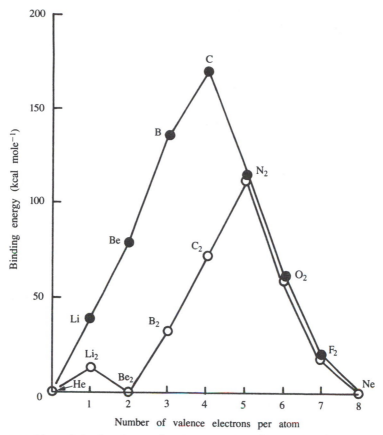

Figure 7-38 Plot of the bond energies (expressed as $\frac{1}{2}A_2 \rightarrow A$) of gaseous homo-nuclear diatomic molecules and the corresponding ΔH_{atom} values of solid state substances as a function of the number of valence electrons.

and the band model—all of which have been applied to describe this solid.

29. Gaseous diatomic molecules have a maximum in the plot of bond energy versus n (number of valence electrons) at $n = 5$, corresponding to N_2. The corresponding binding energy (ΔH_{atom}) plot for elemental solids maximizes at $n = 4$, or $C(s)$ (Figure 7-38). Explain the difference.

30. Explain why the gaseous diatomic molecule Be_2 is unstable, whereas solid Be is stable with $\Delta H_{atom} = 79.5$ kcal mole⁻¹.

31. Explain why the binding energy of solid potassium is much smaller than that of solid vanadium $[\Delta H_{atom}(K) = 26.35$ kcal mole⁻¹, $\Delta H_{atom}(V)$

= 121.9 kcal mole^{-1}]. What is the binding energy per valence electron for potassium? For vanadium?

32. Draw the predicted molecular structure for the hydrogen-bonded dimer of ammonia, $(NH_3)_2$. Label the electron-donor ammonia molecule a and the other b. What angle do you predict for N_b----H_a—N_a? What angle do you predict for H_b—N_b----H_a? Discuss the expected relative importance of the four factors contributing to the energy of the hydrogen bond in $(NH_3)_2$.

33. Draw the predicted molecular shape for the two hydrogen-bonded isomers of hydrogen fluoride with water: H_2O—HF and HF—H_2O.

Suggestions for Further Reading

Brewer, L., "Bonding and Structures of Transition Metals," *Science* 161: 115 (1968).

Cotton, F.A., and Wilkinson, G., *Advanced Inorganic Chemistry,* 3rd ed., New York: Wiley-Interscience, 1972.

Cottrell, A.H., "The Nature of Metals," *Scientific American,* September, 1967.

Dasent, W.E., *Inorganic Energetics,* Middlesex, England: Penguin, 1970.

Deer, W.A., Howie, R.A., and Zussman, J., *Rock Forming Minerals,* vols. 1–5, New York: Wiley, 1962.

Dickerson, R.E., Gray, H.B., and Haight, G.P., Jr., *Chemical Principles,* 3rd ed., Menlo Park, Calif.: Benjamin, 1979.

Gray, H.B., *Chemical Bonds,* Menlo Park, Calif.: Benjamin, 1973.

Hill, T.L., *Matter and Equilibrium,* Menlo Park, Calif.: Benjamin, 1966.

Huheey, J.E., *Inorganic Chemistry: Principles of Structure and Reactivity,* 2nd ed., New York: Harper and Row, 1978.

Ladd, M.F.C., *Structure and Bonding in Solid State Chemistry,* New York: Wiley, 1979.

Moore, W.J., *Seven Solid States,* Menlo Park, Calif.: Benjamin, 1967.

Mott, N., "The Solid State," *Scientific American,* September, 1967.

Murrell, J.N., Kettle, S.F.A., and Tedder, J.M., *The Chemical Bond,* New York: Wiley, 1978.

Phillips, C.S.G., and Williams, R.J.P., *Inorganic Chemistry,* vol. 1, Oxford: Clarendon Press, 1965.

Pimentel, G.C., and Spratley, R.D., *Understanding Chemistry,* San Francisco: Holden-Day, 1971.

Reiss, H., "Chemical Properties of Materials," *Scientific American,* September, 1967.

Umeyama, H., and Morokuma, K., "The Origin of Hydrogen Bonding. An Energy Decomposition Study," *J. Amer. Chem. Soc.* 99: 1316 (1977).

Wells, A.F., *Structural Inorganic Chemistry,* 4th ed., New York: Oxford, 1975.

Appendix A
Physical Constants and Conversion
Factors Used in Text

Constant	Symbol and Value*
Bohr radius	$a_0 = 0.52917706(44) \times 10^{-8}$ cm
Electron mass	$m_e = 9.109534(47) \times 10^{-28}$ g
Electron charge	$e = 4.803242(14) \times 10^{-10}$ esu
Pi	$\pi = 3.1415927$
Planck's constant	$h = 6.626176(36) \times 10^{-27}$ erg sec
Rydberg constant	$R_\text{H} = 109,678.764$ cm^{-1}
(experimental value)	$R_\infty = 109,737.3177(83)$ cm^{-1}
Velocity of light	$c = 2.997924580(12) \times 10^{10}$ cm sec^{-1}
Avogadro's constant	$N = 6.022045(31) \times 10^{23}$ mole^{-1}

Conversion Factors

$$1 \text{ amu} = 1.6605655(86) \times 10^{-24} \text{ g}$$
$$1 \text{ eV} = 1.6021892(46) \times 10^{-12} \text{ erg}$$
$$1 \text{ eV} = 8065.479(21) \text{ cm}^{-1}$$
$$1 \text{ eV particle}^{-1} = 23.060362(65) \text{ kcal mole}^{-1}$$
$$1 \text{ cal} = 4.184 \text{ joule}$$

*The numbers in parentheses represent the uncertainty in the last two significant figures.

The electron volt (eV) is the energy acquired by an electron when it is accelerated by a potential difference of 1 volt.

The erg is the unit of energy in the centimeter-gram-second system of units; 1 erg = 1 g cm^2 sec^{-2}.

We have used several sources in our effort to compile and use the latest values for physical constants, bond lengths, bond dissociation energies, ionization energies, and electron affinities. The principle references we have used are:

Physical constants.
    Cohen, E.R., "The 1973 Table of the Fundamental Physical Constants," *Atomic Data and Nuclear Data Tables* 18: 587 (1976).

Bond lengths and bond energies for main-group diatomic molecules.
    Bourcier, S., ed., *Constantes sélectionnées données spectroscopiques relatives aux molécules diatomiques,* Elmsford, N.Y.: Pergamon, 1970.
    Gray, H.B., *Chemical Bonds,* Menlo Park, Calif.: Benjamin, 1973.

Bond lengths and bond energies for transition-group diatomic molecules.

Cheetham, C.J., and Barrow, R.F., "The Spectroscopy of Diatomic Transition Element Molecules," in *Advances in High Temperature Chemistry,* vol. 1, edited by L. Eyring, New York: Academic, 1967, p. 7.

Ionization energies.

Moore, C.E., "Ionization Potentials and Ionization Limits Derived from the Analyses of Optical Spectra," NSRDS–NBS 34, National Bureau of Standards, Washington, D.C., 1970.

Martin, W.C., Hagan, L., Reader, J., and Sugar, J., "Ground Levels and Ionization Potentials for Lanthanide and Actinide Atoms and Ions," *J. Phys. Chem. Ref. Data* 3: 771 (1974).

Electron affinities.

Chen, E.M., and Wentworth, W.E., "The Experimental Values of Atomic Electron Affinities," *J. Chem. Educ.* 52: 486 (1975).

Appendix B
The Greek Alphabet

Alpha	A	α
Beta	B	β
Gamma	Γ	γ
Delta	Δ	δ
Epsilon	E	ϵ
Zeta	Z	ζ
Eta	H	η
Theta	Θ	θ
Iota	I	ι
Kappa	K	κ
Lambda	Λ	λ
Mu	M	μ
Nu	N	ν
Xi	Ξ	ξ
Omicron	O	o
Pi	Π	π
Rho	P	ρ
Sigma	Σ	σ
Tau	T	τ
Upsilon	Y	υ
Phi	Φ	ϕ
Chi	X	χ
Psi	Ψ	ψ
Omega	Ω	ω

Appendix C
Answers to Selected
Questions and Problems

Chapter 1

1. A) (d)
   B) (b)
   C) (b)

2. (a) First
   (b) First
   (c) Second
   (d) First
   (e) First

3. $IE = 0.544$ eV

4. $\lambda = 468.65$ nm

5. $\lambda = 450$ nm
   $E = 4.4 \times 10^{-12}$ erg
   $\bar{\nu} = 2.2 \times 10^4$ cm^{-1}

6. 1.89 eV

7. $\nu = 4.57 \times 10^{15}$ sec^{-1}
   $\bar{\nu} = 1.52 \times 10^5$ cm^{-1}

8. (a) $1s^2 2p^1 \rightarrow 1s^2 2s^1$
   (b) $\nu = 4.469 \times 10^{14}$ sec^{-1}
   (c) $\bar{\nu} = 1.491 \times 10^4$ cm^{-1}
   (d) $E = 42.62$ kcal mole^{-1}

9. (a) excited, (b) forbidden, (c) ground, (d) excited, (e) excited, (f) excited

10. $P(Z = 15)$; $1s^2 2s^2 2p^6 3s^2 3p^3$ [the three $3p$ electrons are unpaired]
    $Na(Z = 11)$; $1s^2 2s^2 2p^6 3s^1$
    $As(Z = 33)$; $1s^2 2s^2 2p^6 3s^2 3p^6 4s^2 3d^{10} 4p^3$ [the three $4p$ electrons are unpaired]
    $C^-(Z = 6)$; $1s^2 2s^2 2p^3$ [the three $2p$ electrons are unpaired]
    $O^+(Z = 8)$; $1s^2 2s^2 2p^3$ [the three $2p$ electrons are unpaired]

11. $C(Z = 6)$, 2 unpaired e's
    $F(Z = 9)$, 1 unpaired e
    $Ne(Z = 10)$, no unpaired e's

13. $n = 3$, $l = 2$, $m_l = 2, 1, 0, -1$, or -2

14. $Ca (Z = 20)$; $1s^2 2s^2 2p^6 3s^2 3p^6 4s^2$
    $Mg^{2+} (Z = 12)$; $1s^2 2s^2 2p^6$

15. $n = 7 \rightarrow n = 6$

16. Noble gases: 2, 6, and 12
    Halogens: 5, 11, and 19

18.

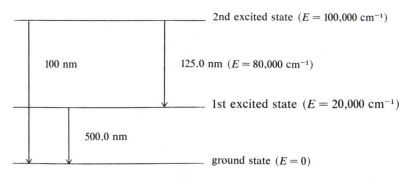

19. In multielectron atoms the effects of shielding and electron-electron repulsion cause the $4s$, $4p$, $4d$, and $4f$ orbitals to have different energies.

20. 5 or higher $\rightarrow$ 4 (Brackett)
    $4 \rightarrow 3$ (Paschen)
    $3 \rightarrow 2$ (Balmer)
    $2 \rightarrow 1$ (Lyman)
    The atom cannot emit successive photons in reverse order, because Lyman emission (2 or higher $\rightarrow$ 1) terminates in the ground state.

21. (a) $\nu = 5.13 \times 10^{15}$ sec^{-1}
    (b) $E = 3.40 \times 10^{-11}$ erg
        $E = 21.2$ eV
    (c) 5.4 eV

22. (a) 9.18 eV
    (b) 8.17 eV
    (c) 4.70 eV
    (d) The $4s$ electron has a lower ionization energy than the $3d$ electron.

23. (a) Si, ^3P
    (b) Mn, ^6S
    (c) Rb, ^2S
    (d) Ni, ^3F

24. The ground state is listed first in each case.
    (a) $2s$; ^2S
    (b) $2p^2 3s$; ^4P, ^2D, ^2S
    (c) $2p3p$; ^3D, ^1D, ^3P, ^1P, ^3S, ^1S
    (d) $3d^2$; ^3F, ^3P, ^1G, ^1D, ^1S
    (e) $2p3d$; ^3F, ^1F, ^3D, ^1D, ^3P, ^1P
    (f) $3d^9$; ^2D
    (g) $2s4f$; ^3F
    (h) $2p^5$; ^2P
    (i) $3d^2 4s$; ^4F, ^4P, ^2G, ^2D, ^2S

Chapter 2

1.   *(unpaired e's)*

Na; $1s^2 2s^2 2p^6 3s$ Na·   (1)

Si; $1s^2 2s^2 2p^6 3s^2 3p^2$:Si·   (2)

P; $1s^2 2s^2 2p^6 3s^2 3p^3$:P̈·   (3)

S; $1s^2 2s^2 2p^6 3s^2 3p^4$:S̈·   (2)

2. Based on the trends in ionization energy observed for the Group V, VI, and VII elements, we expect astatine to have an ionization energy about 1 eV less than iodine. This would place the predicted ionization energy at about 9.5 eV.

3. The electron affinity of silicon is larger than that of phosphorus because the anion of phosphorus is destabilized by *e–e* repulsion in the 3*p* orbital that contains two electrons. The electron affinity of sulfur is larger than that of phosphorus because sulfur has a larger effective nuclear charge than phosphorus.

4. For the *second* ionization energy, the noble gas electronic configuration occurs for Li^+, Na^+, and K^+.

5. 13.598 eV $(8065.479$ cm$^{-1}/1$ eV$) = 109{,}674$ cm^{-1}
    13.598 eV $(23.060$ kcal mole$^{-1}/1$ eV$) = 313.6$ kcal mole^{-1}
    0.754 eV $(8065.479$ cm$^{-1}/1$ eV$) = 6081$ cm^{-1}
    0.754 eV $(23.060$ kcal mole$^{-1}/1$ eV$) = 17.4$ kcal mole^{-1}

6. The relative effective radii are predicted to be $H^- > He > Li^+$. For isoelectronic species the effective radius decreases as Z increases.

7. $1s^2 2s^2 2p^6$. Relative sizes $N^{3-} > O^{2-} > F^- > Ne > Na^+$

8. The 12 values of IE_m/m are:
    7.646, 7.518, 26.71, 27.31, 28.25, 31.08, 32.13, 33.24, 36.44, 36.75, 160.16, 163.55. The values fall into three distinct groups around 7, 30, and 160 eV. The number of ionizations in each group are 2, 8, and 2, respectively. These results are consistent with the known electronic configuration of Mg.

9. (a) Cu (b) K (c) Cl (d) H (e) Ge

11. 218 kcal mole^{-1}

12. BeO : 212 kcal mole^{-1}
    CaO : 164 kcal mole^{-1}

13. Pauling : 2.20
    Mulliken : 2.76

14. According to the Mulliken definition of electronegativity
    $EN = c(IE + EA)$
    For fluorine:
        $IE_F = 17.422$ eV $EA_F = 3.34$ eV
        $IE_F + EA_F = 20.76$ eV
    For neon:
        $IE_{Ne} = 21.564$ eV $EA_{Ne} = 0.0$ eV
        $IE_{Ne} + EA_{Ne} = 21.564$ eV
    Since $(IE_{Ne} + EA_{Ne}) > (IE_F + EA_F)$
    we must have $EN_{Ne} > EN_F$
    The large EN for Ne results from its large IE.

21.

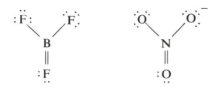

These species are isoelectronic. BF_3 and NO_3^- are predicted to be trigonal planar.

22. $:C\equiv N:$ $:C\equiv O:^+$
C—O bond is shorter in CO than in CO_2 (triple bond *vs.* double bond).

24. H—Ö: H—F̈:
          |
          H
Both molecules have dipole moments.

25. $[:N\equiv C-\ddot{\underset{..}{S}}:]^- \leftrightarrow [:\ddot{\underset{..}{N}}=C=\ddot{S}:]^-$

27. (a) 18 (b) 1 $(5g^{18})$ (c) $5g^7$ and $5g^{11}$ (d) At least 9 $(5g^9)$ and possibly $10(5g^97d^1)$, in either case a new record. (e) 0.544 eV $= IE(5g$ of H). This is smaller than the *IE* of a 5g electron in one of the hypotransition elements, because Z_{eff} will be much greater than the $Z=1$ of hydrogen. (f) Screening effects will lead to an energy order $5s < 5p < 5d < 5f < 5g$.

28. One of the resonance structures for $(HO)_5IO$ will exhibit a double bond for the I-O. bond. The I-OH bonds will consist of single bonds only.

29. (a) CS_2;   linear, nonpolar
     (b) SO_3;   trigonal planar, nonpolar
     (c) ICl_3;   T-shaped, polar
     (d) BF_3;   trigonal-planar, nonpolar
     (e) CBr_4;   tetrahedral, nonpolar
     (f) SiH_4;   tetrahedral, nonpolar
     (g) SF_2;   angular, polar
     (h) SeF_6;   octahedral, nonpolar
     (i) PF_3;   trigonal pyramidal, polar
     (j) ClO_2;   angular, polar
     (k) IF_5;   square pyramidal, polar
     (l) OF_2;   angular, polar
     (m) H_2Te; angular, polar

30. (a) CO_2 NO_2^+ NO_2 NO_2^- SO_2
           linear   linear   angular   angular   angular
           nonpolar   —   polar   —   polar
     (b)   0   0   1   0   0
     (c)   180°   180°   > 120°   < 120°   < 120°
     (d) Relative N–O bond lengths
          $NO_2^- > NO_2 > NO_2^+$

31. XeF_4, square planar
     XeO_4, tetrahedral
     XeF_2, linear
     $XeOF_2$, T shape

32. AsH_3, trigonal pyramidal
     ClF_3, T shape
     $SeCN^-$, linear

33. (a) SO_3^{2-}, trigonal pyramidal
     (b) SO_3, trigonal planar
     (c) BrF_5, square pyramidal
     (d) I_3^-, linear
     (e) CH_3^+, trigonal planar
     (f) CH_3^-, trigonal pyramidal
     (g) PCl_3F_2, trigonal bipyramidal with chlorine in the equatorial positions.

34. BrF_5 only.

35.

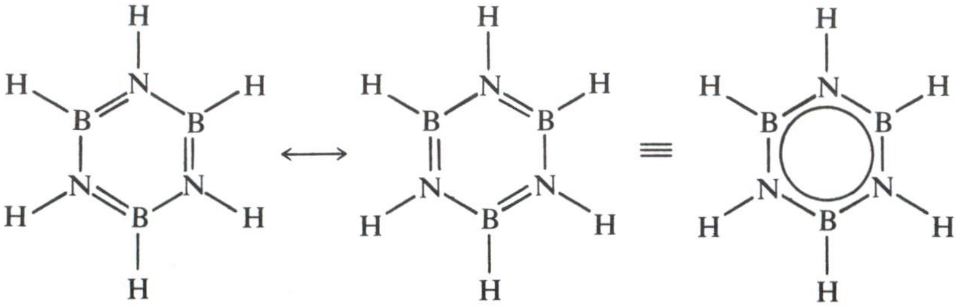

36. An equal mixture of $(NH_4)_3SbCl_6$ and $(NH_4)SbCl_6$ is expected because these compounds contain Sb^{3+} and Sb^{5+}, both of which are expected to result in diamagnetic compounds.

37. Trigonal planar, trigonal pyramidal, and T shape structures are found for three coordinate molecules. These structures have zero, one, and two lone electron pairs on the central atom, respectively.
     Examples: BCl_3, PCl_3, $BrCl_3$
                  nonpolar polar   polar

38. (a) TeF_2
     (b) TeF_3^-
     (c) TeF_3^+
     (d) TeF_4
     (e) TeF_4^{2-}
     (f) TeF_5^-
     (g) TeF_5^+
     (h) TeF_6

39. (a) AlF_6^{3-}, octahedral
     (b) TlI_3^{2-}, VSEPR predicts seesaw – the experimental structure is tetrahedral. This is a good place to discuss the so-called "inert $6s^2$ pair."
     (c) $GaBr_4$, tetrahedral
     (d) NO_3^-, trigonal planar
     (e) NCO^-, linear
     (f) CNO^-, linear
     (g) $SnCl_3^-$, trigonal pyramidal
     (h) $SnCl_6^{2-}$, octahedral

40. The ion XeF_5^+ is predicted to have a square pyramidal structure.

41. (a) OCCCO, linear
     (b) ONF, angular
     (c) NSF, angular
     (d) ClClF⁺, angular

(e)

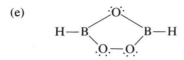

$$H-B \underset{\overset{\cdot\cdot}{\underset{\cdot\cdot}{O}}\:-\:\overset{\cdot\cdot}{\underset{\cdot\cdot}{O}}}{\overset{\overset{\cdot\cdot}{O}\overset{\cdot\cdot}{\cdot}}{\diagdown}} B-H$$

(f)

$$\overset{\cdot\cdot}{\underset{\cdot\cdot}{O}} \diagup \overset{\overset{\cdot\cdot}{O}\overset{\cdot\cdot}{\cdot}}{\diagdown} B \diagdown \overset{\cdot\cdot}{\underset{\cdot\cdot}{O}}$$

(g) $\overset{\cdot\cdot}{\underset{H}{N}}=C=\overset{\cdot\cdot}{O}:$

(h)

$$\overset{\overset{\cdot\cdot}{O}}{\diagdown} \overset{N}{\diagup} \overset{\overset{\cdot\cdot}{O}:}{\cdot} $$
$$:\overset{\cdot\cdot}{\underset{H}{O}} $$

(i)

$$H-\overset{\cdot\cdot}{\underset{\cdot\cdot}{O}} \diagdown \overset{\overset{\cdot\cdot}{O}\overset{\cdot\cdot}{\cdot}}{\diagup} $$
$$:\overset{\cdot\cdot}{\underset{\cdot\cdot}{O}}-N \diagdown \overset{\cdot\cdot}{\underset{\cdot\cdot}{O}}: $$

42. (a) PF_3; E, C_3, $3\sigma_v$
     (b) SO_3; E, C_3, $3C_2$, σ_h, $3\sigma_v$, S_3
     (c) H_2Te; E, C_2, $2\sigma_v$
     (d) IF_5; E, C_4, C_2, $4\sigma_v$
     (e) ClF_3; E, C_2, $2\sigma_v$
     (f) XeF_4; C_4, $5C_2$, i, S_4, σ_h, $4\sigma_v$
     (g) $(BHNH)_3$; E, C_3, $3C_2$, σ_h, $3\sigma_v$, S_3

43. In H_2O the bond dipoles are *enhanced* by the lone pair dipoles as shown in
     Figure 2-16. In F_2O the bond dipoles will point in the opposite direction as for
     H_2O. Consequently, the bond dipoles and lone pair dipoles tend to cancel each
     other. The result is a small dipole moment for F_2O.

44.

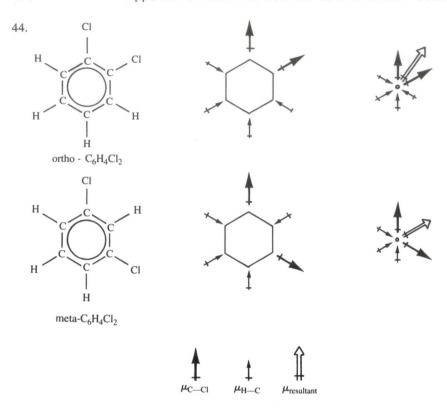

ortho - $C_6H_4Cl_2$

meta-$C_6H_4Cl_2$

μ_{C-Cl} μ_{H-C} $\mu_{resultant}$

$\mu_{ortho} = 2(\mu_{CCl} + \mu_{CH}) \cos 30° = 2.50$ D
Therefore $(\mu_{CCl} + \mu_{CH}) = 2.50/[(2)(0.866)] = 1.44$ D
$\mu_{meta} = 2(\mu_{CCl} + \mu_{CH}) \cos 60°$
$\qquad = 2(1.44$ D$)(0.5) = 1.44$ D
The calculated value of 1.44 D may be compared to the experimental value of 1.72 D.

Chapter 3

1. (a) CS_2, sp
    (b) SO_3, sp^2
    (c) BF_3, sp^2
    (d) CBr_4, sp^3
    (e) SiH_4, sp^3
    (f) SeF_6, d^2sp^3

    (g) SiF_5^-, dsp^3
    (h) AlF_6^{3-}, d^2sp^3
    (i) PF_4^+, sp^3
    (j) IF_6^+, d^2sp^3
    (k) NO_2^+, sp
    (l) NO_3^-, sp^2

2. :N=N=N· , linear. The central nitrogen atom is sp hybridized. There are two pairs of σ bonding electrons and two pairs of π bonding electrons. There is one unpaired electron. The molecule is nonpolar. The N–N bond length in N_2 should be shorter than in N_3 since N_2 has a triple bond.

4. $C_3H_5^+$:

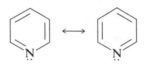

Each C atom is sp^2 hybridized. Two electrons reside in a π orbital. A summary representation for the allyl cation is

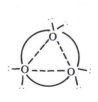

5. (a) $N = \pm(1 + \gamma^2)^{-1/2}$

6. (a) $\phi_a = 0.436\ 2p_z + 0.707\ 2p_y + 0.556\ 2s$
$\phi_b = 0.436\ 2p_z - 0.707\ 2p_y + 0.556\ 2s$
$\phi_c = -0.786\ 2p_z + 0.618\ 2s$
(b) ϕ_a and ϕ_b : 19% $2p_z$, 50% $2p_y$, 31% $2s$
ϕ_c : 62% $2p_z$, 38% $2s$

7.

8. (a) $\phi = 0.830\ 2p + 0.557\ 2s$
(b) 69% $2p$, 31% $2s$
(c) $\phi_{lp} = 0.788\ 2p - 0.616\ 2s$
62% $2p$, 38% $2s$

9. (a) No, because $\cos \theta$ is positive if $\theta < 90°$ and therefore γ would be an imaginary number.
(b)

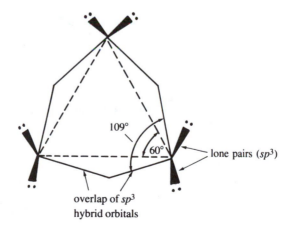

109°

60°

lone pairs (sp^3)

overlap of sp^3
hybrid orbitals

10.

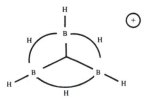

(b) The molecule is predicted to be bent because one of the H atoms would bond to ϕ_+ and one would bond to ϕ_-.

11. $\alpha = \frac{1}{2}(b + h - 3q)$
$\beta = b + q$
For $B_3H_6^+$, $b = 3$, $h = 6$, $q = 1$
$\alpha = \frac{1}{2}(3 + 6 - 3) = 3$
$\beta = 3 + 1 = 4$
Therefore, a suitable structure must have 4 three-center bonds (B-H-B or B-B-B) and 3 two-center bonds (B-H or B-B).

12. For $B_3H_8^-$, $b = 3$, $h = 8$, $q = -1$
$\alpha = \frac{1}{2}(3 + 8 + 3) = 7$
$\beta = 3 - 1 = 2$
Therefore, any suitable structure must have 2 three-center bonds (B-H-B or B-B-B) and 7 two-center bonds (B-H or B-B).
Possible structures:

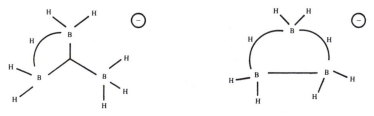

15. $\beta = q$

$\alpha = \frac{1}{2}(4c + h - 3q)$

Chapter 4

1. (c) would be highest, (b) would be lowest (and paramagnetic).

2. $E(\sigma_g^b) = \dfrac{q + \beta}{1 + S}$

$E(\sigma_u^*) = \dfrac{q - \beta}{1 - S}$

The energy of stabilization of the bonding orbital, ΔE_g, can be calculated as follows:

$\Delta E_g = q - \dfrac{q + \beta}{1 + S} = \dfrac{qS - \beta}{1 + S}$

Likewise,

$\Delta E_u = \dfrac{q - \beta}{1 - S} - q = \dfrac{qS - \beta}{1 - S}$

Because S is positive,

$\Delta E_g = \dfrac{qS - \beta}{1 + S} < \dfrac{qS - \beta}{1 - S} = \Delta E_u$

3. Cl_2 : Bond order $= 1$
   Cl_2^+ : Bond order $= 1\frac{1}{2}$

4. $[Hg_1{-}Hg_2]^{2+}$

$\sigma_g^b = \dfrac{1}{\sqrt{2}}(6s_a + 6s_2)$

$\sigma_u^* = \dfrac{1}{\sqrt{2}}(6s_1 - 6s_2)$

Ground state Hg_2^{2+} is $(\sigma_g^b)^2$.
The strong ultraviolet absorption is probably due to the $\sigma_g^b \rightarrow \sigma_u^*$ transition.

6. (a) $^3\Sigma_g^-$, (b) $^1\Sigma_g^+$, (c) $^1\Sigma_g^+$, (d) $^3\Sigma_g^-$, (e) $^2\Sigma$, (f) $^1\Sigma$

7. O_2 has two unpaired valence electrons. In O_2^+, the unpaired core electron can either have its spin parallel or anti-parallel to the unpaired valence electrons.

8. Since C_2 has only $2s$ and $2p$ AO's available for bonding, both $\sigma_g^b(2s)$ and $\sigma_u^*(2s)$ are occupied, resulting in two lone electron pairs. However, Mo_2 has $4d$, $5s$, and $5p$ AO's and no antibonding MO's need to be occupied in the MO scheme.

11. $(1\sigma)^2(2\sigma)^2(3\sigma)^2(1\pi)^1$
    The 1π orbital is localized on the O atom. The lowest electronic transition should be $3\sigma \rightarrow 1\pi$. This transition should be at lower energy than that of OH$^-$, since the $1\pi \rightarrow 4\sigma$ transition in OH$^-$ involves promotion of an electron to the *antibonding* 4σ orbital.

13. The highest occupied molecular orbital in OH and HF is the 1π MO. Since this MO is nonbonding and localized on the O and F atoms, the IE's of the molecules should mimic the VOIE's of the atoms.

15. The highest occupied molecular orbital of SO would be the 2π orbital, so that the molecule should exhibit two unpaired electrons just as does O_2 and S_2.

17. (a) $[:C=\ddot{F}:]^+ \leftrightarrow [:C\equiv F:]^+$
     $:\dot{C}=\ddot{F}:$
     The Lewis dot structures predict a bond order of 2–3 for CF^+ and of 2 for CF.
     (b) $CF^+ : (1\sigma)^2(2\sigma)^2(3\sigma)^2(1\pi)^4(2\pi)^0$
     The CF molecule has one electron in the 2π (antibonding) MO.

19. The highest occupied molecular orbital of BF and CO consists mainly of B and
     C atomic orbital composition, respectively. Therefore, the trends in the first
     ionization energies of the molecules should follow the trends in the atomic
     VOIE's (Table 4-4).

22. The added electron in O_2 and NO must occupy the antibonding 2π molecular
     orbital. However, for CN and C_2 the added electron occupies the bonding 3σ
     molecular orbital.

24. $Cl^{\delta+}F^{\delta-}$

25. The 3σ orbital is antibonding. ScO^+ should have a shorter bond than ScO.

Chapter 5

1. (a) $\dfrac{1}{\sqrt{2}}(1s_b - 1s_c)$
    $\sqrt{\tfrac{2}{3}}(1s_a - \tfrac{1}{2}1s_b - \tfrac{1}{2}1s_c)$
    (b) antibonding
    (c) degenerate, $E = \alpha - \beta$

2. (a) paramagnetic
    (b) σ_u^*
    (c) linear
    (d) $\tfrac{1}{2}$

5. The additional electron in N_3 occupies the π_g^{nb} nonbonding molecular orbital.
    The electron affinity of N_3 is high because the additional electron is able to
    delocalize over both ends of the molecule, thus reducing e-e repulsion.

11. Consider the molecular axis to be the z axis and label $2p_{x_i}$ as x_i, etc.
    (a) We will assume here that the coefficients of the central and terminal orbitals
        in each MO are equal.
        $1\pi_x = \tfrac{1}{2}(x_1 + x_2 + x_3 + x_4)$
        $1\pi_y = \tfrac{1}{2}(y_1 + y_2 + y_3 + y_4)$
        $2\pi_x = \tfrac{1}{2}(x_1 + x_2 - x_3 - x_4)$
        $2\pi_y = \tfrac{1}{2}(y_1 + y_2 - y_3 - y_4)$
        $3\pi_x = \tfrac{1}{2}(x_1 - x_2 - x_3 + x_4)$
        $3\pi_y = \tfrac{1}{2}(y_1 - y_2 - y_3 + y_4)$
        $4\pi_x = \tfrac{1}{2}(x_1 - x_2 + x_3 - x_4)$
        $4\pi_y = \tfrac{1}{2}(y_1 - y_2 + y_3 - y_4)$
    (b) $1\pi_{x,y}$ is ungerade, bonding $2\pi_{x,y}$ is gerade, bonding between C_1—C_2 and
        C_3—C_4 but antibonding between C_2—C_3. The other orbitals can be
        described similarly.
    (c) $C_4 : (1\pi_{x,y})^4(2\pi_x)^1(2\pi_y)^1$ paramagnetic
        $C_2N_2 : (1\pi_{x,y})^4(2\pi_{x,y})^4$ diamagnetic
    (d) Terminal : 2
        Central : $1\tfrac{1}{3}$
    (e) Terminal : $2\tfrac{1}{3}$
        Central : 1

13. (a) With $5s : \frac{1}{2}(1s_a + 1s_b + 1s_c + 1s_d)$

$$5p_x : \frac{1}{\sqrt{2}}(1s_a - 1s_c)$$

$$5p_y : \frac{1}{\sqrt{2}}(1s_b - 1s_d)$$

$$5d_{x^2 - y^2} = \frac{1}{2}(1s_a - 1s_b + 1s_c - 1s_d)$$

15. Ethane should have a higher IE than ethylene because all electrons in ethane are in σ bonds. Ethylene has a π bonding pair which should be more readily ionized.

16. Benzene should absorb at lower energies than ethylene. The π levels are more closely spaced in benzene (the energy of the $\pi \to \pi^*$ promotion is less).

17. Each of these molecules has two types of fluorine atoms so they should exhibit two peaks in the $F(1s)$ photoelectron spectrum. Since the "axial" F atoms are generally considered to be more negative than the equatorial, these $F(1s)$ electrons should ionize at lower energy. The intensity ratio of the peaks should correspond to the relative number of F atoms of each type.

18. In each of these molecules the proton attacks a lone electron pair on the oxygen atom, thus inducing a positive charge on the oxygen atom. Correspondingly, ionization of a core $O(1s)$ electron also induces a positive charge on the oxygen atom. As the binding energy of the oxygen $1s$ orbital decreases, the oxygen atom becomes more negative and hence the proton is more attracted to it.

19. $ClClF^+$

21. (a) The $N(1s)$ photoelectron spectrum should exhibit two peaks with the central nitrogen atom occurring at higher binding energy.
     (b) The $O(1s)$ photoelectron spectrum should exhibit two peaks with the doubly bonded O atom occurring at higher binding energy.

23. All three are forbidden reactions.

25. The PA of CH_4 is much less than that of H_2O because the proton is attacking a bonding orbital in CH_4 and a lone pair orbital in H_2O.

27. (a) CO will be a better electron donor than N_2.
     (b) CO will be a better donor than N_2 since the 3σ orbital of CO is predominantly localized on the C atom. The $2\sigma_g$ orbital of N_2 is delocalized over both atoms.

Chapter 6

1. $O_3CrOCrO_3{}^{2-}$

5. $(CO)_4CoCo(CO)_4$, a metal-metal bond is required.

6. There are numerous possibilities for each of these, some of which are given here.
    (a) $Fe(CO)_4NO^+$
        $Fe(CO)_2(NO)_2$
        $Fe(CO)_3NO^-$
        $Fe_2(CO)_3(NO)_4$ with one bridging CO.
        $Fe_2(CO)_6(NO)_2$ with two bridging CO's.

7. Assume that these atoms lie along the z axis and that the ligand field around each Ru^{4+} ion is essentially octahedral. The Ru^{4+} ions will then each have a $d_{xy}^2 d_{xz}^1 d_{yz}^1$ electronic configuration before reaction with the oxygen. A sigma bond could be formed by the O $2s$ orbital and a d^2sp^3 hybrid on each of the Ru^{4+} ions. Pi bonds can form as follows in the xz plane

    $1\pi_{xz} = xz_1 + O(p_x) + xz_2$, ungerade
    $2\pi_{xz} = xz_1 - xz_2$, gerade

    Each of these orbitals is occupied by 2 electrons, as are the corresponding $1\pi_{yz}$ and $2\pi_{yz}$ orbitals. Thus there are eight π type electrons occupying four π molecular orbitals, and the complex should be diamagnetic. These 8 electrons are obtained from the d_{xz} and d_{yz} orbitals of the Ru^{4+} ions and from the O $2p_x$ and $2p_y$ orbitals.

    A similar bonding description applies to the nitrogen complex.

9. Br^- should displace the halide ions in $PdCl_4^{2-}$ and $PtCl_4^{2-}$.

11. True.

13. $[Cu(NH_3)_{4-y}(Cl)_y][PtCl_{4-y}(NH_3)_y]$
     where $y = 0, 1, 2, 3$, and 4.

17. No d-d transitions are possible for a closed-shell d^{10} complex.

19. $V(CO)_6 : (t_{2g})^5$
     $V(CO)_6^- : (t_{2g})^6$
     Predicted stability $Cr(CO)_6 > V(CO)_6 \gg Mn(CO)_6$ because $Mn(CO)_6$ would be $(t_{2g})^6(e_g^*)^1$.

21. 300 nm light has an energy of 95.3 kcal mole^{-1}; this is the upper limit on the W-CO bond energy. There is no possibility that this wavelength of light could dissociate CO since the CO bond energy is 256 kcal mole^{-1}.

23. (a) $Co(NH_3)_5I^{2+}$, (b) $Co(NH_3)_5Cl^{2+}$, (c) $Pd(NH_3)_4^{2+}$, (d) $Co(CN)_6^{3-}$,
     (e) $Co(CN)_5I^{3-}$, (f) $V(H_2O)_6^{2+}$, (g) $RhCl_6^{3-}$, (h) $Ni(H_2O)_6^{2+}$

25. Charge transfer $CN^- \rightarrow t_{2g}[Fe(CN)_6^{3-}$ is $(t_{2g})^5]$.

26. $M \rightarrow L$ lower in $Cr(CO)_6$ than in $Mn(CO)_6^+$.
     It is easier to excite an electron from Cr^0 than from Mn^{+1}.

28. $(d_{z^2})^2(d_{xz},d_{yz})^4(d_{xy})^2(d_{x^2-y^2})^1$. Three d-d transitions are expected.

29. The CN^- ion has an empty π^* orbital that can bond with the metal d_{xy} orbital (back bonding). This will stabilize the π_{xy}^* ligand-field orbital.

31. Tetrahedral shape with a ground state $(e)^4(t_2)^6$. No d-d transitions are possible for a d^{10} complex.

33.
		unpaired e's
MnO_4^{3-}	tetrahedral	2
$Pd(CN)_4^{2-}$	square planar	0
NiI_4^{2-}	tetrahedral	2
$Ru(NH_3)_6^{3+}$	octahedral	1
$MoCl_6^{3-}$	octahedral	3
$IrCl_6^{2-}$	octahedral	1
$AuCl_4^-$	square planar	0
FeF_6^{3-}	octahedral	5

35. Mn(VII) *vs.* Cr(VI). $L \rightarrow M$ charge transfer decreases in energy as the oxidation no. of M increases.

37. The proton should attack the ligand field t_2 orbital. This orbital is mainly based on the Ni atom.

39. $V(C_6H_6)_2^+$: Zero unpaired electrons.
    $Cr(C_6H_6)_2^+$: One unpaired electron.

41. $Co(C_5H_5)_2$ has an electron count of 19 around the Co atom.

43. Using the results given in Table 6-10:
    $e_{z^2} = 2.75\ \beta S_\sigma^2$
    $e_{xy} = e_{x^2-y^2} = 1.125\ \beta S_\sigma^2$
    $e_{xz} = 0$
    $e_{yz} = 0$

45. $CuCl_6^{4-}$

48. $54.74°$

Chapter 7

1. (a) Electrostatic attraction (in ionic crystals)
   (b) Sharing of electrons (covalent bonds)
   (c) Induced dipole interactions (van der Waals forces)
   (d) Dipolar interactions, an important special case being hydrogen bonds.
   (e) Delocalization of electrons (in metals).

2. Liquids with hydrogen bonding have higher boiling temperatures than would be expected otherwise.

5. The van der Waals radius of hydrogen can be determined by measuring the *intermolecular* H—H distance in solid hydrogen. From a theoretical view-point, the van der Waals radius is determined by the interplay of the attractive and repulsive forces in the van der Waals equation.

6.

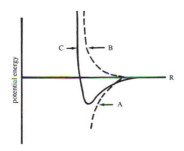

A = potential energy resulting from attractive force
B = potential energy resulting from repulsive force
C = potential energy resulting from net force

7. They are weak, but there are many of them; thus the total effect is significant.

8. The molecular orbitals form bands that are spread over the entire metallic fragment.

9. If the $2s$ and $2p$ molecular-orbital bands of beryllium did not overlap, the $2s$ band would be completely occupied and there would be a forbidden zone before the onset of the $2p$ band.

10. Metals have high coordination numbers (8 to 12) whereas insulators have low coordination numbers (~ 4).

11. Silicon becomes a better conductor if small amounts of boron or phosphorus are added.

16. $E = 8.33 \times 10^{-12}$ erg
     $\nu = 1.3 \times 10^{15}$ sec^{-1}
     $\lambda = 2.3 \times 10^{-5}$ cm
     $\bar{\nu} = 4.3 \times 10^4$ cm^{-1}
     Ultraviolet

18. Van der Waals forces increase from CF_4 to CI_4, thereby increasing the energy required to break intermolecular bonds.

19. $$PE = be^{-ar} - \frac{d}{r^6}$$

     $$= 65.2 \times 10^3 \, e^{-1.61 r} - \frac{125.4 \times 10^3}{r^6}$$

     From a plot of potential energy versus r,
       strength of bond = 335 cal mole^{-1}
         bond distance = 7.6 au $\times$ 0.529 Å au^{-1} = 4.0 Å

20. Van der Waals energy is repulsive (+4.14 kcal mole^{-1}).
     The van der Waals energy contribution reduces the bond energy by about 5%.
     van der Waals energy at 2 Å $\simeq$ 105 kcal mole^{-1}
                          at 1 Å $\simeq$ 360 kcal mole^{-1}
     The calculations show that the closed-shell repulsions are very large at short distances. Therefore the ions cannot be "pushed together."

21. Molecular solids. For BF_3, there is the possibility of forming a partial covalent bond between one electron-deficient BF_3 molecule and an electron pair on a F of another BF_3 molecule. No such possibility exists for NF_3, which already has a lone pair on N. Only weak van der Waals and dipole-dipole forces will operate in solid NF_3. Thus, BF_3 should have a higher melting temperature. It does ($-127°C$); NF_3 has a melting temperature of $-207°C$.

22. In the crystal each cation is surrounded by six anions and each anion is surrounded by six cations. Although the internuclear distance is larger in the crystal, this effect is offset by the fact that there are a larger number of attractions in the crystal than in the gas phase.

23. (a) $U_{r_0} = \dfrac{AN(Z^+)(Z^-)e^2}{r_0}\left(1 - \dfrac{1}{n}\right)$

     (b) $U_{r_0} = -180.5$ kcal mole^{-1}
     (c) The value for U_{r_0} obtained using Equation 7-10 is -180.8 kcal mole^{-1}.

25. (a) $U_{r_0} = -923.7$ kcal mole^{-1}
     (b) EA$(O^{2-}) = -8.5$ eV. The negative value indicates that the second electron in O^{2-} is unbound by 8.5 eV.

26. (a) nonmetallic covalent
     (b) van der Waals
     (c) Graphite is soft because of the weak van der Waals bonds that exist between layers. The melting point of graphite is very high because covalent bonds are broken within layers in going from the solid to the liquid state.
     (d)

(e) The π electrons are responsible for the conduction of electric current in graphite.

27. The lattice energy will be much larger for MgO than for NaF. For MgO $Z^+ = +2$ and $Z^- = -2$ whereas for NaF $Z^+ = +1$ and $Z^- = -1$.

28. The wurtzite structure of ZnS has a lattice structure similar to that of diamond and hence in this respect it could be described in terms of a nonmetallic network structure. The bonding in ZnS could also be described in terms of the ionic model $Zn^{2+}S^{2-}$. The band gap in ZnS is intermediate between that of carbon and silicon.

30. In gaseous Be_2 the electronic configuration is $(\sigma_g^b)^2(\sigma_u^*)^2$ so that the net bond order is zero. In the solid, the $2s$ and $2p$ electronic bands overlap so that the antibonding $2s$ levels are not completely occupied.

31. Potassium has only one valence electron available to participate in the metallic bonding. Vanadium has five valence electrons. The binding energy per valence electron for K is 26.35 kcal mole^{-1} and for V it is 24.38 kcal mole^{-1}.

32.

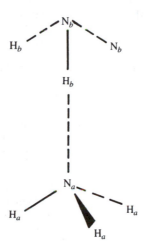

33.

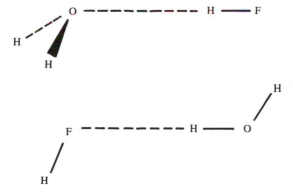

Index